AF356602

Urban Ecosystem Services II: Toward a Sustainable Future

Urban Ecosystem Services II: Toward a Sustainable Future

Editors

Alessio Russo
Giuseppe T. Cirella

MDPI • Basel • Beijing • Wuhan • Barcelona • Belgrade • Manchester • Tokyo • Cluj • Tianjin

Editors

Alessio Russo
School of Arts
University of Gloucestershire
Cheltenham
United Kingdom

Giuseppe T. Cirella
Faculty of Economics
University of Gdansk
Sopot
Poland

Editorial Office
MDPI
St. Alban-Anlage 66
4052 Basel, Switzerland

This is a reprint of articles from the Special Issue published online in the open access journal *Land* (ISSN 2073-445X) (available at: www.mdpi.com/journal/land/special_issues/2ndEd_Urban_Ecosystem_Services).

For citation purposes, cite each article independently as indicated on the article page online and as indicated below:

LastName, A.A.; LastName, B.B.; LastName, C.C. Article Title. *Journal Name* **Year**, *Volume Number*, Page Range.

ISBN 978-3-0365-1838-1 (Hbk)
ISBN 978-3-0365-1837-4 (PDF)

Contents

About the Editors

Alessio Russo

Alessio Russo is a senior lecturer and academic course leader for the Master of Landscape Architecture at the University of Gloucestershire, Cheltenham, United Kingdom. Before joining the University of Gloucestershire, he worked in Russia as an associate professor at RUDN University in Moscow and as a professor and head of the Laboratory of Urban and Landscape Design at Far Eastern Federal University in Vladivostok. He holds a Bachelor in Science in Plant Production from the University of Naples, a Post-Graduate Specialization in Healing Garden Design from the University of Milan, and a Master in Science in Landscape Design and Planning from the University of Pisa. He received his Ph.D. in Urban Forestry from the University of Bologna. Outside of academia, Dr. Russo has worked as a landscape architect in the United Kingdom, Italy, and the United Arab Emirates, dealing with sustainable design and planning.

Giuseppe T. Cirella

Giuseppe T. Cirella is a professor of Human Geography at the Faculty of Economics, University of Gdansk, Sopot, Poland. He received his Ph.D. in Environmental Engineering (specialization: Sustainability) from Griffith University, Australia. He is the founder of the Polo Centre of Sustainability and is the director and head of research. He has acted as a principal investigator and coordinator in a number of international projects and is a reviewer and member of the editorial board of several reputed international journals on sustainability and the environment. He has extensive interdisciplinary and cross-cultural experience in socioeconomics as well as expertise in landscape architecture, urban planning, and societal development.

Preface to "Urban Ecosystem Services II: Toward a Sustainable Future"

Twenty years have passed since the Millennium Ecosystem Assessment was launched in 2001 and published in 2005, yet ecosystem services in the urban environment are more essential than ever for the long sustainability of cities. The long-term sustainability of cities is also dependent on how we plan cities and how ecosystem services become an integral part of spatial planning and design. Twenty years have also passed since the death of Professor Ian McHarg who taught us that we need to design with nature and that it is important for politicians, designers, and all people involved in the city-design process to understand its role as well as any associated ecosystem services. In this regard, the COVID-19 pandemic has presented an important integrative view in which urban economies, urban social structures, and the urban-to-environment outlook have all been shocked. This shock effect has produced novel research reflected in this book. The two editors strongly believe in the sustainability of future cities and decided to publish a second book on urban ecosystem services with the focus on building a sustainable future. Specifically, this second book provides updates on the scientific literature by collecting eleven peer-reviewed articles published in the scientific journal *Land*.

Alessio Russo, Giuseppe T. Cirella
Editors

Editorial

Urban Ecosystem Services: Current Knowledge, Gaps, and Future Research

Alessio Russo [1],* and **Giuseppe T. Cirella [2]**

1   School of Arts, Francis Close Hall Campus, University of Gloucestershire, Swindon Road, Cheltenham GL50 4AZ, UK

2   Faculty of Economics, University of Gdansk, 81-824 Sopot, Poland; gt.cirella@ug.edu.pl

*   Correspondence: arusso@glos.ac.uk; Tel.: +44-(0)12-4271-4557

Citation: Russo, A.; Cirella, G.T. Urban Ecosystem Services: Current Knowledge, Gaps, and Future Research. *Land* **2021**, *10*, 811. https://doi.org/10.3390/land10080811

Received: 28 July 2021
Accepted: 30 July 2021
Published: 1 August 2021

Publisher's Note: MDPI stays neutral with regard to jurisdictional claims in published maps and institutional affiliations.

The term ecosystem services was coined to describe the societal benefit that natural ecosystems provide, as well as to raise awareness about biodiversity and ecosystem conservation [1]. Nowadays, with most people living in cities (i.e., over 50%) and the challenges that come with it, such as the urban heat island effect, food security, floods, pollution, and so on, the concept of ecosystem services is linked to the built environment's sustainability and livability [2,3]. Cities have long been one of the least researched ecosystems in this regard [4]. Consequently, there is a growing interest in urban ecology as well as in urban ecosystem services (UES) that present solutions to the above issues within the environment of the cityscape [2,4]. The need to study these "novel" ecosystems can mitigate the effects of the growing urban population and rural-to-urban transition. In this context, following the first special issue on UES [5], this second special issue aims to update existing knowledge and identify gaps and potential areas for future research. This second issue, in particular, has 10 peer-reviewed papers authored by scholars from all over the world, spanning five continents (Figure 1).

The structure of this article is to look at the current knowledge-base, point out gaps, and summarize future research findings within the dimension of UES. Lourdes et al. [6] point out that research on UES in the Global South has not been extensively examined as it has been in the Global North. To address this issue, they conducted a systematic literature review of UES in Southeast Asia over a two-decade period [6]. Their findings emphasize the region's unequal distribution of UES research and highlight common services, scales, and characteristics examined, as well as methodologies used [6]. They identified that while most research analyze regulatory and cultural UES at a landscape scale, few studies looked at interconnections between services by evaluating synergies and trade-offs [6]. Their results also suggest the urgent need for multitemporal and scenario-based research on the resilience of UES provision [6]. The researchers concluded that more research is needed to incorporate a variety of monetary and non-monetary valuations, as well as increased stakeholder involvement in UES assessments, so that the valuation of UES can encourage more transparent tradeoff assessments to support sustainable city planning [6]. In the Global South, there is also little evidence available on how people perceive the benefits and costs of urban green spaces [7]. To fill this gap, Pineda-Guerrero et al. [7] used semi-structured surveys, statistical analyses, and econometrics to investigate user perceptions of governance and the benefits and costs, i.e., ecosystem services and ecosystem disservices, provided by neotropical green areas in Bogota, Colombia, as well as their willingness to invest in their conservation. Despite the sub-severe watershed's stormwater runoff concerns, their modeling reveals that air quality and biodiversity are very important advantages while water control is not [7]. In terms of costs, inadequate levels of maintenance and infrastructure in the investigated green areas were linked to a sense of insecurity due to crime. The community's unwillingness to invest (UTI) in green space protection was impacted by their perceptions of government openness, corruption, and performance [7]. The findings indicate that socioeconomic backgrounds, government performance, and environmental education all influence the value or priority individuals have on the benefits, costs, and UTI of conservation efforts in urban green spaces [7].

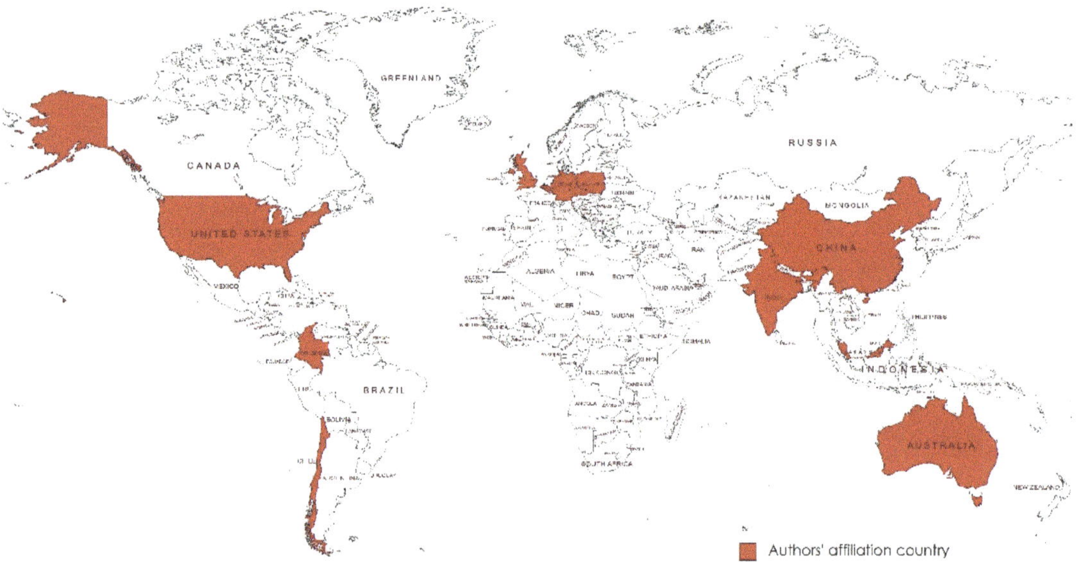

Figure 1. Authors' affiliation country. Source: Created using MapChart.net.

Animals play an important role in providing ecosystem services in cities. Conserving animal populations that offer faunal ecosystem services is critical to ensuring that ecosystems continue to operate properly and provide ecosystem services where there is a need for them [8]. For example, birds provide a variety of ecosystem services and are some of the most successful urban adapters [9]. Based on a literature review of research conducted between 1979 and 2020, Patankar et al. [9] reviewed research studies on traits that influence the survival and response of birds to urbanization, identifying the state of current knowledge, highlighting key knowledge gaps, and identifying geographic and species-specific disparities in research that require specific focus. There is still a gap in the scientific literature on how people perceive urban forestry and related ecosystem services [10]. The research of Alves Carvalho Nascimento and Shandas [10] intended to address this gap by looking at how neighborhood trees and socioeconomic factors affect public perceptions of ecosystem service availability. They studied socioeconomic factors, land cover statistics, and public opinions of neighborhood trees in Portland, OR. The findings showed a strong link between tree canopy, resident income, and sense of responsibility for urban forestry, based on over 2500 survey responses [10]. While a body of literature highlights the importance of trees in delivering UES (e.g., pollination, air quality, carbon storage and sequestration, temperature mitigation, etc.) [11–13], their survey findings show that management and cultural ecosystems services are important to respondents [10]. This finding, while seemingly insignificant, is relevant for several reasons, including the fact that respondents appear to be aware of the financial and maintenance costs associated with trees [10].

Across the globe, land-based financing is becoming more widely recognized as a mechanism for developing infrastructure and supplying ecosystem services [14]. One of the main drivers of urban ecosystem growth in many nations has been a fast increase in land and property values. This occurrence has given project proponents and policymakers with a plethora of alternatives and problems, prompting them to construct or incorporate land-based finance components into their policies and regulations. In particular, the Indian government and state governments have attempted to monetize land through a variety of methods in order to improve the financial sustainability of infrastructure and area development projects. In terms of land monetization approaches, Tirumala and Tiwari [14] examined Indian central and state infrastructure policies and relating acts. Key factors of a successful strategy that captures a rise in land values are highlighted and reported on in their study [14]. Zepp and Inostroza [15] created an ad-hoc assessment to evaluate a typical environmental compensation technique using ecosystem services with an actual planning and development scenario involving a planned route to a restructured former industrial site in Bochum, Germany. The researchers used both techniques to assess the impact of the proposed construction choices [15]. In a subsequent phase, they chose the alternative with the lowest effect and calculated the ecosystem service losses from the compensatory measures [15]. The findings demonstrate that an ecosystem services evaluation offers a sound foundation for selecting development alternatives, identifying compensation areas, and estimating compensation amounts, with the added advantage of enhancing the environmental quality of the impacted areas [15]. When utilizing Zepp and Inostroza's method, there are two main limitations to consider. The first limitation is that they utilized a broad outline of the proposed roadways. In a more comprehensive analysis, the exact demarcation as described in engineering drawings should be used [15]. The second limitation is that they did not examine the area compensated in terms of urban structural subtypes in great depth. In general, adding nature-based solutions to existing urban settings may always enhance the provision of ecosystem services [15]. However, depending on the morphology of the city, a more precise estimate is required [15].

Greening tram lanes in cities appear to be an important strategy for the development of green corridors as well as enhancing UES [16]. Łukaszkiewicz et al. [16] demonstrated how to revitalize the Warsaw cityscape by converting existing tram lines where viable and developing new ones from a "green perspective." Green tram lanes reduce the noise level during tram operation, improve the aesthetic experience of city streets, and, when skillfully-applied, allow the sound level from traffic to be reduced by 10.0 to 15.0 dB [16]. Their research exemplifies future UES-based thinking and moves the bar on how infrastructure, design, and planning come together in a 21st century city. Expanding on this unity-concept, Wang et al. [17] used multiperiod datasets from the Land Use and Land Cover of China databases to study farmland loss owing to urbanization in China's Guangdong–Hong Kong–Macao Greater Bay Area from 1980 to 2018. Then, using valuation methodologies, they produced agricultural ecosystem service values (ESVs) to quantify the ecosystem service changes induced by urbanization in the research region [17]. The findings revealed that over the last 38 years, urbanization has resulted in a total area of farmland loss of 3711.3 km^2, resulting in a direct decrease in total ESVs of 5036.7 million yuan [17]. A sense of urbanization urgency pinpoints the need for better UES knowledge and how the cityscape is growing and changing. Due to this urban upsurge, landscape architects are under increasing strain and require practical knowledge, skills, and methodologies to support their designs [18]. Cultural ecosystem services (CES) have been linked to landscape architecture research, and the findings of CES assessments have the potential to help landscape architecture practice [18]. However, there have been few attempts to investigate CES in landscape architecture research in a systematic manner [18]. Furthermore, how CES assessments are carried out in in landscape architecture studies are rarely investigated. The goal of Cheng et al.'s [18] study points out some of these challenges and recommends employing CES assessments in landscape architecture practice, with an emphasis on landscape architecture design [18]. In the last paper, Russo et al. [19] used Bristol City Centre as an example to illustrate two free user-friendly web resources (i.e., i-Tree Canopy and the Office for National Statistics). They showed that both tools are simple to use and effectively convey ecosystem services and monetary values. Their research has updated the literature on the evaluation of green infrastructure tools in the United Kingdom, as well as identified topics for further research. As such, important UES gains have been made in the past year in which researchers have had to ponder and work within the new COVID-19 era. These abrupt changes have resulted in some UES rethinking, as demonstrated in this article, and from the new work environments city planners and urbanists have faced.

Author Contributions: Visualization and writing—original draft preparation, A.R.; writing—review and editing, A.R. and G.T.C. Both authors have read and agreed to the published version of the manuscript.

Funding: This research received no external funding.

Institutional Review Board Statement: Not applicable.

Informed Consent Statement: Not applicable.

Data Availability Statement: Not applicable.

Acknowledgments: We are grateful to the MDPI Land team of academic editors and reviewers for assisting with the Special Issue's academic excellence. MapChart.net was used to create the authors' affiliation country.

Conflicts of Interest: The authors declare no conflict of interest.

References

1. Birkhofer, K.; Diehl, E.; Andersson, J.; Ekroos, J.; Früh-Müller, A.; Machnikowski, F.; Mader, V.L.; Nilsson, L.; Sasaki, K.; Rundlöf, M.; et al. Ecosystem services—Current challenges and opportunities for ecological research. *Front. Ecol. Evol.* **2015**, *2*, 87. [CrossRef]
2. Russo, A.; Cirella, G.T. Urban Sustainability: Integrating Ecology in City Design and Planning. In *Sustainable Human—Nature Relations: Environmental Scholarship, Economic Evaluation, Urban Strategies*; Cirella, G.T., Ed.; Springer Singapore: Singapore, 2020; pp. 187–204. ISBN 978-981-15-3049-4.
3. Russo, A.; Cirella, G.T. Edible urbanism 5.0. *Palgrave Commun.* **2019**, *5*, 1–9. [CrossRef]
4. Stott, I.; Soga, M.; Inger, R.; Gaston, K.J. Land sparing is crucial for urban ecosystem services. *Front. Ecol. Environ.* **2015**, *13*, 387–393. [CrossRef]
5. Russo, A.; Cirella, G.T. Urban ecosystem services: New findings for landscape architects, urban planners, and policymakers. *Land* **2021**, *10*, 88. [CrossRef]
6. Lourdes, K.; Gibbins, C.; Hamel, P.; Sanusi, R.; Azhar, B.; Lechner, A. A review of urban ecosystem services research in southeast asia. *Land* **2021**, *10*, 40. [CrossRef]
7. Pineda-Guerrero, A.; Escobedo, F.J.; Carriazo, F. Governance, nature's contributions to people, and investing in conservation influence the valuation of urban green areas. *Land* **2020**, *10*, 14. [CrossRef]
8. Gutierrez-Arellano, C.; Mulligan, M. A review of regulation ecosystem services and disservices from faunal populations and potential impacts of agriculturalisation on their provision, globally. *Nat. Conserv.* **2018**, *30*, 1–39. [CrossRef]
9. Patankar, S.; Jambhekar, R.; Suryawanshi, K.R.; Nagendra, H. Which traits influence bird survival in the city? A review. *Land* **2021**, *10*, 92. [CrossRef]
10. Alves Carvalho Nascimento, L.; Shandas, V. Integrating diverse perspectives for managing neighborhood trees and urban ecosystem services in portland, OR (US). *Land* **2021**, *10*, 48. [CrossRef]
11. Roy, S.; Byrne, J.; Pickering, C. A systematic quantitative review of urban tree benefits, costs, and assessment methods across cities in different climatic zones. *Urban. For. Urban. Green.* **2012**, *11*, 351–363. [CrossRef]
12. Speak, A.; Escobedo, F.J.; Russo, A.; Zerbe, S. Total urban tree carbon storage and waste management emissions estimated using a combination of LiDAR, field measurements and an end-of-life wood approach. *J. Clean. Prod.* **2020**, *256*, 120420. [CrossRef]
13. Salmond, J.A.; Tadaki, M.; Vardoulakis, S.; Arbuthnott, K.; Coutts, A.; Demuzere, M.; Dirks, K.N.; Heaviside, C.; Lim, S.; Macintyre, H.; et al. Health and climate related ecosystem services provided by street trees in the urban environment. *Environ. Health* **2016**, *15*, S36. [CrossRef] [PubMed]
14. Tirumala, R.D.; Tiwari, P. Land-based financing elements in infrastructure policy formulation: A case of india. *Land* **2021**, *10*, 133. [CrossRef]
15. Zepp, H.; Inostroza, L. Who pays the bill? Assessing ecosystem services losses in an urban planning context. *Land* **2021**, *10*, 369. [CrossRef]
16. Łukaszkiewicz, J.; Fortuna-Antoszkiewicz, B.; Oleszczuk, Ł.; Fialová, J. The potential of tram networks in the revitalization of the warsaw landscape. *Land* **2021**, *10*, 375. [CrossRef]
17. Wang, X.; Yan, F.; Zeng, Y.; Chen, M.; He, B.; Kang, L.; Su, F. Ecosystem Services changes on farmland in response to urbanization in the Guangdong–Hong Kong–Macao greater bay area of china. *Land* **2021**, *10*, 501. [CrossRef]
18. Cheng, X.; Van Damme, S.; Uyttenhove, P. Applying the evaluation of cultural ecosystem services in landscape architecture design: Challenges and opportunities. *Land* **2021**, *10*, 665. [CrossRef]
19. Russo, A.; Chan, W.T.; Cirella, G.T. Estimating air pollution removal and monetary value for urban green infrastructure strategies using web-based applications. *Land* **2021**, *10*, 788. [CrossRef]

 *land*

Review

A Review of Urban Ecosystem Services Research in Southeast Asia

Karen T. Lourdes [1], Chris N. Gibbins [2], Perrine Hamel [3,4], Ruzana Sanusi [5,6], Badrul Azhar [5] and Alex M. Lechner [1,7,*]

1 Landscape Ecology and Conservation Lab, School of Environmental and Geographical Sciences, University of Nottingham Malaysia, Semenyih 43500, Malaysia; hgxkl1@nottingham.edu.my
2 School of Environmental and Geographical Sciences, University of Nottingham Malaysia, Semenyih 43500, Malaysia; christopher.gibbins@nottingham.edu.my
3 Asian School of the Environment, Nanyang Technological University, 50 Nanyang Avenue, Singapore 639798, Singapore; perrine.hamel@ntu.edu.sg
4 Natural Capital Project, Woods Institute for the Environment, Stanford University, 371 Serra Mall, Stanford, CA 94305, USA
5 Department of Forestry Science and Biodiversity, Faculty of Forestry and Environment, Universiti Putra Malaysia, Serdang 43400, Malaysia; ruzanasanusi@upm.edu.my (R.S.); b_azhar@upm.edu.my (B.A.)
6 Laboratory of Sustainable Resources Management (BIOREM), Institute of Tropical Forestry and Forest Products, Universiti Putra Malaysia, Serdang 43400, Malaysia
7 Lincoln Centre for Water and Planetary Health, School of Geography, University of Lincoln, Lincoln LN6 7TS, UK
* Correspondence: alechner@lincoln.ac.uk

Abstract: Urban blue-green spaces hold immense potential for supporting the sustainability and liveability of cities through the provision of urban ecosystem services (UES). However, research on UES in the Global South has not been reviewed as systematically as in the Global North. In Southeast Asia, the nature and extent of the biases, imbalances and gaps in UES research are unclear. We address this issue by conducting a systematic review of UES research in Southeast Asia over the last twenty years. Our findings draw attention to the unequal distribution of UES research within the region, and highlight common services, scales and features studied, as well as methods undertaken in UES research. We found that while studies tend to assess regulating and cultural UES at a landscape scale, few studies examined interactions between services by assessing synergies and tradeoffs. Moreover, the bias in research towards megacities in the region may overlook less-developed nations, rural areas, and peri-urban regions and their unique perspectives and preferences towards UES management. We discuss the challenges and considerations for integrating and conducting research on UES in Southeast Asia based on its unique and diverse socio-cultural characteristics. We conclude our review by highlighting aspects of UES research that need more attention in order to support land use planning and decision-making in Southeast Asia.

Keywords: natural capital; blue-green infrastructure; urban environmental challenges; Global South; tropical cities

Citation: Lourdes, K.T.; Gibbins, C.N.; Hamel, P.; Sanusi, R.; Azhar, B.; Lechner, A.M. A Review of Urban Ecosystem Services Research in Southeast Asia. *Land* **2021**, *10*, 40. https://doi.org/10.3390/land10010040

Received: 10 December 2020
Accepted: 28 December 2020
Published: 5 January 2021

Publisher's Note: MDPI stays neutral with regard to jurisdictional claims in published maps and institutional affiliations.

1. Introduction

The global urban population has grown rapidly in the last few decades, with over 70% of the population in the Global North now residing in urban areas [1]. Similar trends are evident in the Global South and while developed regions may be better equipped to manage urban transformations [2], cities in developing regions such as Southeast Asia face increasing environmental pressures. In 2018, an estimated 320 million people lived in the urban areas of Southeast Asia (49% of the region's total population), and this figure is expected to increase to 66% of the total population by 2050 [1,3]. This rapid urbanisation has been accompanied by a range of environmental problems, including the urban heat island effect, floods, poor air quality and noise pollution, all of which directly impact the health of urban residents [4–11]. These issues are expected to be further exacerbated by the general vulnerability of the region to climate change impacts [12–14]. Moreover, countries within Southeast Asia have extremely diverse biophysical, cultural, socio-economic and political characteristics (Table 1). Levels of urbanisation range from 23% (Cambodia) to 100% (Singapore) and gross national incomes range from the 11th (Singapore) to the 162nd (Myanmar) rank globally. Efforts to mitigate urban environmental challenges should take into consideration these characteristics, in so doing provide context-specific solutions [15–17].

Table 1. Basic statistics that characterise the diverse biophysical, socio-economic and political backgrounds of Southeast Asian countries for the year 2019.

Country	Land Area (km²) [18]	Population Size (mil) [19]	Percentage of Urban Population (%) [20]	Average Annual Population Growth Rate (%) * [19]	Per Capita Gross National Income Ranking [21]	CO_2 Emissions (Million Metric Tons) [†] [20]	Governance (−2.5 TO +2.5)	
							Government Effectiveness [22]	Voice and Accountability [22]
Singapore	720	5.80	100.0	0.90	11	37.5	2.22	−0.18
Brunei Darussalam	5765	0.433	77.1	1.06	31	7.6	1.32	−0.95
Malaysia	331,388	31.94	76.2	1.33	69	248.2	1.0	−0.04
Indonesia	1,916,862	270.62	56.0	1.14	118	512.7	0.18	0.16
Thailand	513,140	69.62	53.6	0.31	86	283.7	0.36	−0.83
Philippines	300,000	108.11	47.1	1.41	123	122.2	0.05	0.03
Lao PDR	236,800	7.16	35.6	1.52	139	17.7	−0.78	−1.80
Vietnam	331,230	96.46	35.0	0.98	141	192.6	0.04	−1.38
Timor Leste	14,870	1.29	30.9	1.94	151	0.49	−0.88	0.38
Myanmar	676,576	54.04	30.0	0.64	162	25.2	−1.15	−0.84
Cambodia	181,035	16.48	23.8	1.48	158	9.9	−0.58	−1.20

Notes: * from 2015 to 2020; [†] for the year 2016.

Planning and designing cities to incorporate blue-green spaces is vital for mitigating socio-environmental problems affecting health and well-being [23–25]. Urban blue-green spaces promote greater resilience, sustainability and liveability in cities through the provision of services such as shading and cooling, carbon sequestration, stormwater management, noise attenuation, habitat for biodiversity and recreational opportunities [26–30]. These services, termed 'urban ecosystem services' (UES), capture the role of water (blue) (i.e., lakes and wetlands) and vegetation (green) (i.e., parks and urban forests) in or near the built environment at different spatial scales (streets, buildings, cities, regions) [31–33]. Generated through the functions and processes of blue-green structures, UES can alleviate the environmental pressures of urbanisation and enhance the wellbeing of urban residents [34–39].

The complex pathways through which UES are delivered can be analysed by the relationships between (i) structures (e.g., mangrove forests), (ii) their biophysical processes and functions (e.g., wave attenuation), and (iii) the derived services that deliver goods and benefits to humans (e.g., coastal flood protection) [40]. The interactions between these different components can be illustrated through frameworks such as the cascade model, which acts as a communication tool between experts and local stakeholders to help support UES assessments for urban planning [41]. Moreover, incorporating UES into urban planning requires an understanding of the various interactions between services, which are linked to one another as they stem from the same structures and functions of a particular ecosystem [42,43]. These interactions include synergies and tradeoffs, described respectively as positive-positive or positive-negative relationships between two or more services [44,45].

Research on UES can also be undertaken from various perspectives, given the interdisciplinary nature of the concept. The field has gained prominence for its ability to integrate natural and social sciences, communicating the dependence of society on ecological structures [43]. A wide range of methods have been used to characterise UES and assess their value to humans. These methods range from biophysical modelling to social surveys applied at various scales (e.g., landscape scale, site-based scale), with benefits valued biophysically (e.g., tonnes of carbon sequestered per year), economically (e.g., $500 per hectare per year) and socio-culturally (e.g., sense of place) [46–48]. UES hold diverse values to various communities and the valuation of UES is necessary to understand local demands or benefits [30,49]. Valuations should be supported by the involvement of stakeholders to further deepen the understanding of local UES needs, while promoting the consideration of alternative management options [50,51].

Previous reviews of global UES research by Haase [28] and Luederitz [41] highlight that research has mostly been undertaken in Europe and North America, with research in Asia dominated by China. Although these reviews have explored the scope and nature of research on a global scale, they lack the finer resolution needed to understand patterns and traits of research in any one region. Despite the rapid economic growth and urbanisation in Southeast Asian countries, UES research across this region has not been reviewed. Hence, a systematic review of UES is timely, to assess the nature and extent of research on UES in Southeast Asia.

This review covers the last 20 years, the period within which the global UES literature has burgeoned. Inspired by Luederitz [41], we address four specific research questions: (1) How is UES research distributed across Southeast Asia and at what scale(s) are UES analysed? (2) Does UES research focus on single or multiple services and what type of blue-green structure are assessed? (3) Which components of the 'cascade' are assessed, and how are the interactions between UES conceptualised and stakeholders involved? (4) What research perspectives, and data collection and analytical methods are used to assess UES? Upon reviewing the current state of research in the region, we discuss the challenges and considerations for integrating UES research given the unique context of Southeast Asia. We conclude our review with recommendations for UES research in order to support planning in the region.

2. Methods

The search string composed terms that expressed the geographical area of interest ('Southeast Asia' and all the countries within the region), the topic of interest ('ecosystem service', its alternative term 'natural capital' and, to capture studies that did not explicitly refer to these two phrases, we included the keywords 'human', 'environment' and 'benefit') as well as terms that further specified the subtopic of interest (i.e., the urban environment). The search was applied to publication Titles, Abstracts and Keywords in the Scopus and Web of Science database as shown below:

(TITLE-ABS-KEY (("Southeast Asia" OR "South East Asia" OR "Indonesia" OR "Vietnam" OR "Thailand" OR "Malaysia" OR "Singapore" OR "Philippines" OR "Cambodia" OR "Laos" OR "Myanmar" OR "Brunei" OR "Timor-Leste"))) AND (TITLE-ABS-KEY (("ecosystem service*" OR "natural capital" OR ("human" AND "environment" AND "benefit*")))) AND (TITLE-ABS-KEY ("urban" OR "city" OR "cities"))

The initial search return was refined to include only journal articles, book chapters and conference papers (see Supplementary Material for complete search string). This search returned a total of 255 unique articles published in the English language. The abstracts of the returned articles were screened manually to include publications within the scope of this review based on the following guiding criteria:

- Studies conducted in urban or peri-urban areas in Southeast Asian countries.
- Focuses on ecosystem services or benefits provided to an urban population.
- Explicitly includes the phrase 'ecosystem services' or 'natural capital', otherwise describes the link between the environment and the benefits provided to urban populations.

The final list comprised 149 empirical articles, assessing one or more ecosystem services in urban Southeast Asia (see Table S1 in Supplementary Material). Studies that investigated multiple urban areas within and outside of Southeast Asia were included in the review, if at least one study site was located within Southeast Asia. Each article was classified to identify information relevant to the four research questions, as described in the sections which follow. Refer to Table S2 in Supplementary Material for further details on definitions and classification protocol.

(1) How is UES research distributed across Southeast Asia and at what scale(s) have they been analysed?

Following the TEEB classification for ecosystem services [52], we classified the ecosystem services studied into four main categories: (i) provisioning, (ii) regulating, (iii) supporting and (iv) cultural. These four categories will be hereafter referred to as 'ecosystem service domains'. We chose the TEEB classification of ecosystem services over the two other common approaches to classifying ecosystem services—the Millennium Ecosystem Assessment (MEA) and Common International Classification of Ecosystem Services (CICES). TEEB is well-known in the context of environmental economics and provides a robust framework for applications in urban planning and policies [53]. Moreover, the TEEB framework emphasises the need for valuing ecosystem services such that the wide range of benefits of ecosystems and biodiversity is recognised by decision-makers [54]. We also recorded the location (e.g., city and country) of studies and quantified the number of times ecosystem service domains were assessed for each country. To analyse the scale of UES assessment, we recorded the population size and area of study sites, scale of assessment as well as distinguished between 'urban' and/or 'peri-urban' areas.

(2) Does UES research focus on single or multiple services and what type of blue-green structure have been assessed?

We recorded the ecosystem services assessed as one of the 17 ecosystem services defined by the TEEB framework [52]. As studies can mention more ecosystem services than those that were empirically assessed, we only classified ecosystem services that were explicitly investigated. We evaluated the number of services assessed in each study and whether the services belonged to the same ecosystem service domain. The ecological structures that provide the investigated ecosystem service(s) were classified as either vegetative (green) or water (blue) structures. We identified 12 categories of blue-green structures, comprising four blue structures (coastal lands, wetlands, rivers, lakes) and eight green structures (urban forests, parks, street greenery, gardens, rooftops, green walls, cultivated lands and grasslands; refer to Table S3 in Supplementary Material for detailed definitions of blue-green structures).

(3) Which components of the 'cascade' have been assessed, how are the interactions between UES conceptualised and stakeholders involved?

The components of ecosystem service 'cascade' were assessed according to structure, function, services, and each linkage between structure-function-services (Figure 1). We described studies as either reviewing one of these six components, all of the six components or none, if studies did not assess any of the components in depth. We recorded the explicit assessment of synergies and tradeoffs in studies as well as the involvement of stakeholders in supporting UES assessments. The latter was defined as the feedback or involvement of external parties, aside from the researcher, in assessing UES. Studies involving surveys and interviews were considered to have involved stakeholders.

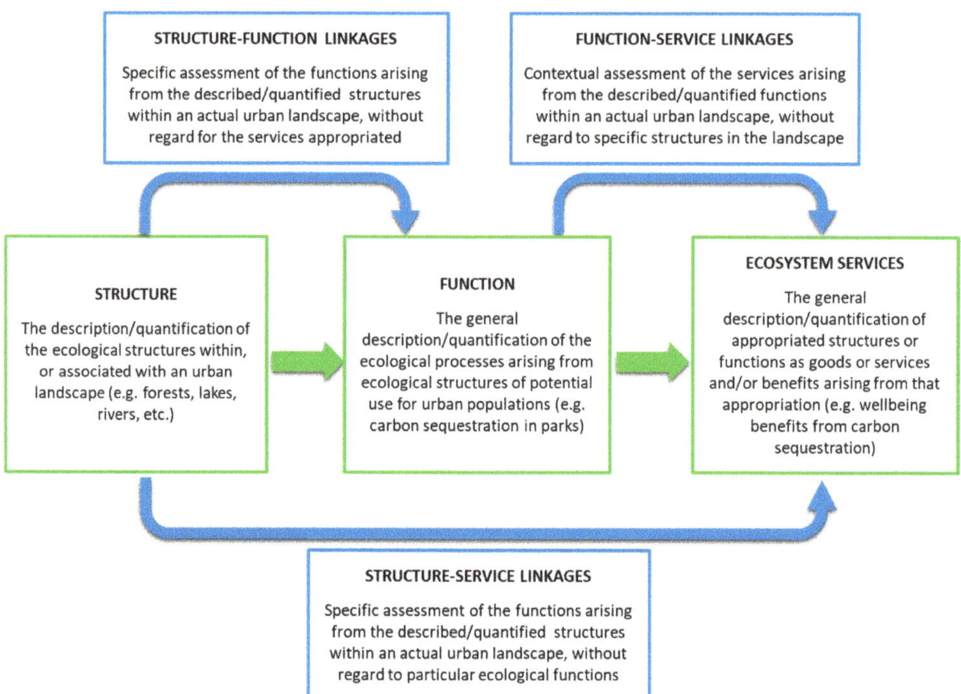

Figure 1. The components and definitions of the cascade model used to classify UES studies [41].

(4) What research perspectives, and data collection and analytical methods are used to assess UES?

We assigned each study one of the following six research perspectives: (i) ecology, (ii) social, (iii) planning, (iv) governance, (v) economic and (vi) methods [41]. The definitions for this classification are available in Table S2 in Supplementary Material. Although a single study can be undertaken with more than one perspective, we classified each study by its most dominant research perspective.

Data collection methods were classified into four categories: (i) 'field-based empirical', (ii) 'biophysical modelling' which is sub-divided into 'process/mechanistic modelling' and 'land cover proxy' (e.g., remote sensing of land cover), and (iii) 'social surveys' and (iv) case studies. We also recorded the type of data collected (i.e., 'quantitative', 'qualitative', or 'both') and the temporal focus of the study. Studies were also reviewed for the valuation of UES and where valuations were conducted, we distinguished between 'monetary valuation' (i.e., economic) and/or 'non-monetary valuation' (i.e., social or biophysical).

3. Results

3.1. Distribution and Scale of UES Assessment across Southeast Asia

Of the 149 studies reviewed in Southeast Asia, 29% were conducted in Singapore (n = 44), followed by 22% in Indonesia (n = 33). Myanmar (n = 4) and Timor Leste (n = 1) had very few studies, while no published studies from Brunei were returned in the search (Figure 2). About 64% of studies had authors with their primary research institution in Southeast Asia (n = 95), with 59% of studies conducted in the country where their primary research institution was located. As for studies with authors' primary research institutions located outside of Southeast Asia, 16% of authors were based in Europe (n = 24), 11% in other parts of Asia (n = 17) (mainly East Asia), 5% in North America (n = 8) and 3% in Australia (n = 5). Note the possibility of some bias in this analysis due to the inclusion of only studies published in English.

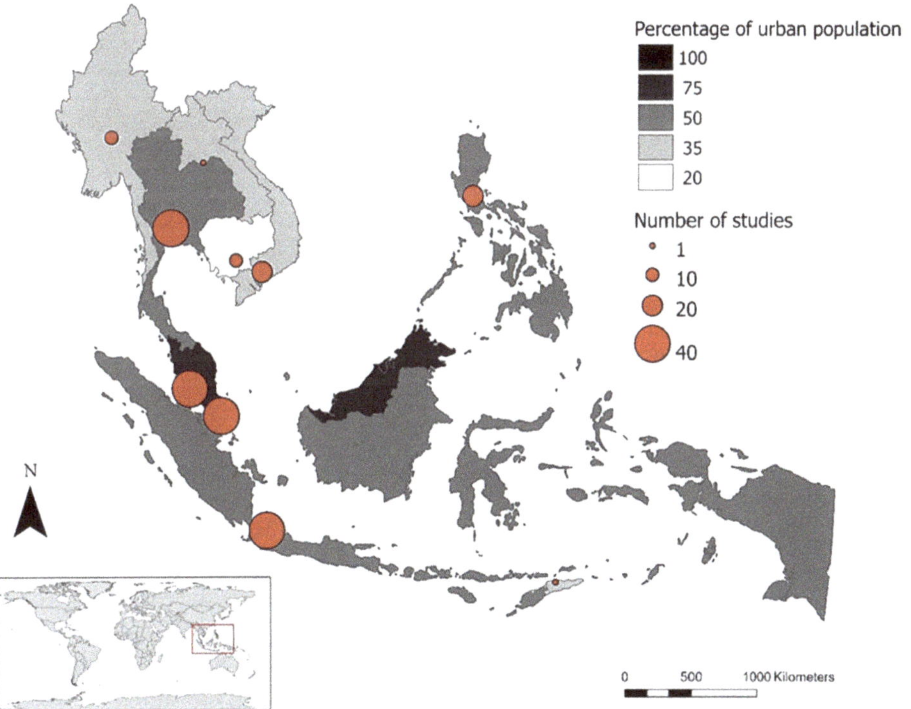

Figure 2. Number of UES studies in Southeast Asia and percentage of urban population in each country. Not visible in the map is the percentage of urban population in Singapore, which is 100%. No studies from Brunei were reviewed.

77% of studies are concentrated in four cities; the city-state of Singapore was most frequently studied (n = 44), followed by the metropolitan capital cities of Bangkok in Thailand (n = 13), Jakarta in Indonesia (n = 12) and Kuala Lumpur in Malaysia (n = 8).

In terms of the type of urban area assessed, 76% of studies were conducted in fully urban areas (n = 113), 15% in peri-urban areas (n = 23) and 8% of studies spanned both urban and peri-urban areas (n = 13). Around 43% of studies were conducted at a 'single-city scale' (n = 64), followed by 32% at the 'sites within cities' scale (n = 48) (Figure 3). Only 17% of studies assessed multiple cities (n = 26) and 7% of studies were conducted at scales larger than cities (i.e., regional or continental scales) (n = 11). Study area sizes varied markedly, extending from a few square kilometres to tens of millions of square kilometres. Similarly, population sizes within the study areas differed greatly, from 750 (Botoc village, Philippines) to 9.6 million (Jakarta metropolitan).

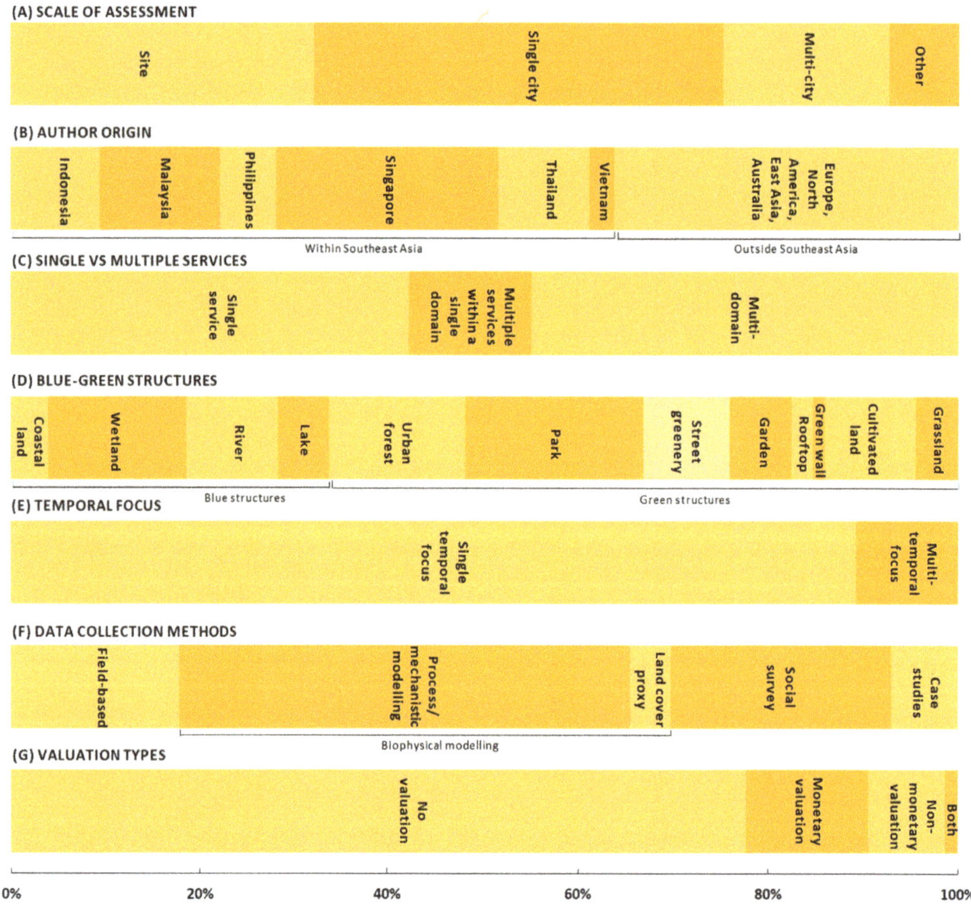

Figure 3. The general characteristics of UES research. (**A**) The various scales at which UES were assessed; (**B**) The percentage of studies conducted by authors based in and outside of Southeast Asia (denoted by the country in which the primary institution of the first author is located); (**C**) The assessment of UES within and/or across domains; (**D**) The types blue-green structures assessed; (**E**) The temporal focus of the study where 'single temporal focus' represents studies that examined UES at one point/period in time and 'multitemporal' represents studies that compared UES across time; (**F**) The methods of data collection and analysis of UES; and (**G**) The types of UES valuations conducted by studies.

3.2. Services and Blue-Green Structures Assessed

Of the four domains, regulating (36%) and cultural (26%) services were most assessed. Most countries had studies encompassing services across all four domains; exceptions were Laos and Timor Leste (Figure 4a). Studies comprised all 17 ecosystem services across all domains, with the 'recreation and mental and physical health' as the most frequently studied service (n = 54), followed by the 'moderation of extreme events' service (n = 51). The 'medicinal resources' and 'biological control' services were least assessed, with only 5 and 4 studies respectively (Table 2). Most studies took a multi-domain approach (n = 67) by investigating ecosystem services across multiple domains. Around 42% of studies assessed a single ecosystem service (n = 63), while 13% studied multiple services from a single domain (n = 19).

60% of studies assessed a single blue-green structure (n = 90) while the remaining studies assessed two or more structures; the maximum was seven structures (n = 2) (Figure 4b). Of the 12 blue-green structures, parks were most frequently studied (n = 57), followed by wetlands (n = 45) and urban forests (n = 44) (note: values differ from the number of times each structure was studied under ecosystem service domains, see Figure 4b). All 12 structures were studied across the four ecosystem services domains except for green walls, which were not studied for provisioning services. Rivers, urban forests and cultivated lands were most commonly studied for provisioning services, while street greenery and wetlands were commonly studied for regulating services. Parks were almost equally studied for regulating (n = 39) and cultural services (n = 36), although studies of cultural services pre-dominantly assessed parks in comparison to all other structures (23%).

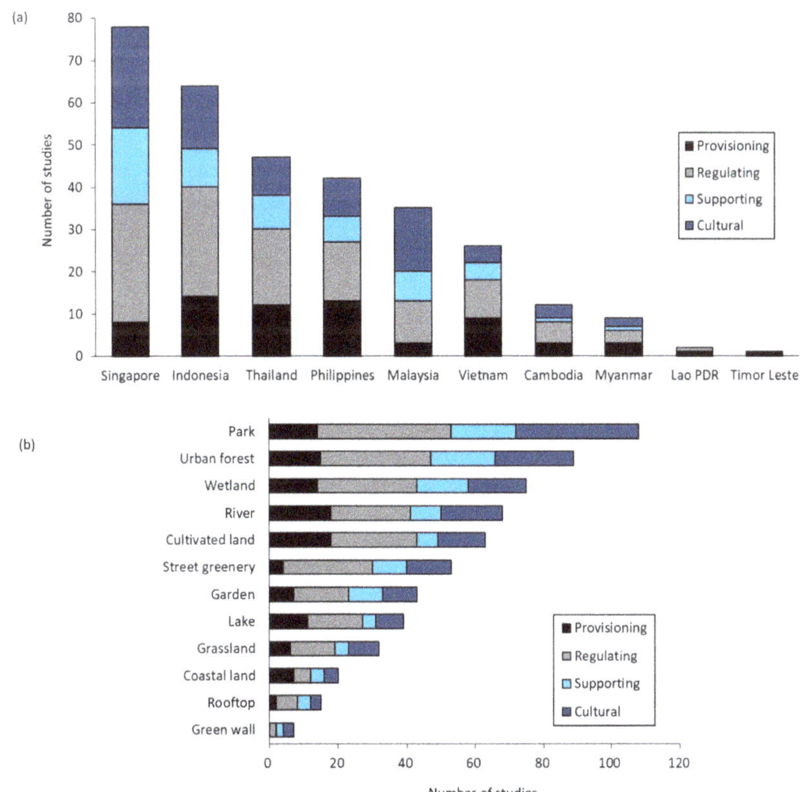

Figure 4. Ecosystem service domains assessed according to (**a**) countries and, (**b**) blue-green structures.

Table 2. The number of studies that assessed each ecosystem service according to the TEEB classification system [52]. Note that some studies assessed multiple ecosystem services; thus, the total number of ecosystem services assessed is greater than the 149 publications reviewed.

Domain	Ecosystem Service	Number of Studies
Provisioning	Food	40
	Raw materials	29
	Fresh water	18
	Medicinal resources	4
Regulating	Local climate and air quality	44
	Carbon sequestration and storage	27
	Moderation of extreme events	51
	Wastewater treatment	11
	Erosion prevention and maintenance of soil fertility	15
	Pollination	9
	Biological control	5
Supporting	Habitats for species	43
	Maintenance of genetic diversity	8
Cultural	Recreation and mental and physical health	54
	Tourism	19
	Aesthetic appreciation and inspiration for culture, art and design	44
	Spiritual experience and sense of place	25

3.3. Components of the 'Cascade' and Stakeholder Involvement

Only 2% of studies (n = 3) did not assess any component of the cascade in depth. These studies were mainly on the management of ecosystem services using frameworks that did not focus on any specific component of the cascade (e.g., [55,56]). Conversely, 16% of studies (n = 24) assessed all three components (Table 3). For instance, Remondi [57] simulated changes to land use surrounding rivers in Jakarta under different urbanisation scenarios. The study modelled the capacity of the river (structure) to retain water (function), in providing fresh water and flood protection services to the local population (services and benefits). The most studied component was the structure-function linkage (26% of studies; n = 38), while the function component was least assessed, with only 4% of studies (n = 6). Only 4% (n = 6) of the 149 studies had explicitly investigated ecosystem service interactions such as synergies and tradeoffs. The majority of the studies (56%) did not involve stakeholders either through surveys, interviews or expert input. Of those that did, most assessed cultural services (n = 46). Links between UES and climate change were only assessed by 3% of studies (n = 5).

Table 3. Distribution of the number of studies assessing various components of the ecosystem services cascade.

Cascade Component	Number of Studies
Structure	11
Structure-function	38
Function	6
Function-benefit	14
Benefit	29
Structure-benefit	24
All	24
None	3

3.4. Research Perspectives and Methods of UES Assessment

The number of studies in the region has increased across all ecosystem service domains (Figure 5a), particularly over the last decade; the review only yielded three studies prior to 2011 (Figure 5b), with more than 89% being published post-2014 (n = 133). The highest annual number of studies was in the year 2018, although bearing mind that for 2020 the review only included studies published between January and August, this year also saw a relatively high number of papers published.

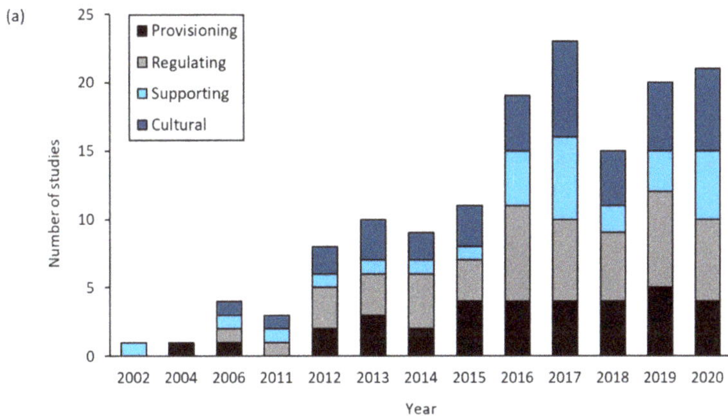

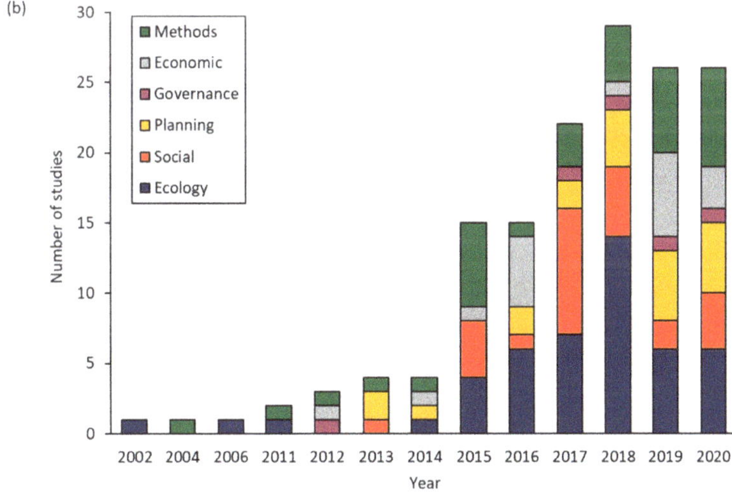

Figure 5. The (**a**) ecosystem service domains and (**b**) research perspectives, undertaken by studies over time. Note: A study may have assessed more than one ecosystem service domain but only one research perspective. Hence, **b** represents the actual number of studies reviewed over time.

Only papers published between 2018 and 2020 encompass all six research perspectives. Of the 149 studies, 32% were dominated by an ecological perspective (n = 47), while studies undertaken with a governance perspective were least common (3%; n = 5). Since 2013, more studies have been undertaken with social and planning perspectives, while governance-based research has received more attention since 2017. Studies with an ecological perspective were conducted in all countries except Timor Leste, which had the least number of studies in the region (Figure 6a). Singapore, Indonesia, Thailand and Vietnam had studies comprising all six perspectives. About 34% (n = 15) of the 44 studies conducted in Singapore had an ecological perspective, while only 5% (n = 2) had an economic perspective and one study had a governance perspective. There were no studies with an economic or governance perspective in Malaysia, although the social perspective comprised 43% (n = 10) of studies in this country.

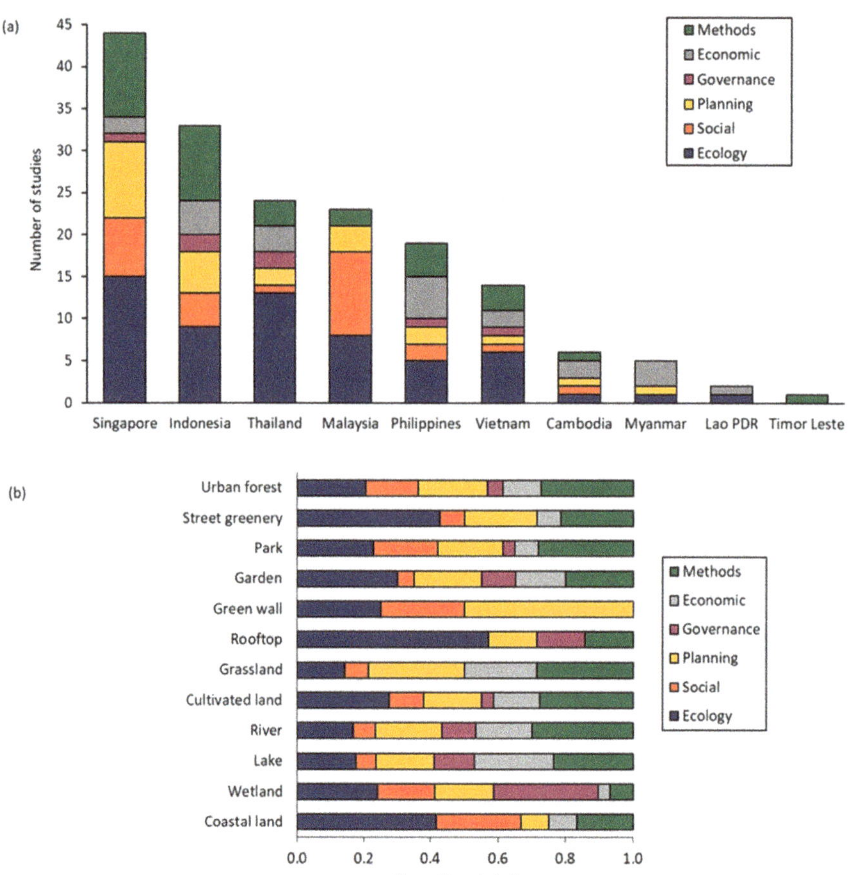

Figure 6. Research perspectives undertaken by studies across (**a**) countries and (**b**) blue-green structures.

Street greenery, gardens and rooftops were mainly studied from an ecological perspective, while parks, urban forests and rivers were predominantly assessed from a methods perspective (Figure 6b). Wetlands were most studied under the governance perspective (n = 9) and two of four studies of green walls had a planning perspective. Cultivated lands were equally studied using methods and ecological perspectives.

Over 89% of studies (n = 133) examined UES in a single time period or duration, with only 16 studies comparing services over two or more points in time. With respect to the type of data collected, 65% of studies (n = 97) collected only quantitative data, while only 5% of studies (n = 8) examined qualitative data. The remaining 44 studies examined both qualitative and quantitative data. Process and/or mechanistic models were the most utilised method of data collection and analysis in the region, comprising 88 studies (note: studies can utilise more than one method). Social surveys were the next most common method (n = 43), followed by field sampling (n = 33). 13 studies used case studies to assess UES and 8 studies used landcover proxies. Only 23% (n = 34) of studies conducted valuations, of which over half were monetary (n = 19). Two studies conducted both monetary and non-monetary valuations, while the remaining studies (n = 12) conducted non-monetary valuations.

4. Discussion

4.1. Current State of Research

Our review found that the there was a growing body of research on UES in the Southeast Asia, particularly in the last five years. The research was biased towards more developed countries, in particular the city-state of Singapore, where about one third (29%) of published research was conducted. Previous reviews have also found that UES research tends to focus on highly developed and urbanised countries [28,41]. It is also apparent that little research has been conducted in less developed countries such as Myanmar, Cambodia and Laos. While most papers were authored by researchers based in Southeast Asia, there were no clear differences between the research foci of authors based in Southeast Asia and those based outside of the region.

Studies in Southeast Asia provided sufficient contextual information in their assessment, contrary to the findings of Luederitz [41] in their global review. Studies provided detailed descriptions of the boundary of respective the study areas, population size, location of ecological structures and type of structures studied. Of the four ecosystem service domains, in Southeast Asia, regulating and cultural services were predominantly assessed (62% of all studies). The two most commonly assessed services were recreation, mental and physical health (n = 54) and moderation of extreme events services (n = 51). Parks were the most assessed blue-green structure, while there were few studies focused on coastal areas (n = 12), rooftops (n = 7) and green walls (n = 4).

Over half the studies examined multiple ecosystem services, within and across domains, and mostly at a landscape scale (i.e., city scale or larger). Studies also assessed multiple components of the cascade, although there is room for a more holistic research approach, as interactions, such as synergies and tradeoffs between services were rarely examined (4%). There was also a lack of studies with a multitemporal focus (11%). Process/mechanistic modelling was the dominant method of UES assessment [58–60], although valuations of services were lacking.

Stakeholder involvement was higher in studies that examined regulating and cultural services. Many studies that involved stakeholders also had social or planning research perspectives suggesting a strong applied focus on managing UES. There were few studies with a dominant governance perspective and this finding is not unique to Southeast Asia, as global reviews by Haase [28] and Luederitz [41] also report the lack of governance discourse on UES research. While the nature of UES research within the region may have some commonalities with its global counterpart, we highlight aspects of research that are specific to Southeast Asia, discussing considerations and opportunities for integrating UES in the region below.

4.2. Specificity of Research in Southeast Asia

The transferability of research may be limited due to the diverse characteristics of Southeast Asian countries—in particular economic power and government effectiveness (see Table 1 and Figure 2) [17]. Furthermore, even within countries there can be diversity in values. There is diversity in environmental conditions as well as the nature of urbanisation and cultural perspectives and values. For example, Hassan [61] highlighted substantial differences in wetland management preferences between urban and rural areas in Malaysia. While, in Singapore, contrary to popular assumptions around the desire for natural green spaces, some urban residents do not favour high conservation value vegetation and unmanaged secondary forests due to perceived wildlife threats and poor aesthetics [62,63].

If regional uniformity is assumed in how services are perceived and valued, the specific preferences and/or needs of minority groups may be overlooked when managing UES.

Considering that countries in Southeast Asia are renowned agro-industrial producers and exporters [64], provisioning services and services from agricultural landscapes (e.g., oil palm) were fairly understudied in the region. Although it is generally expected that highly urbanised areas are less likely to include productive areas, agricultural landscapes can be commonly found within the urban matrix of Southeast Asia [65,66]. While this adds to the uniqueness of urban-scapes in the region, the interactions between provisioning services and other service types, as well as implications for different stakeholders is yet to be fully understood.

Much remains to be learnt about biodiversity and UES in Southeast Asia. As Mammides [67] reported, despite most of the world's biodiversity being concentrated in the tropics and the imminent threats it faces, research on tropical conservation is largely underrepresented. In our review, the initial search string, which contained only UES related terms, returned only 48 relevant publications. It was only through expanding our search string with more general keywords that we were able to increase the number of publications. Like most other Global South regions, the underrepresentation of research could be attributed to Southeast Asia being data poor [68,69], which was noted in a number of studies [62,70,71].

Limitations in the quality, availability and access to data pose major challenges to UES research in the region. Databases and organisations that collect and provide open-access regional environmental data are few to none, compared to those in North America or Europe (e.g., United States Geological Survey, European Soil Data Centre, National Biodiversity Network, Biodiversity Information System for Europe, European Environment Agency). This was reported in several studies such as Balmford [72] who used global environmental data in their assessment of road networks in the Greater Mekong subregion, as finer scale, regional data was not available. Estoque [70] also utilised global ecosystem service values reported by Costanza [73] due to the limited availability of local data in Baguio, Philippines. Estoque [74] highlighted the need for available and accessible city-scale data across Philippines for conducting heat vulnerability assessments, while Belcher and Chisholm [62] reported that in Singapore LULC data is not publicly available. This limitation significantly affects research outputs as collection and generation of high-resolution regional data requires important human and time resources.

5. Conclusions: Research Needs to Move Forward

As Southeast Asian cities grow and the population density in urban areas rise, demand for ecosystem services will become increasing important [75]. The recognised importance of UES is also seen with the increased number of UES assessments in the region over the last decade (Figure 5b). Increasing urbanisation and urban sprawls in Southeast Asia often result in the loss of natural ecosystems due to the infrastructure demands of growing urban populations [17]. Conserving nature and supporting the provision of UES is often more cost effective and practical than restoring degraded ecosystems [76,77], so a worthwhile objective for cities in the region is avoiding the loss of natural ecosystems through the consideration of UES in planning.

The prevalence of certain services within UES research suggests some UES are considered to be more important than others, from a research perspective, in the Southeast Asian context. For instance, the preservation of cultural services such as recreation services (n = 54) and aesthetic appreciation (n = 44), which are strongly associated with green spaces [33,78], may be of high interest to urban residents, as these areas are being rapidly lost to high density development patterns, characteristic of urbanisation in Southeast Asian cities [79,80]. Similarly, climate regulating services (n = 44) appear to be valued for their role in reducing urban heat island effects, which is a common issue in the region's densely urbanised tropical cities, with high average temperatures [81–84]. These UES, which have been the focus of research, may be valued for their direct contribution to the wellbeing of urban populations and liveability of cities [33,85,86].

Recent research on the nexus between urban challenges, UES and Nature-based Solutions [17,87,88], highlights the role of UES in improving the liveability, resilience and sustainability of cities [15,89]. However, the future availability of UES is determined by land use decisions made in urban planning [90], which need to be supported by exhaustive assessments of UES. Thus, we highlight the following research areas, based on our review, that need further attention in order for UES research to wholly support land use planning and decision-making in the region (Figure 7).

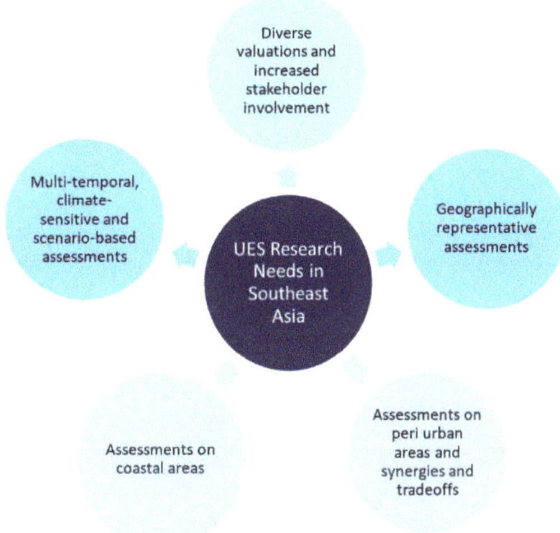

Figure 7. A diagram summarising the needs of UES research in Southeast Asia to support planning.

(i)   Geographically representative assessments

UES research is biased towards specific countries, regions and cities (Figure 2; Figure 4a). Aside from socioeconomic characteristics (Table 1), there are major biophysical differences between locations with maritime, continental and island climates in the region. This means that research conducted in Singapore may not necessarily be applicable in Cambodia or Myanmar. UES research needs to be context specific in order to purposefully address local needs. We therefore stress the need for a diverse range of UES assessments in countries that have very low representation of research such as Myanmar (n = 9), Laos (n = 1), Timor-Leste (n = 1) and Brunei (n = 0). The focus of assessments should also expand from megacities to secondary cities that are underrepresented (77% of studies concentrated on only four megacities—Singapore, Bangkok, Jakarta and Kuala Lumpur).

(ii)   Assessments on peri-urban areas and synergies and tradeoffs

As cities expand, peri-urban areas experience rapid land use change. However, only 15% of assessments examined peri-urban areas, consistent with Richards [91] and Wangai [75] that highlight peri-urban areas as being understudied globally. We encourage UES research in peri-urban areas, as these areas are where the intensity of development is the greatest and UES are being lost or degraded, and therefore where planning is mostly urgently needed.

It is especially important to investigate the synergies and tradeoffs of UES in urban and peri-urban areas so the consequence of planning decisions can be considered systematically [92,93]. Although 58% of studies in the region assessed multiple ecosystem services, only 4% dealt with synergies and tradeoffs. A clear understanding of the complex interactions between UES, as well as UES and land use management, is particularly important in rapidly developing peri-urban areas.

We also highlight the need for synergy and tradeoffs assessments between urbanisation and provisioning services. The spatial expansion of cities has negative impacts on urban/peri-urban agriculture [94], which is commonplace in Southeast Asia [66,95]. Given the importance of agricultural production to local livelihood in the region [96], the sustainability and multifunctional capacity of urban agricultural landscapes needs to be better understood [97,98]. Careful management of land use as peri-urban areas develop can yield more sustainable UES provision, than attempts to retrofit restoration efforts in the future.

(iii)   Assessments on coastal areas

Many of Southeast Asia's densely populated cities are located along the coastlines (e.g., Greater Jakarta, Singapore, Ho Chi Minh, Bangkok, Manila), yet few studies examined UES in coastal areas. Coastal cities are particularly vulnerable to coastal and riverine floods, coastal erosion, storm surges, monsoons and tsunamis [99,100], all of which bring adverse health risks to the urban population [101,102]. Moreover, many of these extreme events are expected to increase in frequency in Southeast Asia because of climate change effects [14,99]. Thus, we bring to attention the exigency of assessments of coastal structures as a Nature-based Solution in coastal cities. Research should also focus on opportunities to support the resilience of urban communities through the sustainable provision of UES [103].

(iv)   Multi-temporal, climate-sensitive and scenario-based assessments

Few studies (11%) have conducted temporal assessments of UES, which are key to understanding changes in service provision and demand [104]. This is challenging in practice as there is limited information on how UES change over time and/or under different future scenarios [57,105]. Moreover, Southeast Asian cities are seen to be highly vulnerable to climate change effects [17,106,107], yet few studies have examined the link between UES and climate resilience (n = 5). Assessments of changes in UES can be used to identify areas vulnerable to weather-related disaster risks and/or support decisions on appropriate land use management strategies [14]. Our review highlights a pressing need for multitemporal and/or scenario-based research on the resilience of UES provision. Research should also address the increased risk of diseases in tropical ecosystems due to the effects of climate change [108,109], as well as the consequent impacts to UES, particularly provisioning services [110,111].

(v)   Diverse valuations and increased stakeholder involvement

Literature supporting the valuation of ecosystem services is abundant [30,112–114], with recent research emphasising diverse perspectives in valuations through value pluralism [49]. However, our review found that only 23% of studies in the region conducted valuations. Valuations support decision-making by providing explicit quantification of UES demand, which can be in monetary or non-monetary terms [46]. The involvement of stakeholders (44%) in UES assessments can support valuations by identifying context-specific demands and preferences of the people appropriating the services [41,49].

In line with TEEB and the Intergovernmental Panel on Biodiversity and Ecosystem Services (IPBES) [49,115,116], we urge research to incorporate diverse valuations, monetary and non-monetary, as well as increase stakeholder involvement in UES assessments. Although contentious [117], the comprehensive representation of UES through valuations has been proven to be effective in influencing decision-makers towards planning agendas [118]. This is because valuations can be used as a tool to demonstrate the cost of restoring ecosystems or the critical importance of alternative land use options objectively to decision-makers [119–121]. As the invisibility of nature in economic choices often drives its depletion [53], valuation of UES can encourage more transparent assessments of tradeoffs to support the planning of sustainable cities.

Supplementary Materials: The following are available online at https://www.mdpi.com/2073-445X/10/1/40/s1, Search strings for Scopus and Web of Knowledge databases; Table S1: Bibliography of the 149 studies analysed in this review; Table S2: Review categories, description and classification method; Table S3: Ecological structures and definitions.

Author Contributions: This review is part of a PhD research by K.T.L. The idea for the paper was conceived by K.T.L., A.M.L. and C.N.G., K.T.L. performed the literature search, data analysis and drafted the original manuscript. A.M.L., C.N.G., P.H., R.S. and B.A. provided critical feedback and edited the manuscript. All authors have read and agreed to the published version of the manuscript.

Funding: This work is part of a PhD research that is funded by the Faculty of Science and Engineering and the Landscape Ecology and Conservation Laboratory, School of Environmental and Geographical Sciences at the University of Nottingham Malaysia. We are also thankful for the funding received from the International Research Collaboration Award from the Research Knowledge and Exchange Hub, University of Nottingham Malaysia, Singapore's National Research Foundation (NRFF12-2020-0009) and the Putra Grant-Putra Young Initiative (IPM) from Universiti Putra Malaysia (Grant No: GP-IPM/2018/9637800).

Institutional Review Board Statement: Not applicable.

Informed Consent Statement: Not applicable.

Data Availability Statement: The data presented in this study are available within the article and supplementary materials.

Acknowledgments: We thank the reviewers for their kind and constructive feedback which has helped improve the manuscript.

Conflicts of Interest: The authors declare no conflict of interest.

References

1. United Nations. *World Urbanization Prospects: The 2014 Revision, Highlights (ST/ESA/SER.A/352)*; Department of Economic and Social Affairs, Population Division: New York, NY, USA, 2014.
2. Nagendra, H.; Bai, X.; Brondizio, E.S.; Lwasa, S. The urban south and the predicament of global sustainability. *Nat. Sustain.* **2018**, *1*, 341–349. [CrossRef]
3. Yuen, B.; Kong, L. Climate change and urban planning in Southeast Asia. *SAPIENS* **2009**, 2. Available online: http://journals.openedition.org/sapiens/881 (accessed on 10 December 2020).
4. Jacobs, A.; Appleyard, D. Toward an Urban Design Manifesto Toward an Urban Design Manifesto. *J. Am. Plan. Assoc.* **1987**, *53*, 37–41. [CrossRef]
5. Jones, G.W. Urbanisation and development in Southeast Asia. *Malaysian J. Econ. Stud.* **2014**, *51*, 103–120.
6. Mahmoudi, M.; Ahmad, F.; Abbasi, B. Livable streets: The effects of physical problems on the quality and livability of Kuala Lumpur streets. *Cities* **2015**, *43*, 104–114. [CrossRef]
7. Morillas, J.B.; Gozalo, G.R.; González, D.M.; Moraga, P.A.; Vílchez-Gómez, R. Noise Pollution and Urban Planning. *Curr. Pollut. Rep.* **2018**, *4*, 208–219. [CrossRef]
8. Harun, Z.; Reda, E.; Abdulrazzaq, A.; Abbas, A.A.; Yusup, Y.; Zaki, S.A. Urban heat island in the modern tropical Kuala Lumpur: Comparative weight of the different parameters. *Alex. Eng. J.* **2020**, *59*, 4475–4489. [CrossRef]
9. Jusuf, S.K.; Wong, N.H.; Hagen, E.; Anggoro, R.; Hong, Y. The influence of land use on the urban heat island in Singapore. *Habitat Int.* **2007**, *31*, 232–242. [CrossRef]
10. Laeni, N.; Brink, M.V.D.; Arts, J. Is Bangkok becoming more resilient to flooding? A framing analysis of Bangkok's flood resilience policy combining insights from both insiders and outsiders. *Cities* **2019**, *90*, 157–167. [CrossRef]
11. Padawangi, R.; Douglass, M. Water, Water Everywhere: Toward Participatory Solutions to Chronic Urban Flooding in Jakarta. *Pac. Aff.* **2015**, *88*, 517–550. [CrossRef]
12. Li, L.; Cao, R.; Wei, K.; Wang, W.; Chen, L. Adapting climate change challenge: A new vulnerability assessment framework from the global perspective. *J. Clean. Prod.* **2019**, *217*, 216–224. [CrossRef]
13. ADB. *The Economics of Climate Change in Southeast Asia: A Regional Review*; Asian Development Bank: Mandaluyong, Philippines, 2009.
14. Yusuf, A.A.; Francisco, H. *Hotspots! Mapping Climate Change Vulnerability in Southeast Asia*; Worldfish Philippine Country Office: Laguna, Philippines, 2010.
15. Savage, V.R. Ecology matters: Sustainable development in Southeast Asia. *Sustain. Sci.* **2006**, *1*, 37–63. [CrossRef]
16. Arfanuzzaman, M.; Dahiya, B. Sustainable urbanization in Southeast Asia and beyond: Challenges of population growth, land use change, and environmental health. *Growth Chang.* **2019**, *50*, 725–744. [CrossRef]

17. Lechner, A.M.; Gomes, R.L.; Rodrigues, L.; Ashfold, M.J.; Selvam, S.B.; Wong, E.P.; Raymond, C.M.; Zieritz, A.; Sing, K.W.; Moug, P.; et al. Challenges and considerations of applying nature-based solutions in low- and middle-income countries in Southeast and East Asia. *Blue Green Syst.* **2020**, *2*, 331–351. [CrossRef]
18. ASEANStats. Indicators ASEAN Member States 2020. Available online: https://data.aseanstats.org/ (accessed on 2 December 2020).
19. UNdata. Statistics. 2020. Available online: https://www.data.un.org (accessed on 1 December 2020).
20. Asian Development Bank. Key Indicators Database. 2020. Available online: https://kidb.adb.org/ (accessed on 2 December 2020).
21. WorldBank. DataBank. 2020. Available online: https://databank.worldbank.org/databases/page/1/orderby/popularity/direction/desc?qterm=grossnationalincomerank (accessed on 2 December 2020).
22. WGI. Worldwide Governance Indicators, Interactive Data Access, 2020. Available online: https://info.worldbank.org/governance/wgi/Home/Reports (accessed on 2 December 2020).
23. El-Baghdadi, O.; Desha, C. Conceptualising a biophilic services model for urban areas. *Urban For. Urban Green.* **2017**, *27*, 399–408. [CrossRef]
24. Xue, F.; Gou, Z.; Lau, S. The green open space development model and associated use behaviors in dense urban settings: Lessons from Hong Kong and Singapore. *Urban Des. Int.* **2017**, *22*, 287–302. [CrossRef]
25. Kabisch, N.; Qureshi, S.; Haase, D. Human–environment interactions in urban green spaces-A systematic review of contemporary issues and prospects for future research. *Environ. Impact Assess. Rev.* **2015**, *50*, 25–34. [CrossRef]
26. Baharuddin, Z.M.; Rusli, N.; Ramli, L.; Othman, R.; Yaman, M. The Diversity of Birds and Frogs Species at Perdana Botanical Lake Garden, Kuala Lumpur, Malaysia. *Adv. Sci. Lett.* **2017**, *23*, 6256–6260. [CrossRef]
27. Cortinovis, C.; Geneletti, D. A performance-based planning approach integrating supply and demand of urban ecosystem services. *Landsc. Urban Plan.* **2020**, *201*, 103842. [CrossRef]
28. Haase, D.; Larondelle, N.; Andersson, E.; Artmann, M.; Borgström, S.; Breuste, J.; Gomez-Baggethun, E.; Gren, Å.; Hamstead, Z.; Hansen, R.; et al. A Quantitative Review of Urban Ecosystem Service Assessments: Concepts, Models, and Implementation. *Ambio* **2014**, *43*, 413–433. [CrossRef]
29. Kanniah, K.D.; Siong, H.C. Tree canopy cover and its potential to reduce CO_2 in South of Peninsular Malaysia. *Chem. Eng. Trans.* **2018**, *63*, 13–18.
30. Keeler, B.L.; Hamel, P.; McPhearson, T.; Hamann, M.H.; Donahue, M.L.; Prado, K.A.M.; Arkema, K.K.; Bratman, G.N.; Brauman, K.A.; Finlay, J.C.; et al. Social-ecological and technological factors moderate the value of urban nature. *Nat. Sustain.* **2019**, *2*, 29–38. [CrossRef]
31. Bolund, P.; Hunhammar, S. Ecosystem services in urban areas. *Ecol. Econ.* **1999**, *29*, 293–301. [CrossRef]
32. Gomez-Baggethun, E.; Gren, Å.; Barton, D.N.; Langemeyer, J.; McPhearson, T.; O'Farrell, P.; Andersson, E.; Hamstead, Z.; Kremer, P. Urban Ecosystem Services. In *Urbanization, Biodiversity and Ecosystem Services: Challenges and Opportunities*; Springer Science and Business Media LLC: Dordrecht, The Netherlands, 2013; pp. 175–251.
33. Braat, L.C.; de Groot, R. The ecosystem services agenda:bridging the worlds of natural science and economics, conservation and development, and public and private policy. *Ecosyst. Serv.* **2012**, *1*, 4–15. [CrossRef]
34. Burls, A. People and green spaces: Promoting public health and mental well-being through ecotherapy. *J. Public Ment. Health* **2007**, *6*, 24–39. [CrossRef]
35. Gascon, M.; Triguero-Mas, M.; Martínez, D.; Dadvand, P.; Forns, J.; Plasència, A.; Nieuwenhuijsen, M.J. Mental Health Benefits of Long-Term Exposure to Residential Green and Blue Spaces: A Systematic Review. *Int. J. Environ. Res. Public Health* **2015**, *12*, 4354–4379. [CrossRef]
36. Müller-Riemenschneider, F.; Petrunoff, N.; Sia, A.; Ramiah, A.; Ng, A.; Han, J.; Wong, M.; Tai, B.C.; Uijtdewilligen, L. Prescribing Physical Activity in Parks to Improve Health and Wellbeing: Protocol of the Park Prescription Randomized Controlled Trial. *Int. J. Environ. Res. Public Health* **2018**, *15*, 1154. [CrossRef] [PubMed]
37. Cilliers, S. Social Aspects of Urban Biodiversity: An Overview. In *Urban Biodiversity and Design*; Müller, N., Werner, P., Kelcey, J.G., Eds.; Zoological Society of London: London, UK, 2010; p. 626.
38. Nath, T.K.; Han, S.S.Z.; Lechner, A.M. Urban green space and well-being in Kuala Lumpur, Malaysia. *Urban For. Urban Green.* **2018**, *36*, 34–41. [CrossRef]
39. Sandifer, P.A.; Sutton-Grier, A.E.; Ward, B.P. Exploring connections among nature, biodiversity, ecosystem services, and human health and well-being: Opportunities to enhance health and biodiversity conservation. *Ecosyst. Serv.* **2015**, *12*, 1–15. [CrossRef]
40. Potschin, M.B.; Haines-Young, R.H. Ecosystem services. *Prog. Phys. Geogr. Earth Environ.* **2011**, *35*, 575–594. [CrossRef]
41. Luederitz, C.; Brink, E.; Gralla, F.; Hermelingmeier, V.; Meyer, M.; Niven, L.; Panzer, L.; Partelow, S.; Rau, A.-L.; Sasaki, R.; et al. A review of urban ecosystem services: Six key challenges for future research. *Ecosyst. Serv.* **2015**, *14*, 98–112. [CrossRef]
42. De Groot, R.S.; Wilson, M.A.; Boumans, R.M. A typology for the classification, description and valuation of ecosystem functions, goods and services. *Ecol. Econ.* **2002**, *41*, 393–408. [CrossRef]
43. Martínez, A.J.C.; García-Llorente, M.; Martín-López, B.; Palomo, I.; Iniesta-Arandia, I. Multidimensional approaches in ecosystem services assessment. *Earth Obs. Ecosyst. Serv.* **2013**, 441–468.
44. Bennett, E.M.; Peterson, G.D.; Gordon, L.J. Understanding relationships among multiple ecosystem services. *Ecol. Lett.* **2009**, *12*, 1394–1404. [CrossRef]

45. La Notte, A.; D'Amato, D.; Mäkinen, H.; Paracchini, M.L.; Liquete, C.; Egoh, B.; Geneletti, D.; Crossman, N.D. Ecosystem services classification: A systems ecology perspective of the cascade framework. *Ecol. Indic.* **2017**, *74*, 392–402. [CrossRef] [PubMed]

46. Conte, M.N. Valuing Ecosystem Services. *Encycl. Biodivers.* **2013**, *7*, 314–326. [CrossRef]

47. Costanza, R.; de Groot, R.; Braat, L.; Kubiszewski, I.; Fioramonti, L.; Sutton, P.; Farber, S.; Grasso, M. Twenty years of ecosystem services: How far have we come and how far do we still need to go? *Ecosyst. Serv.* **2017**, *28*, 1–16. [CrossRef]

48. Gomez-Baggethun, E.; Gren, Å.; Barton, D.N.; Langemeyer, J.; McPhearson, T.; O'Farrell, P.; Andersson, E.; Hamstead, Z.; Kremer, P. Urban Ecosystem Services. *Urban. Biodivers. Ecosyst. Serv. Chall. Oppor.* **2013**, 175–251.

49. Pascual, U.; Balvanera, P.; Díaz, S.; Pataki, G.; Roth, E.; Stenseke, M.; Watson, R.T.; Dessane, E.B.; Islar, M.; Kelemen, E.; et al. Valuing nature's contributions to people: The IPBES approach. *Curr. Opin. Environ. Sustain.* **2017**, *26–27*, 7–16. [CrossRef]

50. Brown, G.; Stricklandmunro, J.K.; Kobryn, H.; Moore, S.A. Stakeholder analysis for marine conservation planning using public participation GIS. *Appl. Geogr.* **2016**, *67*, 77–93. [CrossRef]

51. Zoderer, B.M.; Tasser, E.; Carver, S.; Tappeiner, U. An integrated method for the mapping of landscape preferences at the regional scale. *Ecol. Indic.* **2019**, *106*, 105430. [CrossRef]

52. TEEB. *The Economics of Ecosystems and Biodiversity Ecological and Economic Foundations*; Earthscan: London, UK; Washington, DC, USA, 2010.

53. TEEB. TEEB Manual for Cities: Ecosystem services in Urban Management. TEEB Man. *Cities Ecosyst. Serv. Urban Manag.* **2011**, 2–41.

54. TEEB. The Economics of Ecosystems and Biodiversity Challenges and responses. *Nat. Balanc. Econ. Biodivers.* **2014**, 135–152.

55. Pierce, J.R.; Barton, M.A.; Tan, M.M.J.; Oertel, G.; Halder, M.D.; Lopez-Guijosa, P.A.; Nuttall, R. Actions, indicators, and outputs in urban biodiversity plans: A multinational analysis of city practice. *PLoS ONE* **2020**, *15*, e0235773. [CrossRef]

56. Warner, K.; Zommers, Z.; Wreford, A.; Hurlbert, M.; Viner, D.; Scantlan, J.; Halsey, K.; Halsey, K.; Tamang, C. Characteristics of Transformational Adaptation in Climate-Land-Society Interactions. *Sustainability* **2019**, *11*, 356. [CrossRef]

57. Remondi, F.; Burlando, P.; Vollmer, D. Exploring the hydrological impact of increasing urbanisation on a tropical river catchment of the metropolitan Jakarta, Indonesia. *Sustain. Cities Soc.* **2016**, *20*, 210–221. [CrossRef]

58. Achmad, A.; Ramli, I.; Irwansyah, M. The impacts of land use and cover changes on ecosystem services value in urban highland areas. *IOP Conf. Series Earth Environ. Sci.* **2020**, *447*. [CrossRef]

59. Nguyen, L.D.; Nguyen, C.T.; Le, H.S.; Tran, B.Q. Mangrove Mapping and Above-Ground Biomass Change Detection using Satellite Images in Coastal Areas of Thai Binh Province, Vietnam. *For. Soc.* **2019**, *3*, 248–261. [CrossRef]

60. Srichaichana, J.; Trisurat, Y.; Ongsomwang, S. Land Use and Land Cover Scenarios for Optimum Water Yield and Sediment Retention Ecosystem Services in Klong U-Tapao Watershed, Songkhla, Thailand. *Sustainability* **2019**, *11*, 2895. [CrossRef]

61. Hassan, S.; Olsen, S.B.; Thorsen, B.J. Urban-rural divides in preferences for wetland conservation in Malaysia. *Land Use Policy* **2019**, *84*, 226–237. [CrossRef]

62. Belcher, R.N.; Chisholm, R.A. Tropical Vegetation and Residential Property Value: A Hedonic Pricing Analysis in Singapore. *Ecol. Econ.* **2018**, *149*, 149–159. [CrossRef]

63. Richards, D.; Fung, T.K.; Leong, R.A.T.; Sachidhanandam, U.; Drillet, Z.; Edwards, P.J. Demographic biases in engagement with nature in a tropical Asian city. *PLoS ONE* **2020**, *15*, e0231576. [CrossRef]

64. Kontgis, C.; Schneider, A.; Ozdogan, M.; Kucharik, C.; Tri, V.P.D.; Duc, N.H.; Schatz, J. Climate change impacts on rice productivity in the Mekong River Delta. *Appl. Geogr.* **2019**, *102*, 71–83. [CrossRef]

65. Budidarsono, S.; Susanti, A.; Zoomers, E.B. *Biofuels-Economy, Environment and Sustainability*; Zhen, F., Ed.; IntechOpen: London, UK, 2013. Available online: https://www.intechopen.com/books/biofuels-economy-environment-and-sustainability/oil-palm-plantations-in-indonesia-the-implications-for-migration-settlement-resettlement-and-local-e (accessed on 10 December 2020). [CrossRef]

66. Shevade, V.S.; Loboda, T.V. Oil palm plantations in Peninsular Malaysia: Determinants and constraints on expansion. *PLoS ONE* **2019**, *14*, e0210628. [CrossRef]

67. Mammides, C.; Goodale, U.M.; Corlett, R.T.; Chen, J.; Bawa, K.S.; Hariya, H.; Jarrad, F.; Primack, R.B.; Ewing, H.; Xia, X.; et al. Increasing geographic diversity in the international conservation literature: A stalled process? *Biol. Conserv.* **2016**, *198*, 78–83. [CrossRef]

68. Perez, L.P.; Santos, R.; De Almeida, G.M.J.A.; Carvalho, G.C. Spatial data in the Global South: A case study of alternative land management tools for cities with limited resources. In Proceedings of the 2017 IEEE Global Humanitarian Technology Conference (GHTC), San Jose, CA, USA, 19–22 October 2017; pp. 1–10.

69. Karnad, D.; Martin, K.S. Assembling marine spatial planning in the global south: International agencies and the fate of fishing communities in India. *Marit. Stud.* **2020**, *19*, 375–387. [CrossRef]

70. Estoque, R.C.; Murayama, Y. Examining the potential impact of land use/cover changes on the ecosystem services of Baguio city, the Philippines: A scenario-based analysis. *Appl. Geogr.* **2012**, *35*, 316–326. [CrossRef]

71. Estoque, R.C.; Murayama, Y. Quantifying landscape pattern and ecosystem service value changes in four rapidly urbanizing hill stations of Southeast Asia. *Landsc. Ecol.* **2016**, *31*, 1481–1507. [CrossRef]

72. Balmford, A.; Chen, H.; Phalan, B.; Wang, M.; O'Connell, C.; Tayleur, C.; Xu, J.-C. Getting Road Expansion on the Right Track: A Framework for Smart Infrastructure Planning in the Mekong. *PLoS Biol.* **2016**, *14*, e2000266. [CrossRef]

73. Costanza, R.; d'Arge, R.; de Groot, R.; Farber, S.; Grasso, M.; Hannon, B.; Limburg, K.; Naeem, S.; O'Neill, R.V.; Paruelo, J.; et al. The value of the world's ecosystem services and natural capital. *Nature* **1997**, *387*, 253–260. [CrossRef]

74. Estoque, R.C.; Ooba, M.; Seposo, X.T.; Togawa, T.; Hijioka, Y.; Takahashi, K.; Nakamura, S. Heat health risk assessment in Philippine cities using remotely sensed data and social-ecological indicators. *Nat. Commun.* **2020**, *11*, 1–12. [CrossRef] [PubMed]

75. Wangai, P.W.; Burkhard, B.; Müller, F. A review of studies on ecosystem services in Africa. *Int. J. Sustain. Built Environ.* **2016**, *5*, 225–245. [CrossRef]

76. Holl, K.D.; Howarth, R. Paying for Restoration. *Restor. Ecol.* **2000**, *8*, 260–267. [CrossRef]

77. Loomis, J.; Kent, P.; Strange, L.; Fausch, K.; Covich, A. Measuring the total economic value of restoring ecosystem services in an impaired river basin: Results from a contingent valuation survey. *Ecol. Econ.* **2000**, *33*, 103–117. [CrossRef]

78. Liu, H.; Remme, R.P.; Hamel, P.; Nong, H.; Ren, H. Supply and demand assessment of urban recreation service and its implication for greenspace planning-A case study on Guangzhou. *Landsc. Urban Plan.* **2020**, *203*, 103898. [CrossRef]

79. Thiagarajah, J.; Wong, S.K.M.; Richards, D.R.; Friess, D.A. Historical and contemporary cultural ecosystem service values in the rapidly urbanizing city state of Singapore. *Ambio* **2015**, *44*, 666–677. [CrossRef]

80. Friess, D.A. Singapore as a long-term case study for tropical urban ecosystem services. *Urban Ecosyst.* **2016**, *20*, 277–291. [CrossRef]

81. Buyadi, S.; Wan Mohd, W.; Misni, A. Quantifying Green Space Cooling Effects on the Urban Microclimate using Remote Sensing and GIS Techniques. In Proceedings of the XXV International Federation of Surveyors, Kuala Lumpur, Malaysia, 16–21 June 2014; pp. 1–16.

82. Richards, D.; Edwards, P.J. Quantifying street tree regulating ecosystem services using Google Street View. *Ecol. Indic.* **2017**, *77*, 31–40. [CrossRef]

83. Gunawardena, K.; Wells, M.; Kershaw, T. Utilising green and bluespace to mitigate urban heat island intensity. *Sci. Total Environ.* **2017**, *584*, 1040–1055. [CrossRef]

84. Heng, S.L.; Chow, W.T. How 'hot' is too hot? Evaluating acceptable outdoor thermal comfort ranges in an equatorial urban park. *Int. J. Biometeorol.* **2019**, *63*, 801–816. [CrossRef]

85. Beck, H. Linking the quality of public spaces to quality of life. *J. Place Manag. Dev.* **2009**, *2*, 240–248. [CrossRef]

86. Yap, K.S.; Thuzar, M. *Urbanization in Southeast Asia: Issues & Impacts*; Institute of Southeast Asian Studies: Singapore, 2012.

87. Lafortezza, R.; Chen, J.; van den Bosch, C.K.; Randrup, T.B. Nature-based solutions for resilient landscapes and cities. *Environ. Res.* **2018**, *165*, 431–441. [CrossRef] [PubMed]

88. Almenar, J.B.; Elliot, T.; Rugani, B.; Philippe, B.; Gutierrez, T.N.; Sonnemann, G.; Geneletti, D. Nexus between nature-based solutions, ecosystem services and urban challenges. *Land Use Policy* **2021**, *100*, 104898. [CrossRef]

89. United Nations. *The 2030 Agenda for Sustainable Development, A/RES/70/1*; United Nations: New York, NY, USA, 2015; pp. 13–14.

90. Geneletti, D.; Cortinovis, C.; Zardo, L.; Esmail, B.A. Planning for Ecosystem Services in Cities. In *Environmental Education and Ecotourism*; Springer International Publishing: Berlin/Heidelberg, Germany, 2020.

91. Richards, D.; Masoudi, M.; Oh, R.R.Y.; Yando, E.S.; Zhang, J.; Friess, D.A.; Grêt-Regamey, A.; Tan, P.Y.; Edwards, P. Global Variation in Climate, Human Development, and Population Density Has Implications for Urban Ecosystem Services. *Sustainability* **2019**, *11*, 6200. [CrossRef]

92. Mokondoko, P.; Manson, R.H.; Ricketts, T.H.; Geissert, D. Spatial analysis of ecosystem service relationships to improve targeting of payments for hydrological services. *PLoS ONE* **2018**, *13*, e0192560. [CrossRef]

93. Holt, A.R.; Mears, M.; Maltby, L.; Warren, P.H. Understanding spatial patterns in the production of multiple urban ecosystem services. *Ecosyst. Serv.* **2015**, *16*, 33–46. [CrossRef]

94. Artmann, M.; Sartison, K. The Role of Urban Agriculture as a Nature-Based Solution: A Review for Developing a Systemic Assessment Framework. *Sustainability* **2018**, *10*, 1937. [CrossRef]

95. Arif, S.; Ae, A.; Hezri, A.A. From Forest Landscape to Agricultural Landscape in the Developing Tropical Country of Malaysia: Pattern, Process, and Their Significance on Policy. *Environ. Manag.* **2008**, *42*, 907–917.

96. Gasparatos, A.; Helen. Ecosystem Services Provision from Urban Farms in a Secondary City of Myanmar, Pyin Oo Lwin. *Agriculture* **2020**, *10*, 140. [CrossRef]

97. Aerts, R.; Dewaelheyns, V.; Achten, W. Potential ecosystem services of urban agriculture: A review. *PeerJ Prepr.* **2016**, *4*, e2286v1.

98. Dressler, W.H.; Wilson, D.; Clendenning, J.; Cramb, R.A.; Keenan, R.; Mahanty, S.; Bruun, T.B.; Mertz, O.; Lasco, R.D. The impact of swidden decline on livelihoods and ecosystem services in Southeast Asia: A review of the evidence from 1990 to 2015. *Ambio* **2017**, *46*, 291–310. [CrossRef] [PubMed]

99. Eckstein, D.; Künzel, V.S.; Schäfer, L. Global Climate Risk Index 2020 Germanwatch e.V. 2020. Available online: https://www.germanwatch.org/en/17307 (accessed on 10 December 2020).

100. Hallegatte, S.; Green, C.; Nicholls, R.J.; Corfee-Morlot, J. Future flood losses in major coastal cities. *Nat. Clim. Chang.* **2013**, *3*, 802–806. [CrossRef]

101. Wells, J.A.; Wilson, K.A.; Abram, N.K.; Nunn, M.; Gaveau, D.L.A.; Runting, R.K.; Tarniati, N.; Mengersen, K.L.; Meijaard, E. Rising floodwaters: Mapping impacts and perceptions of flooding in Indonesian Borneo. *Environ. Res. Lett.* **2016**, *11*, 064016. [CrossRef]

102. Arifin, H.S.; Nakagoshi, N. Landscape ecology and urban biodiversity in tropical Indonesian cities. *Landsc. Ecol. Eng.* **2011**, *7*, 33–43. [CrossRef]

103. Uy, N.; Shaw, R. Ecosystem resilience and community values: Implications to ecosystem-based adaptation. *J. Disaster Res.* **2013**, *8*, 201–202.

104. Willemen, L. It's about time: Advancing spatial analyses of ecosystem services and their application. *Ecosyst. Serv.* **2020**, *44*, 101125. [CrossRef]

105. McDougall, R.; Kristiansen, P.; Rader, R. Small-scale urban agriculture results in high yields but requires judicious management of inputs to achieve sustainability. *Proc. Natl. Acad. Sci. USA* **2019**, *116*, 129–134. [CrossRef]

106. Matthews, T.; Wilby, R.L.; Murphy, C. Communicating the deadly consequences of global warming for human heat stress. *Proc. Natl. Acad. Sci. USA* **2017**, *114*, 3861–3866. [CrossRef]

107. Mora, C.; Dousset, B.; Caldwell, I.R.; Powell, F.E.; Geronimo, R.C.; Bielecki, C.R.; Counsell, C.W.W.; Dietrich, B.S.; Johnston, E.T.; Louis, L.V.; et al. Global risk of deadly heat. *Nat. Clim. Chang.* **2017**, *7*, 501–506. [CrossRef]

108. Bowen, K.J.; Ebi, K.L. Health risks of climate change in the World Health Organization South-East Asia Region. *WHO South-East Asia J. Public Health* **2017**, *6*, 3–8. [CrossRef]

109. Saulnier, D.D.; Ribacke, K.B.; Von Schreeb, J. No Calm After the Storm: A Systematic Review of Human Health Following Flood and Storm Disasters. *Prehospital Disaster Med.* **2017**, *32*, 568–579. [CrossRef]

110. Bito-onon, J.B. Climate risk vulnerability assessment: Basis for decision making support for the agriculture sector in the province of Iloilo. *Int. J. Innov. Creat. Chang.* **2020**, *13*, 186–202.

111. Van Noordwijk, M.; Kim, Y.-S.; Leimona, B.; Hairiah, K.; Fisher, L.A. Metrics of water security, adaptive capacity, and agroforestry in Indonesia. *Curr. Opin. Environ. Sustain.* **2016**, *21*, 1–8. [CrossRef]

112. Sun, Y.; Xie, S.; Zhao, S.Q. Valuing urban green spaces in mitigating climate change: A city-wide estimate of aboveground carbon stored in urban green spaces of China's Capital. *Glob. Chang. Biol.* **2019**, *25*, 1717–1732. [CrossRef]

113. Hermes, J.; Tabrizian, P.; Burkhard, B.; Plieninger, T.; Fagerholm, N.; Von Haaren, C.; Albert, C. Assessment and valuation of recreational ecosystem services of landscapes. *Ecosyst. Serv.* **2018**, *31*, 289–295. [CrossRef]

114. Gómez-Baggethun, E.; Barton, D.N. Classifying and valuing ecosystem services for urban planning. *Ecol. Econ.* **2013**, *86*, 235–245. [CrossRef]

115. TEEB. *A Quick Guide to the Economics of Ecosystems and Biodiversity for Local and Regional Policy Makers*; Earthscan: London, UK, 2010.

116. Díaz, S.; Pascual, U.; Stenseke, M.; Martín-López, B.; Watson, R.T.; Molnár, Z.; Hill, R.; Chan, K.M.A.; Baste, I.A.; Brauman, K.A.; et al. Assessing nature's contributions to people: Recognizing culture, and diverse sources of knowledge, can improve assessments. *Science* **2018**, *359*, 270–272. [CrossRef]

117. Heikkinen, A.; Mäkelä, H.; Kujala, J.; Nieminen, J.; Jokinen, A.; Rekola, H. 7 Urban ecosystem services and stakeholders: Towards a sustainable capability approach. In *Strongly Sustainable Societies: Organising Human Activities on a Hot and Full Earth*; Bonnedahl, K.J., Heikkurinen, P., Eds.; Routledge: London, UK; New York, NY, USA, 2019; pp. 116–133.

118. Kenter, J.O.; Hyde, T.; Christie, M.; Fazey, I. The importance of deliberation in valuing ecosystem services in developing countries—Evidence from the Solomon Islands. *Glob. Environ. Chang.* **2011**, *21*, 505–521. [CrossRef]

119. Griffin, R.G.; Chaumont, N.; Denu, D.; Guerry, A.D.; Kim, C.-K.; Ruckelshaus, M.H. Incorporating the visibility of coastal energy infrastructure into multi-criteria siting decisions. *Mar. Policy* **2015**, *62*, 218–223. [CrossRef]

120. Brown, G.; Donovan, S.M.; Pullar, D.; Pocewicz, A.; Toohey, R.; Ballesteros-Lopez, R. An empirical evaluation of workshop versus survey PPGIS methods. *Appl. Geogr.* **2014**, *48*, 42–51. [CrossRef]

121. Rambonilaza, T.; Neang, M. Exploring the potential of local market in remunerating water ecosystem services in Cambodia: An application for endogenous attribute non-attendance modelling. *Water Resour. Econ.* **2019**, *25*, 14–26. [CrossRef]

land

Article

Governance, Nature's Contributions to People, and Investing in Conservation Influence the Valuation of Urban Green Areas

Alexandra Pineda-Guerrero [1], Francisco J. Escobedo [2,*] and Fernando Carriazo [3]

[1] Nelson Institute for Environmental Studies, University of Wisconsin Madison,550 North Park Street, Madison, WI 53706, USA; Pinedaguerre@wisc.edu

[2] US Forest Service, Pacific Southwest Research Station, 4955 Canyon Crest Dr., Riverside, CA 92507, USA

[3] Urban Management and Development Program, School of Political Science, Government and International Relations, Universidad del Rosario, Cra 6 No. 12C-13, Bogotá 111711, Colombia; fernando.carriazo@urosario.edu.co

* Correspondence: Francisco.Escobedo@usda.gov; Tel.: +1-951-680-1544

Abstract: There is little information concerning how people in the Global South perceive the benefits and costs associated with urban green areas. There is even less information on how governance influences the way people value these highly complex socio-ecological systems. We used semi-structured surveys, statistical analyses, and econometrics to explore the perceptions of users regarding governance and the benefits and costs, or Ecosystem Services (ES) and Ecosystem Disservices (ED), provided by Neotropical green areas and their willingness to invest, or not, for their conservation. The study area was the El Salitre sub-watershed in Bogota, Colombia, and 10 different sites representative of its wetlands, parks, green areas, and socioeconomic contexts. Using a context-specific approach and methods, we identified the most important benefits and costs of green areas and the influence of governance on how people valued these. Our modelling shows that air quality and biodiversity were highly important benefits, while water regulation was the least important; despite the sub-watershed's acute problems with stormwater runoff. In terms of costs, the feeling of insecurity due to crime was related to poor levels of maintenance and infrastructure in the studied green areas. Perceived transparency, corruption, and performance of government institutions influenced people's Unwillingness to Invest (UTI) in green space conservation. Results show that socioeconomic backgrounds, government performance, and environmental education will play a role in the value or importance people place on the benefits, costs, and UTI in conservation efforts in urban green areas. Similarly, care is warranted when directly applying frameworks and typologies developed in high income countries (i.e., ES) to the unique realities of cities in the Global South. Accordingly, alternative frameworks such as Nature's Contributions to People is promising.

Keywords: urban biodiversity; urban watersheds; Bogota Colombia; corruption; Unwillingness to Invest

check for
updates

Citation: Pineda-Guerrero, A.; Escobedo, F.J.; Carriazo, F. Governance, Nature's Contributions to People, and Investing in Conservation Influence the Valuation of Urban Green Areas. *Land* **2021**, *10*, 14. https://dx.doi.org/10.3390/land10010014

Received: 30 November 2020
Accepted: 23 December 2020
Published: 27 December 2020

Publisher's Note: MDPI stays neutral with regard to jurisdictional claims in published maps and institutional affiliations.

1. Introduction

The link between human well-being and urban green areas, forests, parks, wetlands, and other natural and semi-natural ecosystems in cities has been well established [1,2] Several studies have valued multiple benefits using a diverse set of case studies, methods, and ecosystem service frameworks like the Millennium Ecosystem Assessment, The Economics of Ecosystem and Biodiversity, and others [3]. These have classified and defined urban ecosystem services as well as reviewed methods for their valuation. A similar body of literature has also discussed ecosystem disservices, or the social, environmental, and economic costs that these detrimental ecosystem functions have on people's well-being [4–6].

This urban ecosystem service–disservice literature has primarily used case studies in contexts such those of Europe and the United States to explore these functions as well as the links between citizens and the benefits from green spaces [7–9] Similarly, several urban ecological functions result in a suite of disservices and costs including: Human injuries and infrastructure damage from vegetation debris and growth, wildlife nuisance, allergies, and maintenance costs, among others [6,10,11]. More recently, because cities are complex socio-ecological systems, other socioeconomic functions—in addition to ecological ones—are being included in the assessment of urban ecosystem disservices and can include: Fear of crime and tree fall, unpleasing aesthetics, diseases from remnant natural areas (i.e., wetlands) and foregone property premiums to name a few [6,7,12]. Despite this, there is much less information on how people in the Global South perceive benefits [13–16], and even less so on ecosystem disservices and if conventional urban ES/ED typologies are relevant for green areas of the Global South [17,18].

More recently, the Intergovernmental Panel on Biodiversity and Ecosystem Services has proposed the Nature's Contributions to People (NCP) framework or "the positive contributions, or benefits, and occasionally negative contributions, losses or detriments, that people obtain from nature" to complement the ecosystem service framework; particularly in places like the Global South [16]. Although NCP "goes further by explicitly embracing concepts associated with other worldviews on human–nature relations and knowledge systems" [16], the concept has sparked a lively debate in the ecosystem service community; see [18] and responses therein to the article. Despite this recent NCP versus ecosystem services controversy, the terms "benefits" and "costs" as related to urban green areas have a long history of use and application dating back to the early 1990s [1] before the advent and frequent use of these other metaphors [17].

Other studies regularly use geospatial and statistical methods to understand the supply of these ecosystem services and benefits in cities [19,20]. Surveys are also regularly used to better understand the perception residents have towards urban ecosystem services [6,15]. Some of these studies use psychometric scales and methods [15], as well as stated preferences and econometrics, to determine value [6,21]. Fewer studies have, however, measured the role that governance, perceived corruption, and policy processes have in influencing people's willingness to pay to conserve the ecosystems providing such benefits [22,23]. Similarly, the realities of inequity, weak governance, perceived corruptions, and lack of resources is systematically omitted in stated preference studies in low–middle income countries [21].

These processes and dynamics between actors or stakeholders, governments, and the management and planning of these benefits are key elements that link the supply and demand for benefits [24]. These policy processes, or governance, of ecosystem benefits has been looked at using several lenses including: Political ecology [25,26], urban and rural forest management [27,28], biodiversity [22,24], program evaluation, and the urban ecosystem services framework [8,9,11,29]. However, most of these urban context studies are predominantly from high income countries such as those of North America, Europe, and Australia [17].

The concept of governance has many definitions and applications, and has been described as "an emergent, often complex decision making process" [30]. Huang, C.W. et al. [24] define effective governance as a process that "facilitates the development and implementation of law, regulations, and institutions that have a role in the management of land resources". Although generally used as a means to describe the processes used by governments to include the governed or society in the decision making process (state-centered), it can also include community and market sectors and situations where actors take a prominent role in the co-management of ecosystems (society-centered; [25,27]). Governance as such includes processes and interactions that organize power relations, influences, interests, and government performance and transparency into the decision making process in order to determine socioeconomic and environmental benefits [28].

Lawrence, A., et al. [28] and Launay, G.C. et al. [31] emphasize the role of measuring these processes, their applications, and outcomes in terms of evaluating the effectiveness in assuring good governance. Kenward, R.E. et al. [22] and Huang, C.W. et al. [24] investigated the performance of governance strategies and context in achieving successful biodiversity conservation outcomes and supplies of ecosystem services. However, Turnhout, E. et al. [26] argue that an increased focus on measuring transparency, efficiency, and effectiveness can lead to an "impoverished understanding of biodiversity itself". Examples of frameworks, models, and discourses related to governance in regards to urban and peri-urban forests are discussed in detail in [28,32].

However, few studies discuss what good governance is in regards to urban ecosystems in cities of the Global South [13,20,33]. Lockwood [34] defines "good governance", which encompasses: Legitimacy, transparency, accountability, inclusiveness, fairness, connectivity, and resilience, while [31] promoting criteria such as transparency, corruption, and government performance when evaluating proper governance in Latin American countries. According to Barrett, C.B. et al. [35], the inverse of transparency, or corruption, in regards to natural resources is regularly used as an explanation for environmental degradation. Indeed, in low and middle income countries, perceptions of corruption influence how people value and access environmental benefits in both urban and rural settings [21]. Yet, corruption comes in many forms, levels, and scales, and the causal effects and relationships between corruption and natural resource use and condition can be complex [35].

The above studies document how people perceive and value urban ecosystem services in several cities [1,12], and the role of good governance and willingness to pay for conserving biodiversity and ecosystems [9,24,33,36]. However, as previously mentioned, less known is how people in places such as Latin America perceive this urban benefit-cost bundle and how context-specific realities such as lack of transparency, perceived corruption, and inequity affect people´s willingness to invest in conserving the ecosystems that provide these services [17,20]. Indeed, even the relevance and direct application of the ecosystem service framework in places such as the Global South have recently been questioned [13,18].

This study explores how people perceive the benefits, costs, and the influence of governance on their willingness to invest—or not—to conserve urban green areas. We surveyed representative areas in an urbanized sub-watershed in Bogota, Colombia. Specifically, we have three different study objectives. First, we assess how people perceive urban benefits and costs in an urban sub-watershed in Latin America. Second, we assess how socioeconomic factors affect perceptions. Third, we explore the influence of the different dimensions of governance (e.g., perceived corruption, transparency, government performance) on people´s willingness to invest for the conservation of the green areas and wetlands providing these benefits and costs (i.e., Ecosystem Services–Ecosystem Disservices (ES–ED)). So as to avoid the ecosystem service versus NCP controversy [18], we use the terms urban "benefits" and "costs" as defined by Dwyer et al. [1] in our study and analyses; but we do discuss the relevance of these metaphors (i.e., ES, ED, and NCP) in our Discussion and Conclusion with a focus on the promising use of NCP.

2. Materials and Methods

2.1. Study Area

The study area was the El Salitre urban sub-watershed in Bogotá, Colombia (Figure 1). Bogotá is located at 2600 m in elevation and has a subtropical highland climate temperature that varies between 7–17 degrees C, and total average annual rainfall is about 825 mm [37]. The study´s sub-watershed encompasses the localities of Usaquén, Santa Fé, Chapinero, Teusaquillo, Barrios Unidos, Engativá, and Suba, and within these are 1894 different neighborhoods encompassing 11,791 has. Although the sub-watershed does encompass a large portion of the adjacent Eastern Hills Protected Forest Area (i.e., Reserva Forestal Protectora Bosque Oriental de Bogotá) to the east of Bogota, the study was done entirely in the urban portion. The El Salitre River is mostly channelized in this urban portion and is often referred to as the Arzobispo, Quebrada Molinos, Rio Callejas, and La Sirena streams or drainage channels. There are four officially designated wetlands and approximately 175 different parks and water bodies within the watershed. In all, 3 wetlands, 5 parks, and 2 green areas were selected for sampling the sub-watershed (Table 1).

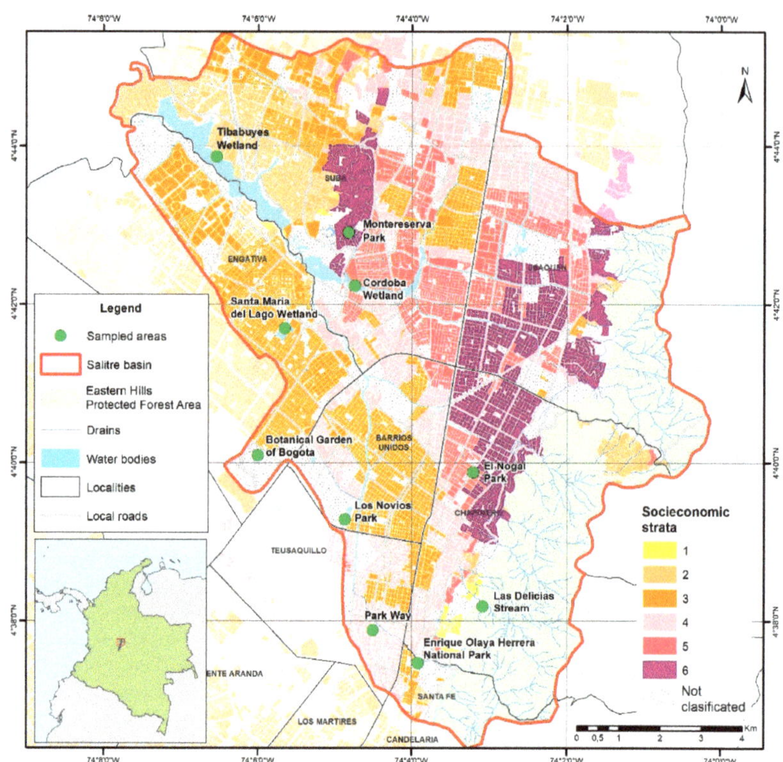

Figure 1. The El Salitre urban sub-watershed in Bogota, Colombia, and its socioeconomic strata and ten sample sites.

Table 1. Ten different sites characterizing three different green area types in the urban sub-watershed of El Salitre, Bogota Colombia.

Site	Socioeconomic Strata	Use
Wetlands		
Tibabuyes—Juan Amarillo	2	Largest wetland in Bogota (223 has)
Córdoba	5	40-hectare wetland with the highest number of registered bird species
Santa María del Lago	3	11-hectare wetland with high visitor use and water quality
Parks		
Parque Nacional Enrique Olaya Herrera	3	283-hectare recreational park bordering an extensive forest reserve to the east of Bogota
Parque de los Novios	3	23-hectare recreational park with a large lake in the center
Parque El Nogal	5	3100 m2 recreational park in a residential and business district
Park Way	4	Linear, recreational park 30 m in width extending for about 9 city blocks
Parque Montereserva	6	Recreational green space adjacent to multi-story residential buildings
Green areas		
Jardín Botánico de Bogotá	3	Municipal botanical garden
Quebrada las delicias	1	Conservation riparian area at the edge of forest reserve to the east of Bogota

The sub-watershed also encompasses all the different sub-neighborhood level socioeconomic strata that characterize Bogota. Bogota is divided into six different designated socioeconomic strata, that were designed to subsidize utility payments and infrastructure based on resident's average income, and thus are a measure of a residential zone´s income status; socioeconomic strata 1 being the lowest income and 6 the highest. These strata are correlated with green space cover and subsequent ecosystem service provision and benefits [38]. Bogota has a high population and building density and as such, the sub-watershed poses several socio-political and environmental realties typical of medium income Latin American cities. For example, sewage pollution discharges into the wetlands and streams in the forest reserve are common [39]. Urbanized areas near wetlands also experience frequent flooding, and informal settlements—both high and low income—are sporadically being established in the foothills and are characterized by high impervious surfaces and inadequate waste management [39].

Air and water pollution concentrations are also high along transportation land uses and stream channels. Streams, as previously mentioned, are channelized and made impervious to deal with excess stormwater and effluents [37,38]. Specific neighborhoods in the study sub-watershed such as Engativá and Suba are characterized by high rates of criminal activity [27,40]. Thus, given the complexity of sampling in densely populated areas with disparate socioeconomic realities and access and safety issues, we were not able to use other standardized methods for selecting urban sites that are frequently used in places such a Europe [41]. Instead, we selected 10 different and representative sites based on safety and access and we use these to represent the sub-watershed´s different socioeconomic strata and land uses (Table 1).

2.2. Survey Instrument

We used a semi-structured, in-person survey consisting of 18 different questions that assessed people´s perception and value for different urban benefits and costs as well as respondent's demographic and socioeconomic backgrounds (Appendix A). Questions were a combination of closed and open-ended items. In the survey, the first part of the questionnaire was about socioeconomic strata, gender, age, and education level. In a second section, we measured people´s awareness about the watershed's ecology in terms of their ability to recognize key ecological information by asking 4 questions about: Ecological health, different species, the existence of established wetlands, and an extensive forest reserve to the east of the city (Figure 1). Accordingly, employees of the Jardin Botanico de Bogota and students from the Universidad del Rosario surveyed people's perception towards climate change and their Willingness to Invest to conserve and restore the ten different green areas and wetlands we used in the study. In all, we surveyed 500 different people, or 50 respondents per site. Approximately 75% of the people who were approached participated in taking the survey and signed an informed consent form.

Based on [1,4,12,16,19,20], we analyzed 8 different benefits and 8 costs that have been reported to influence people´s perception and values regarding urban green and nature in cities. We also include crime-related costs [40] and poverty alleviation–income generating benefits [14] that are not accounted for in ecosystem service typologies [16,18]; but were important to the citizenry. We have found from previous experience that survey respondents in this study area do not distinguish among cognitively complex and technically difficult processes regularly used in ES frameworks such as "ecological functions", "ecosystem services and disservices", and "economic benefits". Rather, they simply recognize them as "benefits and costs". Thus, we also include crime related ED and poverty alleviation–income generating benefits that are not accounted for in typologies developed in high income countries such as those associated with the "ecosystem service cascade" [16]. Thus, we emphasize that the more scientific and technical terms ES and ED were posed as "benefits and costs" in the instrument to better communicate with respondents. However, in the following methods, results, and discussion sections, they are presented as ES, ED, and NCP to better contextualize and discuss relative to other relevant literature and studies.

2.3. Statistical and Econometric Analysis

Survey responses were in the form of dichotomous questions and Likert scales that were used to statistically characterize the survey population and their responses using two approaches (Appendix A; [15]). First, we summarized respondent´s socioeconomic strata, gender, age, and education level. Graphical analyses were used to identify trends and patterns in responses towards benefits and costs. Then an ordinary least squares (OLS) regression with robust standard errors: 1. Explored the variables that can influence perception towards different benefits and costs, and 2. identified variables for a subsequent analysis using a more predictive model. Following the work of [42], the following regression model (Equation (1)) accounted for the social–cultural dimensions of benefits (B_{ij}):

$$Benefits_{ij} = \beta_0 + \beta_1 EA_i + \beta_2 SS_i + \beta_3 AE_i + \beta_k X_{ik} + \varepsilon_i \tag{1}$$

where $Benefits_{ij}$ corresponds to the number of benefits identified by each individual surveyed in three categories: Provision, environmental, and cultural. This variable was modelled according to three types of sites: Parks, wetlands, and green areas. The independent variable EA_i was the environmental awareness of each individual; SS_i is the sense of security; AE_i is a dichotomic variable that takes the value of 1 when the individual has a university degree, and 0 when the individual has a lower degree than a university degree; X_i is a 3 by 1 vector of control variables. This vector includes the following variables: Age, a dichotomous variable if the individual is from Bogotá, and another dichotomous variable if the individual is aware of climate change.

Second, we used survey responses and most of the variables from our OLS model in a logistic regression to econometrically assess the effects of Weak Governance (Weak_governance) on respondent's Unwillingness to Invest (UTI) using Equation (2). Per Colombian program evaluation standards [31], a response of not willing to invest (i.e., UTI) included the following reasons for not doing so: 1. Perception of corruption or that funds will not be used appropriately, 2. conservation of green areas and wetlands is already paid for in taxes, 3. the respondents already pay too much tax, and 4. it is the government´s responsibility to conserve green areas. We recoded people´s Unwillingness to Invest as the dependent variable UTI = 1; conversely, UTI = 0 if people were Willing to Invest.

$$ProbUTI = \beta_0 + \beta_1 \, Weak\_Governance + \beta_k X + \varepsilon_i \tag{2}$$

Thus, based on [31] and [34], the Weak_governance variable in Equation (2) was 1 if there was a perception of weak governance and 0 otherwise. The X in Equation (1) is a vector representing the socioeconomic variables that were used as controls. We used Gender = 1 for female, and Age was a categorical variable. Strata were recorded according to respondent´s socioeconomic strata (1, 2, … .6). In order to avoid perfect collinearity with the intercept of the model, stratum 1 was the omitted variable, as was Salitre if the respondent was from this locality and the city of Bogotá. Five of the 500 respondents did not respond to the UTI question, therefore we used 495 responses for this analysis. Logistic regression estimates were reported using Odds Ratios. Both regression models and all statistical analyses were done using the Stata Version 12 software.

3. Results

3.1. Socioeconomic Characteristics

The majority of the respondents (68%) were between 18–45 years in age, and 55% were male (Table 2). About 32% had a university level education, and only 2.4% had post-graduate studies. Fifty % were in the middle-income strata (Strata 3 and 4), while 20% were in the lower income strata (Strata 1 and 2).

Table 2. Socioeconomic characteristics of respondents in the El Salitre sub-watershed in Bogota, Colombia.

Socioeconomic Characteristics	N	Mean	Std. Dev.
Age (years)			
<18	500	9.60%	0.2949
18–30	500	44.60%	0.4976
31–45	500	24.20%	0.4287
46–60	500	14.40%	0.3514
>60	500	7.20%	0.2587
Gender			
Female	500	44.80%	0.4978
Male	500	55.20%	0.4978
Education			
None	500	0.40%	0.0635
Preschool-Primary	500	11.31%	0.3170
Secondary	500	42.82%	0.4983
Technical	500	10.30%	0.3043
Professional-Technological	500	32.32%	0.4681
Specialization/Masters/ Postgraduate	500	2.42%	0.1539
Doctorate	500	0.2%	0.0449
Socioeconomic Strata			
1-Lower	500	10%	0.3003
2-Upper lower	500	10%	0.3003
3-Lower middle	500	30%	0.4587
4-Middle	500	20%	0.4004
5-Upper middle	500	20%	0.4004
6-Upper class	500	10%	0.3003

N, Number; Std. Dev., Standard Deviation.

3.2. Perceptions of Benefits and Costs

We found that air purification was the most frequently identified benefit as opposed to water regulation and quality (Table 3). Most notably, flooding regulation was the least identified benefit in all three wetlands. In parks and green areas, air purification was the most identified benefit, but respondents did identify provisioning benefits in parks, however, similar to wetlands, flood regulation and water quality were the least mentioned in both parks and green areas (Table 3).

Overall respondents identified 11 different costs in the watershed (Table 4.). Two of the wetlands, Tibabuyes and Córdoba wetland, and two parks, Montereserva and Parque Nacional, had the most costs identified. Respondents in the Cordoba and Tibabuyes wetlands, for example, reported crime and lack of maintenance and drug use. The Las Delicias Riparian area, the Botanical Garden of Bogotá, and the Nogal Park were the areas with least amount of costs reported. Overall, drug use was the main cost reported in parks. Since crime is a frequently reported problem or ecosystem disservice reported in other international literature [1,4,22], we here forth focus on this specific cost in subsequent analyses. The Cordoba (70%) and Santa Maria del Lago (94%) wetlands were reported as the safest places—in regards to crime—as opposed to the Tibabuyes wetland, which was the most insecure, as only 8% of respondents considered it safe (Figure 2).

It appears that respondent's perceptions regarding costs from urban ecosystems do affect their identification of benefits, and thus there appears to be a trade-off in the respondent's sense of wellbeing when they feel unsafe relative to the benefits they perceive. In general, the areas identified as safest were those that had the greatest number of benefits identified. Conversely, in sites that were the least secure, respondents identified the least number of benefits in wetlands, parks, and green areas (Figure 2 and Table 3).

Table 3. Perception towards benefits in 10 different green areas and wetlands in the El Salitre sub-watershed. Note: We integrate both the ecosystem services and nature's contribution to people (NCP) metaphors to better represent Bogota, Colombia's context and what citizens value from ecosystems and nature.

Ecosystem Services/NCP	Benefits	Wetlands			Parks					Green Areas	
		Córdoba	Santa María del Lago	Tibabuyes	Montereserva	Park Way	Los Novios	El Nogal	Parque Nacional	Quebrada las Delicias	Jardín Botánico
Provision	Food	0%	0%	0%	2%	0%	0%	0%	0%	0%	0%
	Shadow	0%	0%	0%	0%	4%	0%	0%	0%	0%	0%
Regulation	Air purification	96%	100%	66%	90%	80%	98%	94%	88%	90%	100%
	Water purification	64%	54%	32%	50%	38%	46%	28%	50%	84%	66%
	Climate regulation	48%	80%	34%	62%	62%	66%	72%	64%	74%	72%
	Flood mitigation	32%	62%	22%	52%	34%	50%	30%	44%	42%	42%
	Biodiversity	54%	90%	48%	68%	66%	78%	82%	84%	84%	90%
Cultural	Sense of well-being	78%	84%	58%	80%	80%	86%	92%	94%	92%	92%
	Aesthetics	72%	82%	76%	78%	80%	90%	84%	94%	90%	94%
	Recreation	88%	80%	62%	66%	64%	92%	90%	90%	86%	88%
	Employment	0%	0%	0%	0%	2%	0%	0%	2%	0%	0%

Table 4. Perception towards costs in 10 different green areas and wetlands in the El Salitre sub-watershed. Note: We integrate both the ecosystem disservices and problems from nature metaphors to better represent Bogota, Colombia's context and what citizens least value from ecosystems and nature.

Ecosystem Disservices	Costs	Wetlands					Parks				Green areas	
		Córdoba	Santa María del Lago	Tibabuyes	Montereserva	Park Way	Los Novios	El Nogal	Parque Nacional	Quebrada las Delicias	Jardín Botánico
Environmental	Falling branches and trees	30%	30%	18%	28%	24%	22%	20%	30%	34%	4%
	Trash and refuse	28%	24%	76%	42%	28%	24%	18%	52%	36%	4%
	Invasive vegetation	0%	0%	0%	0%	0%	0%	0%	0%	4%	0%
	Pet excrement	2%	2%	0%	0%	0%	2%	6%	0%	0%	0%
	Insects or rats	24%	28%	50%	32%	26%	28%	16%	12%	4%	0%
Financial	Maintenance costs	18%	10%	14%	24%	16%	32%	18%	10%	0%	4%
	Lack of maintenance	40%	56%	62%	44%	32%	34%	14%	54%	26%	2%
	Illicit drug sales and use	32%	26%	82%	40%	40%	34%	54%	60%	24%	2%
Social	Crime	44%	12%	84%	34%	52%	10%	14%	76%	56%	2%
	Fear	32%	14%	42%	36%	26%	36%	22%	40%	6%	2%
	Rhinitis or allergies	6%	14%	32%	38%	8%	22%	14%	18%	4%	2%
Other	None	32%	22%	0%	24%	6%	22%	8%	0%	16%	82%

35

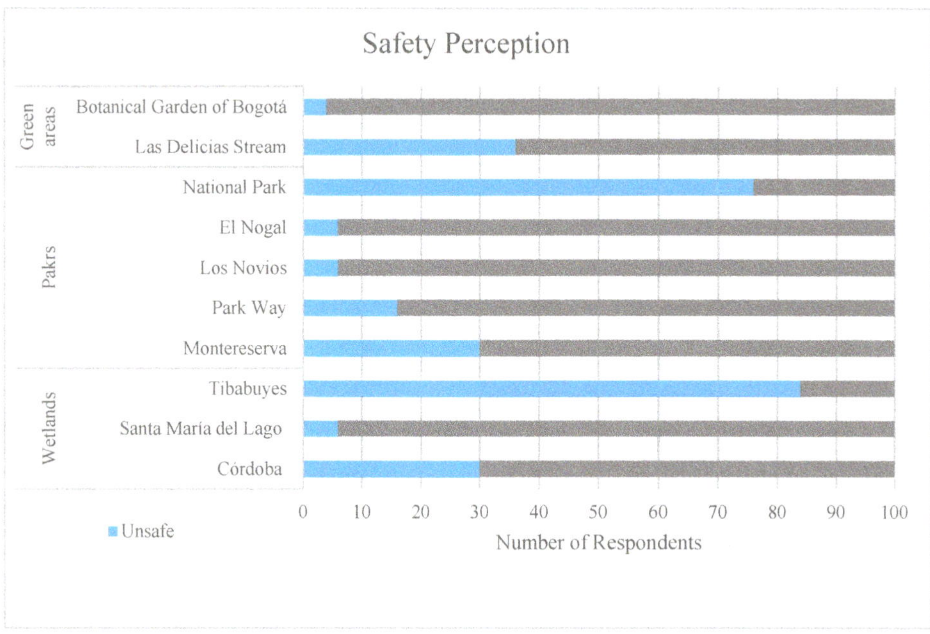

Figure 2. Number of responses towards the perception of levels of a cost in the form of high crime (Unsafe) and low crime (Safe) in 10 different green areas and wetlands in the El Salitre sub-watershed in Bogotá, Colombia.

3.3. Factors Influencing Perceptions towards Benefits

Our OLS model shows that in general, environmental awareness coincided with a recognition of a greater number of environmental and cultural benefits (Table 5). Those with advanced degrees and that felt a greater sense of security, or lack of crime, also identified a greater number of environmental and cultural benefits. There was however no relationship between being born in Bogotá and the ability to identify benefits in the sub-watershed. In terms of respondents surveyed in parks, those with greater levels of education identified a statistically significant greater number of different environmental benefits. In wetlands, respondents: With more education, born in Bogota, and with more awareness about the local environment and climate change effects identified a greater number of environmental and cultural benefits. However, of those surveyed, respondents greater than 60 years old were those that identified the greatest number of environmental benefits (Table 5).

3.4. Factors Influencing the Conservation of Bogota´s Green Areas and Wetlands

Overall, 97% of respondents do consider that conserving the watershed's wetlands and natural areas would improve the wellbeing of people living in their proximity. Forty-seven % felt that air quality improvement was the main benefit, followed by health and wellbeing (27%) and recreational activities (16%). However, when subsequently asked their WTI for an additional fee in their monthly utility bill to "protect and restore Bogotá's natural areas", 57% responded positively while 43% responded no. When asked why they were not willing to invest, 17% said that they already pay taxes and that this is the government´s responsibility, and an additional 16% said that they would not invest because of misuse of funds (i.e., corruption). However, of those that were WTI, 16% was because of improvement in wellbeing and protection of biodiversity (11%).

Table 5. Ordinary least squares regression model for socioeconomic factors that influence the environmental, cultural, and provision benefits identified in the El Salitre watershed (Robust standard errors in parenthesis) *** p < 0.01, ** p < 0.05, * p < 0.1.

Variables	Number of Regulation ES in the Watershed	Number of Cultural ES in the Watershed	Number of Provision ES in the Watershed
	(1)	(2)	(3)
Environmental awareness [a]	0.207 ** (0.0987)	0.177 *** (0.0574)	0.00626 (0.00443)
Sense of security—Crime	0.466 *** (0.152)	0.170 * (0.0985)	0.0116 * (0.00678)
Higher education (> university degree)	0.878 *** (0.317)	0.553 *** (0.0905)	0.00150 (0.00382)
Between 31 and 45 years old	0.0103 (0.156)	−0.140 (0.0947)	−0.00826 (0.00510)
Between 46 and 60 years old	0.329 (0.244)	0.0406 (0.110)	−0.00911 (0.00568)
Greater than 60 years old	0.0714 (0.220)	−0.0875 (0.126)	0.00550 (0.0125)
Recognizes climate change effects	0.903 ** (0.351)	0.900 *** (0.268)	−0.0436 (0.0436)
Born in Bogotá	0.00404 (0.138)	−0.192 ** (0.0785)	−7.29e-05 (0.00628)
Constant	1.739 *** (0.380)	1.481 *** (0.282)	0.0349 (0.0394)
Surveys	499	499	499
R2	0.0634	0.095	0.021
F statistic	3.662	9.306	0.380

[a] The environmental awareness variable consisted of 3 questions: Can you identify 3 trees, palms, or plants in this place? Are you aware of the *Reserva Forestal Protectora Bosque Oriental de Bogotá* (the adjacent large extensive forest reserve to the east of the city)?; and Can you name wetlands that are located in Bogota? (Appendix A).

3.5. The Role of Governance and Unwillingness to Invest for Ecosystem Services

We found that WGOV was positively related and significant at all levels to a respondent's UTI. According to the Odds Ratio, the chance of UTI for conservation was 31 times greater if the respondent perceived WGOV as opposed to strong governance (Table 6). When accounting for other socioeconomic variables, Odds Ratios show that if the respondent perceives WGOV, the UTI for conservation is 38 times greater than when perceiving strong governance (Table 7). The model shows that the control variables are not significant, however, WGOVC was very robust. Gender, age, and being born in Bogota were not significant, and the only significant related variables were for respondents in Strata 2 (upper lower income). That is, the Odds of UTI in strata 2 is 3.1, while for strata 6 (highest income) it is only 0.7.

Table 6. Logistic regression [a] and Odds Ratio of the effects of weak governance (WGOV) on people's Unwillingness to invest (UTI) for urban ecosystem benefits.

UTI	Odds Ratio	Standard Error	Z	P > |z|	[95% Conf. Interval]	
WGOV	31.43382	7.968317	13.6	0	19.1259	51.66216
Constant	0.16	0.027247	−10.76	0	0.114595	0.223396

[a] Number of observations = 495; Log likelihood = −208.44138; LR chi2(1) = 258.53; Prob > chi2 = 0.0000; Pseudo R2 = 0.3828.

Table 7. Logistic regression [a] of the effect of weak governance (WGOV) on people's Unwillingness to Invest (UTI) controlling for gender, age, socio-economic strata, and residence.

UTI	Odds Ratio	Standard Error	Z	P > |z|	[95% Confidence Interval]	
WGOV	37.73149	10.60571	12.92	0	21.74929	65.45804
Gender	0.8305991	0.217637	−0.71	0.479	0.497	1.388119
Age	0.9831168	0.11173	−0.15	0.881	0.7868048	1.22841
Stratum 2	3.155877	1.782857	2.03	0.042	1.042915	9.549736
Stratum 3	1.353498	0.642457	0.64	0.524	0.5338548	3.431566
Stratum 4	1.22499	0.6226557	0.40	0.69	0.452347	3.317368
Stratum 5	1.170744	0.595533	0.31	0.757	0.4319894	3.172857
Stratum 6	0.734595	0.420205	−0.54	0.59	0.23941	2.254003
Resides in Salitre	1.058643	0.294395	0.2	0.838	0.613821	1.825818
Born in Bogota	1.140392	0.322006	0.47	0.642	0.6556999	1.983368
Constant	0.1191804	0.06968	−3.64	0	0.0378913	0.3748613

[a] Number of observations = 495; Logistic Regression chi2(10) = 267.09; Prob > chi2 = 0.0000; Pseudo R2 = 0.3954; Log likelihood = −204.16118.

4. Discussion

Below we will discuss and interpret our findings relative to the relevant literature; thus we will use the term "benefit" and "cost" [1] as synonymous with ecosystem service and disservices, respectively. Later we will discuss the relevance of the NCP framework to our analysis [16,18]. Overall, our findings are similar to those of many other studies in that cultural and environmental benefits were the most identified by respondents, specifically, air quality, aesthetics, well-being, and recreation [43]. Ecosystem service studies in more rural contexts in middle and low-income countries have found that people focus on provisioning ecosystem services or benefits as defined in this study [14,44]. In contrast, in our study these were the least reported. Other studies in natural areas and wetlands in Colombia have also reported that people primarily identify regulation and cultural ecosystem services [45]. However, we do note that water regulation and purification were the least reported benefits in the sub-watershed as opposed to air quality regulation, which was the most frequently reported ES in the study. Shade, as opposed to other studies, was also not a top benefit [6]; but "biodiversity" was also highly valued. Interestingly, we note that these two are considered "ecological structure or functions" in the "ecosystem service cascade" framework—not benefits per se [3,4]. Interestingly, Bogotá's economic activities and topographic, high elevation, and precipitation characteristics, do not make air quality and temperature regulation issues as pressing as other Latin American cities; however, the city suffers regularly from floods and stormwater problems [38].

Overall, respondents from lower socio-economic strata perceived fewer benefits than those from higher income strata. Accordingly, one would surmise that these respondents spend less time in Bogota's green areas, but Scopelliti, M. el al. [15] found that it was low and high socioeconomic strata residents that spent the least time in Bogota´s urban parks. Regardless, our weak governance results seem to indicate that lower income respondents do not feel that they are, or should, participate in the decision-making processes. Our results also show that socioeconomics does play a role in the perception towards benefits. Some specific relationships between socioeconomic factors and the identification of benefits have been reported in [46] and [42].

Although income and education are related, there is an obvious relationship between education and environmental awareness, particularly in regards to the hydrological functions. Allendorf, T. D. et al. [47] found that education level does indeed affect how people perceive ecosystem services. Our results were similar; where people with the highest level of education perceived more benefits than those without. One of the most insightful findings was the relationship between "wellbeing" and ecological structure-function-form in that the better maintained green areas were also those that were "safest" in terms of crime and where the majority of benefits were identified. Similarly, the majority of respondents were aware that there would be a loss in their wellbeing if these green areas were lost to land use change. Other research in Bogota has shown that public areas with increased tree-shrub-palm density and heights as well as tree plantings were related to lower incidences of crime and greater provision of benefits [19,38].

Based on Colombian program evaluation criteria [31] for principles and performance outcomes for good governance [34], our survey and modelling found that the perception of governance regarding public expenditures was playing an influential role in the WTI for benefits. Particularly, perceived weak governance was statistically related to people's UTI in conserving green areas. This finding suggests that transparency, performance, and perceived corruption of government institutions can and will influence buy in and the value citizens place on the benefits from urban ecosystems [33,34]. Furthermore, the effect of governance on the UTI for benefits was not homogeneous across all socio-economic strata. Notably, respondents in Stratum 2 identified a weakness in governance, and this affected their subsequent odds of not investing for conservation–benefits related initiatives; this was on average three times greater than individuals from higher socio-economic strata. Although income and education were related [48], there is an obvious relationship between education and environmental awareness in particular to the hydrological functions and other co-benefits [13,49].

The perception of weak or strong governance on watershed and ecosystem management, conservation, and urban ecosystem services—or benefits as defined in this study—has been previously studied by [21,23,26,28,35]. Based on this literature and our findings, one of the main contributions of this study is that we have identified an influence of weak governance (i.e., lack of transparency, perceived corruption, and poor government performance) on how society values the urban green space benefits from Neotropical urban green areas and watersheds. Such information is key for improved governance, effective conservation, and sustainable provision of benefits to society [24,50].

Our literature review shows how governance is a complex metaphor, and that it has many definitions and is used in the context of urban ES and biodiversity studies; yet few studies have actually attempted to measure it and how it can influence urban benefits-costs, biodiversity, NCP or Nature-based Solutions [24,29]. Thus, to provide for a measure of governance, our study measured governance as consisting of: Transparency, corruption, and government performance. Our measure also incorporated aspects of Launay G.C. et al. [31] and Lockwood, M.'s [34] definitions of good governance as encompassing: Legitimacy, transparency, accountability, inclusiveness, fairness, connectivity, and resilience. Accordingly, we found that our corruption and transparency variables can be used as proxies for perceived corruption, and that funds intended for green area conservation might not be used appropriately as perceived by respondents. Similarly, our tax payment variables accounted for legitimacy, accountability, and government performance (i.e., trust). As such, we found that although residents pay taxes, many respondents are seeing fewer direct benefits (Table 6). Additional variables regarding the government's responsibility for maintaining green areas accounted for the lack of inclusiveness, fairness, and overall transparency (Table 6). We found that even though respondents knew that the public areas they used and pay taxes for—and subsequently received benefits from them—they indicated a lack of confidence and poor government performance that inevitably affected their response.

We recognize that our use of "socio-economic and ecological processes", "ES/ED", and "benefits and costs" does not match the conventional ES typology and framework. Additionally, we note that we did not randomly sample individual sites in the watershed and that our sample size could have been greater. There will also be bias in our results because the surveyor did not randomly survey individuals or specific places in each site. Similarly, concepts such as governance and costs as pointed out by our literature review are complex metaphors to define and measure. Environmental values, resource conflicts, power relationships, green infrastructure types, and structure will also affect how people perceive benefits-cost bundles [20]; these were not studied though. However, given security and access issues and our review of the relevant literature, we feel that our approach and modelling results do provide initial insights into the role of perception and the role of governance on urban green area benefits and costs in a major Latin America city.

Future research could study the perceptions towards benefits and costs and the effect of other field and in situ cognitive measurements of green space and forest structure and composition. As indicated by a reviewer, the opportunity exists for also including other questions as explanatory variables in our models for better understanding citizen opinions of these functions such as perceptions regarding costs, difference dimension of governance, or other environmental and social capital factors and how they affect the valuation of benefits. Indeed, other biophysical factors such as size and density of green areas per neighborhood could also affect human well-being [2]. Similarly, using other methods from environmental psychology and experimental economics could also offer other causal insights as well. That said, the approach used in our study to measure benefits and effects of weak governance and institutional transparency can be used in adopting and improving ecosystem service-governance frameworks such as those proposed by [13,51], and [24] to Latin American socio-ecological systems and their contexts [20]. Implications can also be made regarding the relevance and application of "Nature-Based Solutions" or a more culturally and context relevant approach such as the recently proposed "NCP"; a particularly promising metaphor (Table 3) [18].

Indeed, management and governance of the wetlands and green areas in Bogotá as a public resource does require accounting for such multiple and complex issues and language [3,50]. Our findings, for example, also show the importance of focusing environmental education efforts and topics on certain demographic groups. Although people were knowledgeable about the environment, biodiversity, and nature and its benefits, there was a notable lack of awareness regarding the important hydrological benefits provided by the watershed's green areas. As such, there is an opportunity to educate the public about the existence and benefits of the positive ecological processes provided by wetlands and parks for the protection of lives and property in Bogotá. Finally, based on our study area and findings, in order to get buy in from society in conservation efforts, it is essential that public institutions improve their perceived lack of transparency an performance [31,33–35].

Sarkki, S. et al. [51] proposed different types of "governance services" for different arenas (i.e., policy, markets, society, science) as a way to understand ecosystem service dynamics and incorporate them into management and planning frameworks. Falk, T. et al. [36] also proposed that governance types need to be adapted according to specific benefit types and institutions. Therefore, key to linking well-being with governance and policy uptake is to identify a given society´s perceptions and values concerning the benefits and costs from green areas [24,50]. However, much of this ecosystem services-governance literature presents concepts, conceptual frameworks, and case studies primarily from countries in the Global North and wildland, rural ecosystem-based contexts [16,24,33]. However, our findings add to the emerging ecosystem services, NCP, and governance literature in that we: Addressed socio-ecological dynamics for low-middle income contexts, measured governance using evaluation metrics, and informed how policy processes and incentives can affect citizen´s UTI in green space and wetland conservation.

5. Conclusions

Our findings provide very basic and useful, yet overlooked, information and guidelines for managers, educators, policy makers, and local planners. In particular, urban park and wetland users can identify the various benefits from urban nature. However, as in the case of water regulation, information needs to be targeted and context-specific. That is, in surveyed wetlands, people perceived that air quality was more important than water regulation. Despite the fact that flooding and water quality problems are much more acute problems in the sub-watershed than air quality issues. Additionally, we found that conventional ES typologies and frameworks developed in high income countries need to be adjusted and adapted in order to match the realities on the Global South. The NCP or Nature-Based Solutions approaches as such provide fertile grounds for future use and application in places such as Latin America cities.

Similarly, people's perception of crime and the overall lack of maintenance and infrastructure influenced the overall benefit's they identified from green areas. This was particularly evident with respondents from lower socio-economic neighborhoods. So, people with different education and socioeconomic levels do weigh the tradeoffs of personal security and perceived lack of governance against the conservation and provision benefits of and from these areas in different ways. In particular, costs such as crime and litter did affect them, but detrimental ecosystem structures and their negative functions are easily addressed in a relatively low-cost manner and can effectively influence how people perceive these urban benefits. Thus, it is important for planners to consider safety and maintenance of wetlands and parks because basic simple management, maintenance, and planning activities can play a role in people´s perception and probably in attitudes and investment of public resources towards those areas.

Accordingly, conservation and effective management of wetlands, parks, and green areas is key, and citizens can indeed identify the benefits versus the costs of conserving them. In order to provide long-term benefits from urban ecosystems, effective governance processes and environmental education efforts based on the premise that humans are integral parts of social–ecological systems is key. The perception of good governance is regularly considered important in this link, but the perceived lack of transparency will be a limiting factor in people´s buy in and willingness to invest in and maintain the necessary ecological structure that provides the most benefits and minimizes costs. Accordingly, lack of governance processes and trust towards the institutions developing and implementing them will affect the effectiveness of existing planning and management goals.

The benefits and costs of urban green areas—be they referred to as ecological processes, biodiversity, ES, ED, NCPs, or nature-based solutions, are key in improving the social, economic, and environmental well-being of citizens; regardless of the metaphor used. Participatory management and planning of urban green areas requires information on how the different segments of society perceive both the benefits and costs of urban socio-ecosystem functions. Thus, effective institutional capacities and transparent state and non-state centered processes require information on what the different actors perceive about not only ES/ED, but of the governance and policy processes inherent in delivering the supply of benefits—and the mitigation of costs—from wetlands and green areas. Indeed, where good governance is absent, "bottom up" initiatives by citizens are one means to move towards improving the well-being of people living in urban and peri-urban areas of the Global South; regardless of what metaphors scientists from the Global North desire to use to describe these benefits and costs.

Author Contributions: Conceptualization, A.P.-G. and F.J.E.; methodology, F.J.E. and A.P.-G.; formal analysis, A.P.-G., F.C., and F.J.E.; writing—original draft preparation, A.P.-G. and F.J.E.; writing—review and editing, F.J.E., A.P.-G., and F.C. All authors have read and agreed to the published version of the manuscript.

Funding: This research received no external funding.

Data Availability Statement: The data presented in this study are available on request from A.P.-G. and the corresponding author. The data are not publicly available due to personal data and contact information collected as part of the survey.

Acknowledgments: We thank the Jardin Botánico de Bogotá- José Celestino Mutis for funding and leading this project particularly Carlos Fonseca, José Lopez, Catalina Lopera, and Maribel Vasquez for their assistance and coordination of field crews. We are also grateful to the Universidad del Rosario´s Biology Program's undergraduate Socio-ecological Systems class for their valuable contribution to this work. In particular, we thank Mari Paula Otero, Sara Pedraza, Andrea Aragon, Daniel A Quevedo, and Nubia Vazquez.

Appendix A

Table A1. Table listing response variables, units, and data types for the survey instrument presented in Supplementary Material A. Note: Only analyzed responses and variables are presented.

Question	Response Variable	Data Type/Units	How Data Were Analyzed
Is this place offering any benefits?	Awareness of ecosystem functions	Yes/No	Dichotomous variable 1 = This place is offering benefits. 0 = This place is not offering benefits.
Which of the following benefits is this place offering?	Perceptions of Ecosystem Services: Regulation: • Air purification • Climate Regulation • Flood Mitigation • Water purification • Shade Cultural • Aesthetic values • Recreation • Sense of well-being Provisioning/supporting/Benefits • Food • Biodiversity • Allows informal Job	Categorical	Variables for each ES Category. Regulation: Number of ES identified in this category. Cultural: Number of ES identified in this category. Provisioning: Number of ES identified in this category.
Which are the most important benefits for you?	Items in #2 were assigned an importance 1; being the most important	Ranking	This variable was not used for the statistical analyzes.
Would losing these benefits affect you?	Perceptions of benefits	Yes/No	Dichotomous variable. 1 = People consider that losing of these benefits affect them. 0 = People consider that losing of these benefits affects them.
What problems does this place present?	Perception of Ecosystem Disservices Environmental • Falling branches and trees • Trash and refuse • Invasive vegetation • Pet excrement • Insects or rats Financial • Maintenance costs • Lack of maintenance • Illicit drug sales and use Social • Crime • Fear • Rhinitis or allergies	Categorical	Variables for each EDS
Is this place safe?	Perception of security	Yes/No	Dichotomous variable 1 = Safe 0 = Unsafe

Table A1. *Cont.*

Question	Response Variable	Data Type/Units	How Data Were Analyzed
Do you believe that protecting natural areas improves the wellbeing of people living nearby?	Perceptions of benefits	Yes/No	Dichotomous variable 1 = Improvement of wellbeing 0 = No Improvement
Do you think urbanization affects quality of water and flooding?	Perception of urbanization's effects on water regulation benefits	Yes/No	Dichotomous variable. 1 = Urbanization affects quality of water. 0 = Urbanization has no affects on quality of water.
Can you identify 3 trees, palms, or plants in this place?	Biodiversity awareness	Yes/No	Dichotomous variable. This variable was used to create the environmental awareness variable. We used the polychronic command in Stata to perform the new variable. 1 = At least 1 plant was identified 0 = None
Are you aware of Reserva Forestal Protectora Bosque Oriental de Bogotá?	Policy awareness	Yes/No	Dichotomous variable. This variable was used to create the environmental awareness variable. We used the polychronic command in Stata to perform the new variable. 1 = People are aware of the Reserva Forestal Bosque Oriental 0 = People are not aware of the Reserva Forestal Bosque Oriental.
Is this reserve important for Bogotá's citizens' wellbeing?	Policy awareness	Yes/No	Dichotomous variable 1 = Reserve important for Bogotá's citizens' wellbeing 0 = Reserve is not important
Can you name wetlands that are located in Bogota?	Names of wetlands mentioned by respondents: Humedales de Bogotá Humedal Juan Amarillo o Tibabuyes Humedal Córdoba Humedal La Conejera Humedal de Santa María del lago Humedal El burro Humedal de Jaboque Humedal La vaca Humedal Torca-Guaymaral Humedal Tibanica Humedal de Capellanía Humedal Salitre Humedal La Florida Humedal Chicú Humedal Libélula Humedal de Techo Humedal Laguna de Chinará Humedal Techo Humedales Autopista Norte Complejo de humedales El Tunjo Humedal Capellanía Humedal Chiguasuque Humedal El Salitre Humedal Timiza	Categorical	Dichotomous variable. This variable was used to create the environmental awareness variable. We used the polychronic command in Stata to perform the new variable. 1 = At least 1 wetland is identified 0 = None

Table A1. *Cont.*

Question	Response Variable	Data Type/Units	How Data Were Analyzed
Are you willing to invest an additional amount in monthly utilities for programs that protect and recover natural areas in Bogotá?	Willingness to pay: • Not willing to pay • $1000 to $4000 COP • $5000 to 10,000 COP • More than 10,000 COP • More than 30,000 COP	Categorical Colombian Pesos (COP)	Dichotomous variable that was used to create Unwillingness to Invest (UTI). If UTI = 1; conversely, UTI = 0 if people were willing to pay.
Why are you willing to invest or not?	Governance responses • Taxes are already paid and it's the government's responsibility/It is not my responsibility • Corruption • Lack of resources • It is our responsibility Well-being/other responses • Air purification • I use them • Improves citizens' welfare • Urban biodiversity conservation • I do not know • No answer	Categorical	Variable used to create weak governance if response was: 1. Perception of corruption or funds not used appropriately, 2. Conservation of green areas and wetlands is already paid for in taxes, 3. The respondents already pay too much tax, and 4. It is the government´s responsibility to conserve green areas.
Do you believe that climate change is happening?	Climate change awareness	Yes/No	Dichotomous variable. 1 = Believe in Climate change 2 = Do not believe in Climate change

References

1. Nowak, D.J.; Dwyer, J.F. Understanding the benefits and costs of urban forest ecosystems. In *Urban and Community Forestry in the Northeast*; Springer: Dordrecht, The Netherlands, 2007; pp. 25–46.
2. Valente, D.; Pasimeni, M.R.; Petrosillo, I. The role of green infrastructures in Italian cities by linking natural and social capital. *Ecol. Indic.* **2020**, *108*, 105694. [CrossRef]
3. Haase, D.; Larondelle, N.; Andersson, E.; Artmann, M.; Borgström, S.; Breuste, J.; Gomez-Baggethun, E.; Gren, Å.; Hamstead, Z.; Hansen, R.; et al. A quantitative review of urban ecosystem service assessments: Concepts, models, and implementation. *Ambio* **2014**, *43*, 413–433. [CrossRef] [PubMed]
4. Escobedo, F.J.; Kroeger, T.; Wagner, J.E. Urban forests and pollution mitigation: Analyzing ecosystem services and disservices. *Environ. Pollut.* **2011**, *159*, 2078–2087. [CrossRef] [PubMed]
5. Lyytimäki, J.; Petersen, L.K.; Normander, B.; Bezák, P. Nature as a nuisance? Ecosystem services and disservices to urban lifestyle. *Environ. Sci.* **2008**, *5*, 161–172. [CrossRef]
6. Soto, J.R.; Escobedo, F.J.; Khachatryan, H.; Adams, D.C. Consumer demand for urban forest ecosystem services and disservices: Examining trade-offs using choice experiments and best-worst scaling. *Ecosyst. Serv.* **2018**, *29*, 31–39. [CrossRef]
7. Andersson, E.; Barthel, S.; Ahrné, K. Measuring social–Ecological dynamics behind the generation of ecosystem services. *Ecol. Appl.* **2007**, *17*, 1267–1278. [CrossRef]
8. Ernstson, H.; Barthel, S.; Andersson, E.; Borgström, S.T. Scale-crossing brokers and network governance of urban ecosystem services: The case of stockholm. *Ecol. Soc.* **2010**, *15*, 28. [CrossRef]
9. Kabisch, N. Ecosystem service implementation and governance challenges in urban green space planning—The case of Berlin, Germany. *Land Use Policy* **2015**, *42*, 557–567. [CrossRef]
10. Lyytimäki, J.; Sipilä, M. Hopping on one leg-The challenge of ecosystem disservices for urban green management. *Urban For. Urban Green.* **2009**, *8*, 309–315. [CrossRef]
11. Andersson, E.; McPhearson, T.; Kremer, P.; Gomez-Baggethun, E.; Haase, D.; Tuvendal, M.; Wurster, D. Scale and context dependence of ecosystem service providing units. *Ecosyst. Serv.* **2015**, *12*, 157–164. [CrossRef]
12. Haase, D.; Frantzeskaki, N.; Elmqvist, T. Ecosystem Services in Urban Landscapes: Practical Applications and Governance Implications. *Ambio* **2014**, *43*, 407–412. [CrossRef] [PubMed]
13. Balooni, K.; Gangopadhyay, K.; Kumar, B.M. Governance for private green spaces in a growing Indian city. *Landsc. Urban Plan.* **2014**, *123*, 21–29. [CrossRef]

14. Daw, T.M.; Brown, K.; Rosendo, S.; Pomeroy, R. Applying the ecosystem services concept to poverty alleviation: The need to disaggregate human well-being. *Environ. Conserv.* **2011**, *38*, 370–379. [CrossRef]
15. Scopelliti, M.; Carrus, G.; Adinolfi, C.; Suárez-Cáceres, G.; Colangelo, G.; Lafortezza, R.; Panno, A.; Sanesi, G. Staying in touch with nature and well-being in different income groups: The experience of urban parks in Bogotá. *Landsc. Urban Plan.* **2016**, *148*, 139–148. [CrossRef]
16. Pascual, U.; Balvanera, P.; Díaz, S.; Pataki, G.; Roth, E.; Stenseke, M.; Watson, R.T.; Dessane, E.B.; Islar, M.; Kelemen, E.; et al. Valuing nature's contributions to people: The IPBES approach. *Curr. Opin. Environ. Sustain.* **2017**, *26–27*, 7–16. [CrossRef]
17. Escobedo, F.J.; Giannico, V.; Jim, C.Y.; Sanesi, G.; Lafortezza, R. Urban forests, ecosystem services, green infrastructure and nature-based solutions: Nexus or evolving metaphors? *Urban For. Urban Green.* **2019**, *37*, 3–12. [CrossRef]
18. Dıaz, S.; Pascual, U.M.; Stenseke, B.; Martın-López, R.T.; Watson, Z.; Molnár, R.; Hill, K.M.; Chan, I.A.; Baste, K.A.; Brauman, S.; et al. Assessing nature's contributions to people: Recognizing culture, and diverse sources of knowledge, can improve assessments. *Science* **2018**, *359*, 270–272.
19. Carriazo, F. Arborización y crimen urbano en Bogotá. *IDEAS*. 2017. Documentos CEDE 015286, Universidad de los Andes - CEDE. Available online: https://ideas.repec.org/p/col/000089/015286.html (accessed on 26 December 2020).
20. Dobbs, C.; Escobedo, F.J.; Clerici, N.; De La Barrera, F.; Eleuterio, A.A.; MacGregor-Fors, I.; Reyes-Paecke, S.; Vásquez, A.; Camaño, J.D.Z.; Hernández, H.J. Urban ecosystem Services in Latin America: Mismatch between global concepts and regional realities? *Urban Ecosyst.* **2019**, *22*, 173–187. [CrossRef]
21. Chaudhry, P.; Singh, B.; Tewari, V.P. Non-market economic valuation in developing countries: Role of participant observation method in CVM analysis. *J. For. Econ.* **2007**, *13*, 259–275. [CrossRef]
22. Kenward, R.E.; Whittingham, M.J.; Arampatzis, S.; Manos, B.D.; Hahn, T.; Terry, A.; Simoncini, R.; Alcorn, J.; Bastian, O.; Donlan, M.; et al. Identifying governance strategies that effectively support ecosystem services, resource sustainability, and biodiversity. *Proc. Natl. Acad. Sci. USA* **2011**, *108*, 5308–5312. [CrossRef]
23. Kreye, M.M.; Adams, D.C.; Escobedo, F.J.; Soto, J.R. Does policy process influence public values for forest-water resource protection in Florida? *Ecol. Econ.* **2016**, *129*, 122–131. [CrossRef]
24. Huang, C.-W.; McDonald, R.I.; Seto, K.C. The importance of land governance for biodiversity conservation in an era of global urban expansion. *Landsc. Urban Plan.* **2018**, *173*, 44–50. [CrossRef]
25. Perkins, H.A. Out from the (Green) shadow? Neoliberal hegemony through the market logic of shared urban environmental governance. *Politi Geogr.* **2009**, *28*, 395–405. [CrossRef]
26. Turnhout, E.; Neves, K.; De Lijster, E. 'Measurementality' in Biodiversity Governance: Knowledge, Transparency, and the Intergovernmental Science-Policy Platform on Biodiversity and Ecosystem Services (Ipbes). *Environ. Plan. A Econ. Space* **2014**, *46*, 581–597. [CrossRef]
27. Bell, S.; Hindmoor, A. Governance without government? The case of the forest stewardship council. *Public Adm.* **2011**, *90*, 144–159. [CrossRef]
28. Lawrence, A.; De Vreese, R.; Johnston, M.; Bosch, C.C.K.V.D.; Sanesi, G. Urban forest governance: Towards a framework for comparing approaches. *Urban For. Urban Green.* **2013**, *12*, 464–473. [CrossRef]
29. Connolly, J.J.; Svendsen, E.S.; Fisher, D.R.; Campbell, L.K. Organizing urban ecosystem services through environmental stewardship governance in New York City. *Landsc. Urban Plan.* **2013**, *109*, 76–84. [CrossRef]
30. Van Den Bosch, C.C.K.; Rodbell, P.; Salbitano, F.; Sayers, K.; Villarpando, S.J.; Yokohari, M. The changing governance of urban forests. *Unasylva* **2018**, *69*, 37–42.
31. Launay-Gama, C.; Pachón, M. *Prácticas de evaluación de la gobernanza en América Latina*; Universidad de Los Andes: Bogotá, Colombia, 2011.
32. Atmiş, E.; Batuhan Günşen, H.; Yücedağ, C.; Lise, W. Factors affecting the use of urban forests in Turkey. *Turk. J. For.* **2017**, *18*, 1–10. [CrossRef]
33. Shackleton, R.T.; Angelstam, P.; Van Der Waal, B.; Elbakidze, M. Progress made in managing and valuing ecosystem services: A horizon scan of gaps in research, management and governance. *Ecosyst. Serv.* **2017**, *27*, 232–241. [CrossRef]
34. Lockwood, M. Good governance for terrestrial protected areas: A framework, principles and performance outcomes. *J. Environ. Manag.* **2010**, *91*, 754–766. [CrossRef] [PubMed]
35. Barrett, C.; Gibson, C.; Hoffman, B.; Mccubbins, M. The complex links between governance and biodiversity. *Conserv. Biol.* **2006**, *20*, 1358–1366. [CrossRef] [PubMed]
36. Falk, T.; Spangenberg, J.H.; Siegmund-Schultze, M.; Kobbe, S.; Feike, T.; Kuebler, D.; Settele, J.; Vorlaufer, T. Identifying governance challenges in ecosystem services management—Conceptual considerations and comparison of global forest cases. *Ecosyst. Serv.* **2018**, *32*, 193–203. [CrossRef]
37. Piña, W.H.A.; Martínez, C.I.P. Urban material flow analysis: An approach for Bogotá, Colombia. *Ecol. Indic.* **2014**, *42*, 32–42. [CrossRef]
38. Escobedo, F.J.; Clerici, N.; Staudhammer, C.L.; Corzo, G.T. Socio-ecological dynamics and inequality in Bogotá, Colombia's public urban forests and their ecosystem services. *Urban For. Urban Green.* **2015**, *14*, 1040–1053. [CrossRef]
39. Meza, C.A. Crossroads and Conflict. Urbanization, Conservation and Rurality in the Eastern Mountains of Bogotá. *Rev. Colomb. Antropol.* **2008**, *44*, 439–480. [CrossRef]

40. Escobedo, F.J.; Clerici, N.; Staudhammer, C.L.; Feged-Rivadeneira, A.; Bohorquez, J.C.; Tovar, G. Trees and Crime in Bogota, Colombia: Is the link an ecosystem disservice or service? *Land Use Policy* **2018**, *78*, 583–592. [CrossRef]
41. Wurster, D.; Artmann, M. Development of a Concept for Non-monetary Assessment of Urban Ecosystem Services at the Site Level. *Ambio* **2014**, *43*, 454–465. [CrossRef]
42. Martín-López, B.; Iniesta-Arandia, I.; García-Llorente, M.; Palomo, I.; Casado-Arzuaga, I.; Del Amo, D.G.; Gómez-Baggethun, E.; Oteros-Rozas, E.; Palacios-Agundez, I.; Willaarts, B.; et al. Uncovering Ecosystem Service Bundles through Social Preferences. *PLoS ONE* **2012**, *7*, e38970. [CrossRef]
43. Dobbs, C.; Hernández-Moreno, Á.; Reyes-Paecke, S.; Miranda, M.D. Exploring temporal dynamics of urban ecosystem services in Latin America: The case of Bogota (Colombia) and Santiago (Chile). *Ecol. Indic.* **2018**, *85*, 1068–1080. [CrossRef]
44. Agbenyega, O.; Burgess, P.J.; Cook, M.; Morris, J. Application of an ecosystem function framework to perceptions of community woodlands. *Land Use Policy* **2009**, *26*, 551–557. [CrossRef]
45. Ricaurte, L.F.; Olaya-Rodríguez, M.H.; Cepeda-Valencia, J.; Lara, D.; Arroyave-Suárez, J.; Finlayson, C.M.; Palomo, I. Future impacts of drivers of change on wetland ecosystem services in Colombia. *Glob. Environ. Chang.* **2017**, *44*, 158–169. [CrossRef]
46. Hartter, J.; Goldman, A. Local responses to a forest park in western Uganda: Alternate narratives on fortress conservation. *Oryx* **2010**, *45*, 60–68. [CrossRef]
47. Allendorf, T.D.; Yang, J. The role of ecosystem services in park–people relationships: The case of Gaoligongshan Nature Reserve in southwest China. *Biol. Conserv.* **2013**, *167*, 187–193. [CrossRef]
48. García-Suaza, A.F.; Guataquí, J.C.; Guerra, J.A.; Maldonado, D. Beyond the Mincer equation: The internal rate of return to higher education in Colombia. *Educ Econ.* **2014**, *22*, 328–344. [CrossRef]
49. Kadykalo, A.N.; López-Rodriguez, M.D.; Ainscough, J.; Droste, N.; Ryu, H.; Ávila-Flores, G.; Le Clec'H, S.; Muñoz, M.C.; Nilsson, L.; Rana, S.; et al. Disentangling 'ecosystem services' and 'nature's contributions to people'. *Ecosyst. People* **2019**, *15*, 269–287. [CrossRef]
50. Acey, C. Managing wickedness in the Niger Delta: Can a new approach to multi-stakeholder governance increase voice and sustainability? *Landsc. Urban Plan.* **2016**, *154*, 102–114. [CrossRef]
51. Sarkki, S. Governance services: Co-producing human well-being with ecosystem services. *Ecosyst. Serv.* **2017**, *27*, 82–91. [CrossRef]

 *land*

Review

Which Traits Influence Bird Survival in the City? A Review

Swaroop Patankar [1,*], **Ravi Jambhekar** [1], **Kulbhushansingh Ramesh Suryawanshi** [2,3] **and Harini Nagendra** [1]

1   Center for Climate Change and Sustainability, Azim Premji University, Bengaluru 562125, India; jambhekarm@iisc.ac.in (R.J.); harini.nagendra@apu.edu.in (H.N.)
2   Nature Conservation Foundation, 1311, 'Amritha', 12th A Main, Vijayanagar 1st Stage, Mysore 570017, India; kulbhushan@ncf-india.org
3   Snow Leopard Trust, 4649 Sunnyside Avenue North, Suite 325, Seattle, WA 98103, USA
*   Correspondence: swaroop.patankar@apu.edu.in

Abstract: Urbanization poses a major threat to biodiversity worldwide. We focused on birds as a well-studied taxon of interest, in order to review literature on traits that influence responses to urbanization. We review 226 papers that were published between 1979 and 2020, and aggregate information on five major groups of traits that have been widely studied: ecological traits, life history, physiology, behavior and genetic traits. Some robust findings on trait changes in individual species as well as bird communities emerge. A lack of specific food and shelter resources has led to the urban bird community being dominated by generalist species, while specialist species show decline. Urbanized birds differ in the behavioral traits, showing an increase in song frequency and amplitude, and bolder behavior, as compared to rural populations of the same species. Differential food resources and predatory pressure results in changes in life history traits, including prolonged breeding duration, and increases in clutch and brood size to compensate for lower survival. Other species-specific changes include changes in hormonal state, body state, and genetic differences from rural populations. We identify gaps in research, with a paucity of studies in tropical cities and a need for greater examination of traits that influence persistence and success in native vs. introduced populations.

Keywords: urbanization; birds; ecosystem services; survival; adaptations; traits

Citation: Patankar, S.; Jambhekar, R.; Suryawanshi, K.R.; Nagendra, H. Which Traits Influence Bird Survival in the City? A Review. *Land* **2021**, *10*, 92. https://doi.org/10.3390/land10020092

Academic Editors: Alessio Russo and Giuseppe T. Cirella
Received: 26 November 2020
Accepted: 19 January 2021
Published: 20 January 2021

Publisher's Note: MDPI stays neutral with regard to jurisdictional claims in published maps and institutional affiliations.

1. Introduction

Today, urbanization presents a major threat to biodiversity [1,2]. By 2050, 68 percent of the world's population will live in urban areas [3]. As urban settlements increase, the landscape undergoes drastic change from its pristine state [4]. However, ecologically speaking, urban areas are highly modified and fragmented habitats, in the form of several managed and unmanaged urban green spaces, like public and private garden, nature reserves within the city, vacant plots, to name a few, which are capable of providing resources to a small number of highly adaptable species of fauna [2,4–6]. Not only resident, but several migratory, species make use of this urban habitat [7]. The urban avian community is composed of introduced and invasive species and highly adaptable native species [8,9], although Aronson et al. [8] observed that the community is dominated by locally adapted species, with only five percent non-native species. The urban greenspace along with the biodiversity it harbors, forms a unique ecosystem within the human dominated landscapes, capable of providing several ecosystem services to the cities [2]. Ecosystem services are the services that are beneficial to humans as a result of naturally occurring processes [10], which render them all the more important in human dominated landscapes, like cities.

Birds, which are some of the most successful urban adapters, provide a range of ecosystem services. Birds, in general, are capable of significantly affecting ecosystem processes, due to their specific characters, like flight capability, high metabolic demand, which leads to tolerance to discontinuous food sources; and, flock formation, which significantly affects ecosystems processes, like nutrient cycling [10]. The urban bird community structure also acts as an indicator of the structure and functionality of urban areas as habitats [11]. In this study, it was found that habitat specialists, like woodpeckers and hole-nesting birds, occurred in the peripheral areas of the city, indicating healthier vegetation cover, while habitat generalists increased near the city center and residential areas with less vegetation cover.

Birds have been found to provide all four types of ecosystem services that were mentioned in the UN millennium ecosystem assessment, namely, regulatory, provisioning, supporting, and cultural [10]. Urban birds provide regulatory services by acting as pest control agents when they feed on disease causing insects, prey on rodent population, and scavenge on garbage; provisioning services by acting as seed dispersers when they ingest fruits, which helps in plant regeneration; and, supporting services by acting as nutrient recyclers via their excretory products [12,13]. However, urban birds perhaps play the most crucial role in providing various cultural services. One of the most important services rendered by urban birds is to act as a connecting link between the natural environment and the increasingly nature deprived urban denizens [7]. Birds in residential neighborhoods in cities are generally valued for their color and songs, for providing mental and physical wellbeing, as indicators of seasonal change, for education, and for giving sense of familiarity, by the residents [7,14]. Birds are also an inspiration for art, recreational activities, like birdwatching and wildlife gardening [10]. In India, it is a tradition of grain merchants to give a small portion of their goods to granivorous birds, like house sparrows (*Passer domesticus*), as an offering for prosperity in business. In the city of Delhi in India, it is a common practice of offering meat to scavengers, like black kites (*Milvus migrans*) as religious practice, due to which the black kite population dynamics in the city is significantly affected [15].

It is abundantly clear that the conservation of these ecologically important fauna in urban landscapes is crucial. Many cities are now experimenting with the management of urban green spaces for conservation and, there have been numerous studies on the effect of such management interventions on bird ecosystem services in cities [13]. However, for such management practices to be entirely successful, it is also important to understand the uniqueness of birds who call this urban habitat their home. One is then compelled to ask what characteristics sets apart the birds able to adapt to this unique habitat from those who disappear from it? It would be counter-intuitive to the purpose of conservation if such measures are taken without understanding the changes in "traits" of urban birds.

Studies on trait changes have grown in recent years, as researchers begin to focus on the study of urban areas as modified habitats. It has been suggested that studying the traits of urban biodiversity could be a more effective approach for preserving the ecosystem functions and services in urban areas, ss such data provide authentic and useful information to urban planners [12]. Birds comprise the taxon that is perhaps the best studied in cities. Birds constitute especially useful model species for studying trait changes, as they respond to changes rapidly, are of interest to naturalists and the public at large, are easy to monitor, and they show measurable changes. They have hence been studied widely in the urban context [16,17].

McKinney et al. [1] conducted the last comprehensive review assessing diverse trait changes that were observed in urban birds. In general, only certain species, which are characterized by specific traits or combinations of traits, seem to be more capable of coping with the environmental alterations that are imposed by urbanization [18]. This capability might differ according to season, geographic region, or city structure [9], but such species of urban utilizers can tolerate a wider range of environmental conditions [19] and they appear to exhibit similar characteristics globally [20,21]. The research on bird adaptations to urban habitat has since expanded considerably. Many researchers have carried out reviews and meta-analyses of specific trait changes, including behavioral traits [22,23], community traits [21], and physiological changes [24]. However, there is a need for an updated comprehensive review that collates and synthesizes existing information.

This paper reviews research studies on traits that influence the survival and response of birds to urbanization, based on a literature review of research that was conducted between 1979 and 2020, identifying the state of current knowledge, highlighting key knowledge gaps, and identifying geographic and species-specific disparities in research that require specific focus. The specific aim of this study was to synthesize the information available on studies when comparing urban and rural birds while taking a trait-based approach. Finally, we highlight the gaps in knowledge and pinpoint future directions for research.

2. Materials and Methods

We listed and categorized the traits that are described in birds referring to the review paper by McKinney [1] and making additions of our own. The traits were selected with the knowledge that they were essential for basic survival and reproductive success of birds. To begin this review, we conducted an intensive literature search for each trait, using Google Scholar during August 2019, based on combinations of the following keywords: "urban", "birds", "urbanization", "impacts", "traits", "changes", "effects", "behavioral", "physiological", "functional", "diversity", "global", "avifauna", "song", "life history", "nesting", "foraging", "generalist", and "specialist". Each publication that resulted from this search was reviewed to determine whether it met the criteria for inclusion. The basic criterion was that the study had to be either a comparative field study or an experimental study looking at effects of urbanization in an urban-rural gradient on bird communities or a single bird species. The literature was searched by its importance and relevance to the key words used. Review papers and experimental studies were included as a part of this study. The reviewed papers were grouped into five broad trait groups: ecological traits, which included traits pertaining to ecological niche and community interactions; life history traits, which included traits that are related to reproductive success, breeding cycles, and overall survival; physiological traits, which were related to the physiological, endocrine, and morphological changes in the birds in urban areas; behavioral traits, which were the changes in important behaviors, like singing, fear responses, innovative behavior, to name a few; and finally, genetic traits, which were the trait changes in the genetic structure of urban birds (for complete list, refer Supplementary Materials Table S1). Each paper was read and categorized based on the methodology used, the results, and interpretations of the study. Knowledge gaps and scope for further research were documented.

3. Results

A total of 226 publications were finally included for review (Supplementary Materials Table S2). The earliest paper that we found was published in 1979. The field has grown rapidly since then, with research on several different traits (Figure 1; Table 1). The interdisciplinary nature of the topic has led to an expansion in the number of journals publishing urban ecological research (Supplementary Materials Figure S1). Behavioral Ecology published the most papers on trait adaptations in urban birds, followed by Urban Ecosystems, PLoS ONE, and Scientific Reports (Supplementary Materials Figure S1). Many of the trait changes have been studied on large bird communities at the regional or global scale. Song birds and water birds are well studied, and most of our understanding of impact of urbanization comes from these guilds of birds (Table 2). Additionally, some bird species such as Eurasian blackbird (*Turdus merula*) are widely represented in the literature as compared to others. Studies from temperate countries dominate the literature, with only a few studies from the tropics (Figure 2).

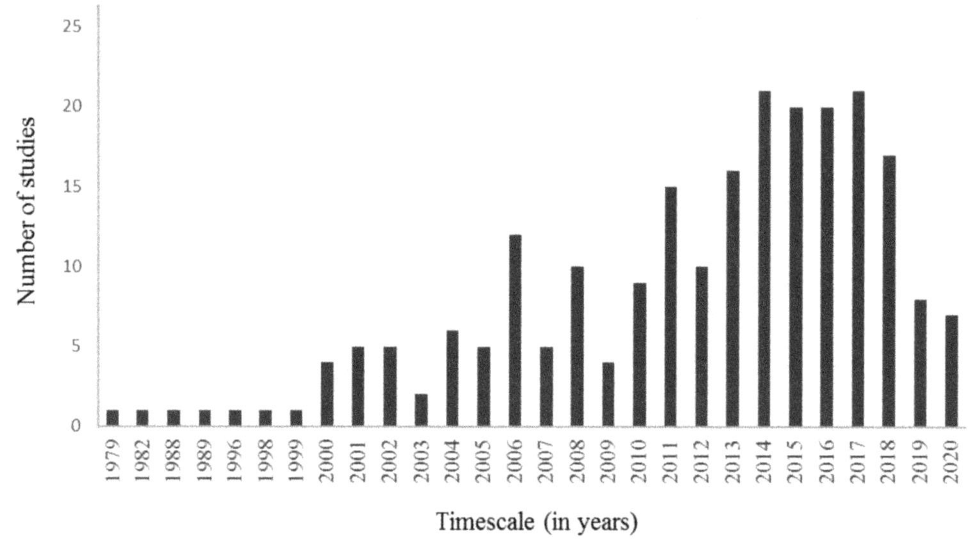

Figure 1. Number of research papers on urban bird traits related to urbanization from 1979 to 2020.

Table 1. Number of published articles on each of the bird traits included for this review along with the number of studies on individual species (<4 species) or large communities (>4 species).

Trait Studied	Number of Studies (<4 Species Studied)	Number of Studies (>4 Species Studied)	Total
Ecological			
Diversity	1	14	15
Nesting spaces	7	9	16
Abundance		8	8
Richness		10	10
Diet breadth	1	9	10
Generalist-specialist	5	7	12
Distribution pattern	1	3	4
Density		3	3
Migration	4	3	7
Habitat dependence	1	4	5
Dispersal		1	1
Sociality and sedentariness		1	1
Community composition		1	1
Habitat preference	2		2
Life history			
Reproduction cycle	3	2	5
Clutch size	5	4	9
Nesting success	4	2	6
Breeding timing and/or performance	2	1	3
Brood size	2	2	4
Fecundity and adult survival	2	1	3
Activity time range	1		1
Migration		1	1
Fitness	1		1
Physiological			
Reproductive physiology, circadian rhythm and migration	12	12	24
Brain size	1	5	6
Body size and/or mass	5	4	9
Inflammatory response, oxidative stress	1	2	3

Table 1. *Cont.*

Trait Studied	Number of Studies (<4 Species Studied)	Number of Studies (>4 Species Studied)	Total
Stress physiology	10	1	11
Endocrine traits		1	1
Bill length	2		2
Plumage coloration	1	1	2
Behavioral			
Fear and stress responses	17	10	27
Foraging behavior innovation	6	6	12
Innovation, learning and problem-solving ability, Neophobia and risk assessment	10	5	15
Entire song structure	7	3	10
Song frequency and/or amplitude	16	1	17
Aggression	11		11
Dawn chorus	1		1
Call frequency, amplitude	2		2
Song frequency, bandwidth	2		2
Song syllable	1		1
Song timing	1		1
Genetic			
Gene expression	2		2
Genetic changes	1	1	2
Genetic divergence	1	1	2
Total	148	139	

Table 2. Bird families represented in the literature studying urbanization induced trait changes. Bird community studies comprise of global reviews or analysis of global datasets, while individual family studies comprise of local datasets and experimental studies.

Bird Family	Common Name	Ecological	Life History	Physiological	Behavioral	Genetic	Total
Bird community (>2 families)		58	7	15	29	2	111
Turdidae	Thrushes	1	1	8	8	2	20
Passerellidae	New world sparrows		1	7	11		19
Paridae	Tits		3		11	2	16
Passeridae	Old world sparrows			3	5		8
Fringillidae	Finches				6		6
Cardinalidae	Cardinals				4		4
Corvidae	Crows	1	1	1	2		5
Strigidae	True owls	3	1		2		6
Zosteropidae	White-eyes				4		4
Accipitridae	Accipiters	2	4	1	4		11
Mimidae	Mimids			2	1		3
Rallidae	Rails				3		3
Emberizidae	American sparrows			2			2
Muscicapidae	Old world flycatchers				2		2
Sturnidae	Starlings				2		2
Troglodytidae	Wrens			1	1		2
Tyrannidae	Tyrants flycatchers		1		1		2
Tytonidae	Barn owls	2	1				3
Artamidae	Crows				1		1
Columbidae	Pigeons and doves				1		1
Corcoracidae	Australian mudnesters	1					1
Falconidae	Falcons and caracaras	2	3		1		6
Icteridae	Songbirds				1		1
Laridae	Gulls			1	5		6
Meliphagidae	Honeyeaters				1		1
Monarchidae	Monarch flycatchers				1		1
Nectariniidae	Sunbirds			1			1
Thraupidae	Tanagers				1		1
Pandionidae	Ospreys				1		1
Haematopodidae	Oystercatchers	1					1
Total		71	23	42	109	6	251

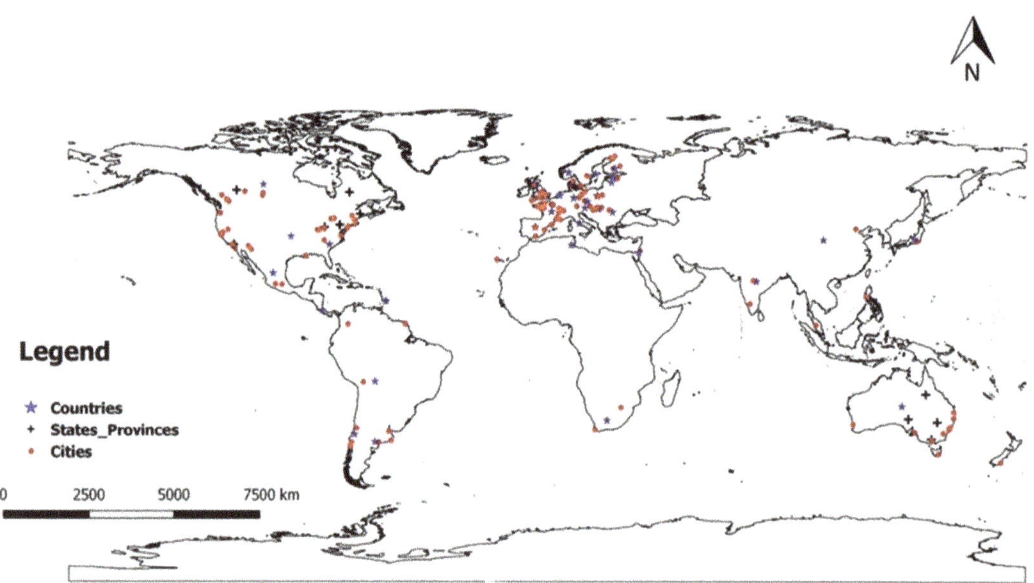

Figure 2. Distribution of research studies on urban bird traits across the world. Studies carried out at state or province level included multiple cities which were not specified in the publication.

3.1. Ecological Traits

3.1.1. Bird Diversity and Abundance

An overall decrease in diversity from rural to urban areas [25–27] as well as an increase in abundance of only a small subset of species are some of the most important ecological effects of urbanization on birds. One study observed a decreased overall abundance of birds in urban areas [28]. Species richness also shows a decreasing trend with an increase in urbanization, a pattern observed globally [29,30]. A study encompassing several cities in Europe observed that urban bird communities also showed lower evolutionary distinctiveness than rural communities [31]. A global review observed that, despite the loss of forest dependent or native species in urban areas, the functional diversity remained the same [32].

The species richness and density of seasonally migrating species also reduces with an increase in built infrastructure [33,34]. Many birds, such as Eurasian blackbird and house sparrow, show a loss in migratory behavior, as there are enough food resources available in urban areas to support them through the winter months. Such a trend may eventually lead to reduced richness in urban areas [35]. Birds with specific dietary requirements may also avoid migrating to cities because of a lack of adequate food resources. For example, insectivorous birds might find difficulty in finding insects in cities, thus reducing their urban population [35,36]. Studies also speculate the indirect effects of urbanization on resident vs. migrant bird dynamics. Resident species may occupy good quality nesting sites in cites before migrants arrive, thus driving migrants away from urban areas by competitive exclusion [37].

3.1.2. Generalism—Specialism

Specialist species are more likely to be negatively affected by change than generalists [38–40]. Devictor et al. [41] found that generalists, which use multiple habitats in the landscape matrix, are less affected by habitat fragmentation than specialists, which are dependent on one or a few habitat types. Generalists should benefit from disturbed landscapes, as there will be reduction in competition from specialists, who do well in stable environments and, hence, will be negatively affected by a degradation in landscape [42,43]. Generalist species usually have large niche breadths, lay multiple clutches, and have broad diets, which makes them more successful in cities [39]. This indicates that, to understand whether a bird can cope with urbanization, we need to consider the effects of multiple traits together.

Whether a species is a generalist or a specialist is largely due to the ability of these birds to adapt to feeding and nesting preferences, as described further below.

3.1.3. Diet Breadth

Diet breadth is one of the most important traits affected by urbanization [44,45]. Bird species that feed on fruits and grains tend to increase in numbers in urbanized areas as compared to insectivorous species [37,46,47]. This might be because cities have a substantial proportion of fruiting trees [48]. Similarly, a review on urban raptors suggested that urban areas are typically inhabited by raptors that are feeding on forest dwelling birds, due to a prevalence of large trees in cities, but open-area raptors feeding on grassland species do not generally inhabit urban areas [49]. Raptors feeding on rodents and scavenging raptors also show similar foraging, depending on the availability of suitable prey or carrion [50–52].

In many countries, there is a culture of putting out bird feeders, supplementing bird diets with seeds, nuts, and grains [53]. Granivorous birds benefit from this and, hence, do well in urbanized areas [27,37]. Callaghan et al. [39] observe that insectivorous and granivorous birds avoid urban areas, a phenomenon that requires further investigation in order to assess whether it is specific to countries where feeding birds is not a popular activity.

Kark et al. [44] report that, in downtown areas, most of the birds were omnivores, especially in temperate countries. Similarly, urban bird composition in Santiago, Chile, predominantly consisted of omnivorous and granivorous species [47]. Omnivorous species, like Eurasian blackbird, great tit (*Parus major*), and house sparrow, are commonly studied in urban avian ecology research (Figure 3). Again, omnivore birds are likely to feed on food scraps, discarded food items and, thus, seem to be possibly making use of the food resources that are discarded by humans. Apart from food preferences, urbanization can negatively affect birds exhibiting solitary feeding behavior [39]. Many birds show subtle changes in foraging behavior in order to adjust to the novel foraging sites. Ground foraging and insectivorous birds forage differentially in suburban remnant patches vs. continuous vegetation [54].

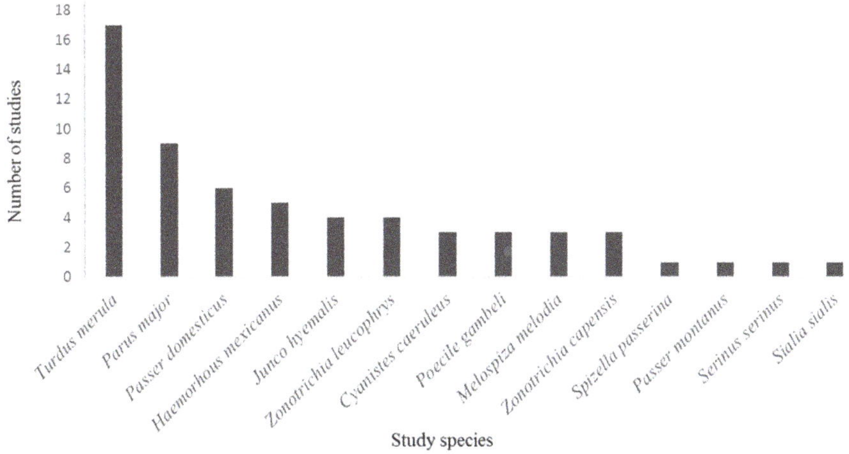

Figure 3. Most commonly studied bird species for urbanization induced trait changes.

3.1.4. Nesting Sites

The utilization of nesting sites is an important factor determining the success of bird species [45]. One of the most consistent effects of urbanization is on ground-nesting birds, whose abundance consistently decreased across most of the studies [16,55]. Small sized ground nesters are most impacted by urbanization in Australia [56]. Birds that nest on high trees and in tree cavities have a better chance of survival in cities [37,57]. Studies speculate that urbanization does not necessarily increase the availability of cavity nesting sites. However, cavity nesters might be less prone to predation, because of their nesting habit, hence surviving better in cities [37]. Alternatively, cavity nesting birds might be using artificial nesting sites, such as nest boxes provided in cities. Cities with good tree cover will benefit birds that nest on trees, as there will not be a scarcity of nesting sites.

Birds that are more adaptable and use a variety of nesting strategies like making use of man-made structures, are more likely to do better in urban areas as compared to birds with specialized nesting preferences [58–60]. Adaptive nesting strategy also ensures better productivity, as is seen in the urban peregrine Falcon (*Falco peregrinus*), as many artificial structures provide better protection against predators or the elements [61]. Consequently, species that are adapted to urban conditions show higher abundance in urbanized areas [8,59,62]. Other birds, such as open-cup nesters, which require trees and shrubs to support their nests, are negatively associated with urbanized areas [59].

3.2. Life History Traits

Life history traits, such as clutch size and brood success, have a considerable positive impact on bird populations in urban areas [37]. Many urban species show an increase in the clutch size and brood size [63]. The increase in number of eggs and chicks helps these birds to overcome the losses that occurred during predation or the effects of urbanization such as mortality caused due to collision with cars or windows [37,61]. Urban birds tend to lay eggs earlier than their rural counterparts [64–68]. This might be because of the improved resource availability in cities. There are exceptions; Chamberlain et al. [69] found that, even though results vary from species to species, most urban bird populations are characterized by slightly greater annual productivity and lower nestling weight. However, certain species do not show such a pattern. For example, Marini et al. [70] find that clutch size does not differ in mountain chickadees (*Poecile gambeli*) as one moves along the urban-rural gradient in North America. Similarly, in magpies (*Pica pica*), a European species, the clutch size does not vary as one moves across the rural to urban gradient [71]. Brood size and nestlings per nesting attempt did not show a consistent pattern of differences between urban and non-urban areas [69]. Some raptor species also fledged fewer offspring in urban areas when there was a lack of prey or excessive human disturbance [68].

3.3. Physiological Traits

3.3.1. Body Mass, Size, and Plumage Coloration

Birds may produce large broods in urban areas, but the condition (average body size, morphological features) of offspring is poor, because there is a high chance of survival, even for the low-quality offspring [72,73]. The change in diet preferences might be associated with higher survival rates, but poor body condition. Birds have ample resources available throughout the year in the city, which is perhaps the reason they do not accumulate more body fat. Another possible reason is that temperatures in cities are higher due to urban heat island effects; hence, birds have smaller sizes: as theory predicts that as temperatures drop, body size increases [73]. Adverse ecological effects may also constrain the body size or condition of offspring [73]. Nestlings in urban habitats are fed less amount of food, or lower quality food, and they reach a lower body mass [74,75]. However, the urban area-smaller mass relation is not observed in all urban birds. A study on silver gulls (*Larus novaehollandiae*) in Tasmania found that adult male gulls in urban areas had greater body mass than adult male gulls in rural areas [76]. Perhaps omnivorous species show an opposite trend in terms of body mass due to the consumption of a wide variety of foods.

Liker et al. [75] found that house sparrows were larger in rural areas as compared to urbanized areas. Sparrows in the Budapest city center were more than 5% lighter than sparrows at the least urbanized locations, and the leanness of urban birds was detectable, even when they compared differently urbanized habitats with similar utilization [75,77].

Although the literature on body size and mass of urban birds is comprehensive, research is only now emerging on changes in plumage coloration in urban birds. A recent study across three cities in Argentina showed that the amount of built area negatively affected the color diversity of bird communities. Hence, the bird community in the city was predominantly composed of grey colored species [78]. Urbanization selects for birds with similar colors, primarily those matching the surrounding habitat. Because plumage coloration is an important trait, not only for the reproductive success of birds, but also as a camouflage to avoid detection, it is important to be studied on a larger spatial scale.

3.3.2. Brain Size

Birds and mammals with relatively larger brain size may be associated with the ability to invade novel habitats [79,80]. Callaghan et al. [39] hypothesize that birds with large brain size might be favored in cities and, hence, such birds should be successful in urban environments. Theory predicts that larger brains might be advantageous to individuals in dealing with altered environments and it might help in innovative behavior and learning [81]. Larger brain size implies that individuals might be more able to explore novel environments and food resources, helping such birds adapt to city life [5]. However, experimental studies suggest that there is a lot of variation in brain size and success in urban environments [82,83]. More work is required in order to definitively declare whether brain size is actually related to success in urban environments.

3.3.3. Stress Response Physiology

Urban birds are exposed to a variety of stressors, like traffic noise [84,85], artificial light pollution [86,87], uneven food distribution, and chemical pollution [85,88]. Stress response to such disturbances is indicated by the level of plasma corticosterone hormone (CORT) secretion [24,84,89–91]. Many studies have revealed changes in the baseline and induced corticosterone levels in birds in response to urban stresses [85,88–90]. The Eurasian blackbird female shows an increased corticosterone secretion in response to artificial light exposure [86], even though Partecke et al. [88] had observed an overall decrease in corticosterone secretion in urban blackbirds. Urban birds also show reduced corticosterone secretion in noisy environments when compared to rural populations, as is displayed by song sparrows (*Melospiza melodia*) [84] and house wrens (*Troglodytes aedon*) [85]. Reduced corticosterone in urban birds is associated with low protein diets [85]. The decrease in CORT secretion in species that have colonized urban areas for a long time could be due to habituation to these stressors [84,86,89]. When male song sparrows from different parts of city were compared for CORT levels, they did not show any difference, which further indicates that these urban song sparrows might be adapted to urban stressors [84]. Bonier [24] reviewed the various endocrine trait changes in urban birds and found no consistent pattern in stress hormone change in all urban species. Even within the populations, the differences fluctuated according to the age and life history stage.

Elevated levels of baseline corticosterone may have a considerable effect on other behavioral and physiological functions. Brain Arginine Vasotocin immunoreactivity, which is associated with various functions, like territoriality and social behavior, differs in response to changed plasma corticosterone in urban curve-billed thrashers (*Toxostoma curvirostre*) [92].

3.3.4. Reproductive Physiology

Living in urban areas influences reproduction in birds [24,93–96]. A global analysis of passerine species found that sexually dichromatic species were less likely to occur in urban areas [97].

Urban populations of Abert's towhees (*Melozone aberti*) are found to have greater plasma luteinizing hormone (hereafter plasma LH) than rural populations [67]. This leads to urban birds developing gonads earlier, starting breeding earlier, and having a prolonged breeding season than rural birds [64,66,67,98–100]. Having a prolonged breeding season might help in the production of more broods and, thus, might be an important trait to have in urban areas to successfully colonize them. Such earlier gonadal development is also prominent in resident rather than migratory birds, as observed in the males of urban Eurasian blackbirds [101] and dark-eyed junkos (*Junco hyemalis*) [101].

Partecke et al. [64] suggest phenotypic plasticity in response to several new conditions, like artificial lights, in urban areas to be the primary reason for this change. Birds living in temperate regions are especially dependent on day length and duration of natural light for their reproductive development [95,98,99]. The presence of artificial light in the city habitat plays a major role in the earlier growth of gonads in male birds [66,87,98,99,102,103]. However, low levels of artificial light inhibit the secretion of plasma LH in male western scrub jays (*Aphelocoma californica*) [91,95]. Apart from artificial light, differential food availability could also affect the pattern of plasma LH physiology [69,91].

Factors, like the vegetation structure, replacement of native plants by exotic species, habitat fragmentation, predatory pressure, and food availability, influence the life history traits of urban birds [69,104]. The reason for lower clutch size in urban compared non-urban areas could be because of lower availability of high quality or specialized food in urban areas, especially during chick development [105]. Such deficiency in good quality food could also lead to increased competition between conspecifics during breeding season, affecting clutch size [69].

Such changes in reproductive physiology could affect life history traits, like fitness levels and nesting success of urban birds [98,99]. A prolonged breeding season could also be favored by urban birds, leading to fewer birds migrating for breeding [64,101].

3.3.5. Immune and Inflammatory Responses, Oxidative Stress Response

Physiological responses to various stressors in cities contribute to oxidative stress in urban birds [106,107]. Increased oxidative stress exposes birds to different types of diseases and organ degeneration [107]. A comparative study of blood and liver transcriptomes for these stress responses revealed that most of the genes for this stress response were expressed in a higher amount in urban birds [108]. An increase in blood antioxidant levels helps urban birds to tolerate such stress [106]. Successful urban colonizers use also genetic and epigenetic mechanisms, DNA methylation, and histone changes to cope with oxidative stress [107]. Urban Eurasian blackbirds developed lower oxidative stress when compared to rural populations, indicating an adaptation to high stress levels [109].

Physiological traits also closely influence behavioral traits: hormone secretion is associated with reproductive behavior, and brain size with innovation.

3.4. Behavioral Traits

3.4.1. Song Structure

One of the major characteristics of urban areas is the increase in low frequency noise levels, such as from traffic [110,111]. Birds in urban areas are impacted by noise in the low frequency range, as it carries over longer distances [111,112]. Low frequency anthropogenic noise tends to mask bird songs, which leads to poor song transmission, and, ultimately, poor reproductive success. Perhaps one of the most widely documented phenomena in behavioral trait changes in birds in response to urbanization is the modification of song and call structure to avoid such masking (Figure 4).

There are thought to be two mechanisms by which birds alter their song structure. One school of thought suggests that birds sing at higher frequencies in areas with high anthropogenic noise levels, as shown by studies on several species of song birds across different continents (e.g., [113–117]). In fact, high frequency songs are one of the selective forces for species to occupy urban habitats [118]. However, this mechanism might not always be effective for reproductive success [119–122]. White-crowned sparrow (*Zonotrichia leucophrys*) change song frequency and bandwidth, which leads to a reduction of vocal performance, which is deleterious for mate attraction and territory defense [120,122].

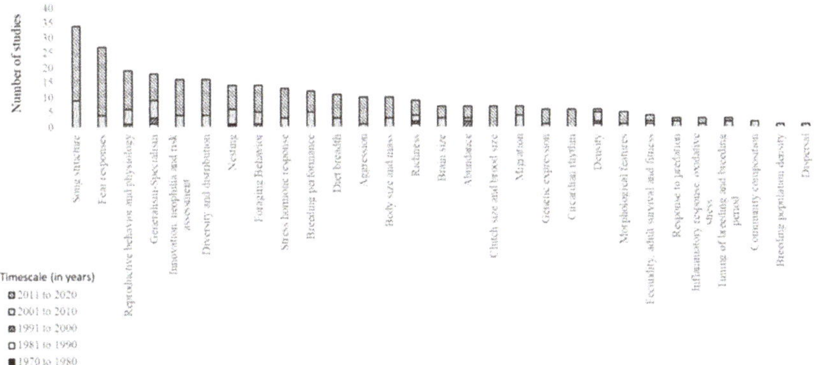

Figure 4. Proportion of studies in each decade from 1970 to 2020 on different trait changes due to urbanization.

Some birds also increase the amplitude of their songs, a phenomenon called "Lombard effect", in order to be heard above the city noises [111,123–126]. Great tits and Eurasian blackbirds are classic examples exhibiting this phenomenon [111]. Changes in bandwidth, trill rate, number of song syllables, and time spent singing have been documented in some birds, for effective song transmission [121,122,127–129]. Urban European robins (*Erithacus rubecula*) sing nocturnally in order to avoid song masking [35,110]. The alarm call structure is also modified due to masking. Urban silvereyes (*Zosterops lateralis*) had lower average, maximum, and minimum frequencies than rural birds [130]. Silvereyes also showed decreased syllable rate in Australia [131].

Apart from traffic noise, reflective structures, vegetation density, ambient temperature, and temporal changes in noise levels due to human activities also affect song communication in birds [132,133]. Male house finches (*Haemorhous mexicanus*) in an urban park in California sing at higher frequencies in areas with higher pedestrian traffic [134]. In the case of the great tit, this phenomenon is attributed to either large scale evolutionary or ontogenetic shift or a local scale song learning from neighboring males [112,113]. Morphological changes, like change in bill structure, due to differential food type in cities, could also change song structure [135].

Urban noise and artificial light levels affect the dawn chorus of urban bird populations [136,137]. Males of four out of five songbird species residing near street lights started singing earlier in the day. This increased their extra pair copulation success, but led to females selecting unsuitable mates [138]. Other studies found that more than artificial light, anthropogenic noise is responsible for temporal shift in the dawn chorus of the study species [117,136,137].

3.4.2. Boldness and Tolerance to Human Presence

Urban habitats have a constant presence of humans and vehicles, altered refuge patches, and differential predator composition [139–144]. Flight initiation distance (FID) is a standard measure for estimating the boldness and tolerance of birds to potential threats in urban areas [23]. Urban birds are observed to be bolder and more tolerant of human and vehicular approach, as they exhibit shorter FIDs [145–152].

A shorter FID reduces the cost associated with flight, and it can help birds exploit novel food sources. Several species of gulls have shown shorter FID in the proximity of human food sources [153–155]. In a study on 39 urbanized species of birds in Europe, urban birds had shorter FIDs than rural counterparts. [156]. This study also showed that the urban bird community had a larger variance in FID when compared to rural bird communities, but this variance decreased with increase in time since urbanization. This means that most of the birds adapted to the threats in urban areas as the time since colonization increased.

Along with FID, urban birds have also adapted to the predation threat of feral animals, like cats and dogs. Urban Eurasian coots (*Fulica atra*) portrayed the same amount of vigilance in the presence of domestic dogs as their natural predators [139].

Behavioral plasticity, which is the inherent ability of an organism to change in response to external stimuli, is one of the main characteristics exhibited by birds who are capable of changing their fear responses. This is thought to be an important adaptation to possess in urban areas. [157–159]. However, it has been argued that habituation induced change is not the reason for the higher tolerance of humans in urban areas. Those individuals who already had a bold personality in their natural habitat were able to colonize urban areas, while others were unsuccessful [82].

3.4.3. Neophobic and Innovative Behavior

Urban areas provide novel types of food resources in the form of artificial feeding or garbage dumps [153,160–162]. Novel urban food sources also come with new types of risks in the form of feral cats and dogs [160,163–165]. Not all native birds have the ability to exploit such sources of food—it seems that bird species with innovative capability and bold personality have an increased capacity to thrive [152,166–168]. Several species of gulls exhibit a great ability to exploit human provided food, for example, by adjusting their foraging time according to peak human activity timing, like school breaks or waste center opening time [169]. Not just foraging innovation, but certain urban species, like Indian house crows (*Corvus splendens*), also show innovative nesting behavior after nesting failure during breeding season [170].

Successful urban colonization requires a balance between neophilia and neophobia [162]. Urban great tits are more tolerant towards a novel object that is placed near their feeders than rural individuals [165]. However, two different studies on house sparrows in Hungary and mountain chikadees in Reno, USA indicated no reduction in object neophobia in urban populations [171,172]. Griffin et al. [162], in a review, suggested that neophilia/neophobia and boldness might be species specific. Corvid species were more neophobic towards novel objects than non-corvid species [173].

Exposure to pollutants and inferior quality food during chick development might also affect the ability to exploit novel food sources as well as learning from parents and conspecifics [162,167]. Research pin pointing the factors influencing the ability to exploit novel urban resources is required.

The ability to innovate different foraging techniques is crucial for successful urban colonization by birds [160,168,174]. The rate of innovation seems to be stronger during early invasion of novel urban areas [5]. The rate of innovativeness also seems to predict the risk–taking ability of the species [168]. The rate of innovation has also been linked to brain size in some urban birds, which also assists in successful urban colonization [175]. Innovative foraging also leads to changes in dominance hierarchy, behavioral strategies that are based on human movement, and introduces new types of competitions [176].

3.4.4. Aggression

Aggressive behavior can be displayed towards competitors [176–179], can be food related [180], due to exposure to chemical pollutants [181], or during nest defense [182]. Urban great tits were more aggressive towards competitors, but they exhibited inconsistent reaction towards a simulated competition when compared to their rural counterparts [179].

Urban great tits also showed greater distress behavior when threatened [183]. In another study on northern mockingbird (*Mimus polyglottos*), urban birds that were exposed to higher amounts of lead were more aggressive simulated competition [181]. In Eurasian coot populations residing in the same urban area, older and more established populations were consistently more territorially aggressive than recently colonized population [184]. Urban sparrowhawks (*Accipiter nisus*) showed more aggressive nest defense than rural sparrowhawks [182]. These changes might not be just behavioral adaptations, but a consequence of micro-evolution over the years in these birds causing changes in the behavior. However, differences in territorial aggression in urban birds appears to be species and situation specific. The species that show increased aggressive behavior in urban areas also generally exhibit bolder personality [155,176].

3.5. Genetic Traits

Although phenotypic trait changes in birds in response to urbanization have been extensively studied, solid evidence for genetic basis for such phenotypes is limited. Studies have focused on finding out the genetic modifications behind observed physiological trait changes like inflammatory and oxidative stress response, morphological trait changes, like change in wing structure, behavioral changes, like risk assessment, migration and urban invasion by certain functional groups [107]. The urban populations of great tits in Europe had elevated gene expression for inflammatory, oxidative stress, and detoxification responses [185]. Similar results were obtained for urban blue tits (*Parus caeruleus*) [186].

Urban Eurasian blackbirds have undergone genetic divergence at a locus coding for risk avoidance [187]. Human induced changes in habitat, along with various stressors, like the presence of novel predators, traffic noise, and pollutants, have led to accelerated changes in genotype in urban populations when compared to rural populations, with the potential to create genetic divergence between urban and non-urban populations [185]. However, an attempt to study the differences in overall genetic composition of Eurasian blackbirds yielded negative results [188]. This suggests that genetic differences between urban and rural populations have only occurred in selected genes, based on adaptive requirements. Urban blackbirds diverged from their rural populations at a single loci coding for risk avoidance [187]. However, there is still little evidence for urbanization-induced micro-evolution.

4. Discussion

Some bird traits are beneficial for their survival in cities, while others could be harmful. This review demonstrates that the impact of urbanization on birds is immense: and yet, our understanding of it is still poor. It is evident that studying one trait might not give a complete picture of how urbanization might affect birds or populations, but looking at a combination of traits would provide more insights.

Behavioral trait changes are dependent on plasticity, individual personalities of the bird populations, and, at least for a few species, the development of separate cognitive skills that are specific to urban needs [22,170]. In particular, as our review demonstrates, behavioral trait changes in urban birds are often a combination of two or more trait changes. Such correlated changes create behavioral syndromes in urban birds [171]. An example of such behavioral syndromes is the occurrence of increased aggression in birds with bolder personalities [176].

The consequences of trait change have not yet been studied in detail, and they require attention. Urban birds could incur energetic and reproductive costs due to changed corticosterone levels or increased aggression in urban areas. The ability to establish a successful breeding population in novel environments also depends on the population size and selection pressure [22]. Earlier and prolonged breeding seasons could have negative effects on reproductive health of urban birds, especially females. Therefore, even though, in the short term, birds might be adapting to urbanization by producing more broods and laying more eggs, we need to understand whether these birds will be successful at surviving in cities in the long term. Additionally, it is important to understand the reasons behind some birds showing a reduction in clutch and brood size in urban areas. Long term data collection is needed for monitoring the population of birds inhabiting the cities. Popular citizen science initiatives can help to answer some of these questions.

In the context of physiological and morphological trait changes, the literature suggests changes in stress hormone levels and reproductive hormone levels. The reason for the lack of consistent pattern in the differences in hormone levels in urban and non-urban populations could be due to the selection of physiologically plastic species in urban areas, which affects the life-history traits and ultimate density of the populations. This could mean potential extinction of species that are currently residing in urban areas, but are not able to adjust their physiology to the novel environment [24]. However, the effect of such differences in hormone levels on the overall fitness of the bird species that are successful in changing their endocrine traits has not yet been investigated in detail.

Trait changes in response to urbanization have the potential for creating genetic divergence between urban and rural populations of a species. The most important trait change contributing to genetic divergence is the change in reproductive behavior. A combination of different traits, like singing behavior, timing of reproduction, and nesting success, play a role in the reproductive pattern of birds. Urban areas modify these traits. Song birds possess high phenotypic plasticity and they have changed song structure in response to urban noise. Trait changes, coupled with isolation due to habitat fragmentation further leads to disconnect between urban and rural populations, leading to differential trait development in both populations. There is already evidence to suggest changes in genetic makeup at certain loci in urban and rural populations. Geographical isolation, coupled with high phenotypic plasticity, has a potential for further genetic divergence in urban birds. Large scale spatial and temporal studies, investigating such microevolution due to urbanization, could help in predicting the course of genetic divergence of urban bird populations in the future. Studying such urbanization induced genetic trait changes could help not only in conserving biodiversity in cities, but could also help to answer some of the basic questions in evolutionary ecology and also help in conserving the ecosystem services in cities [185].

There is a large gap in our knowledge regarding the effects of urbanization on birds in tropics, despite the fact that tropics support many more bird species and they are growing rapidly in terms of urbanization [55]. Future research in this area could also differentiate between introduced species and native species, in order to understand whether the traits discussed in this review contribute to the success of introduced species.

A lack of information regarding how urbanization will affect avifauna in urban tropics can be a major impeding factor while designing conservation strategies to protect the biodiversity in these ever-growing cities. Moreover, a study on southeast Asian cities found that, the wealthier the cities get, the richer they become in terms of urban greenspaces and natural aesthetics [189]. This could lead to potential habitat generation in growing cities, further affecting the bird traits.

Studies have started to emerge on the effect of urbanization on the traits of diurnal and nocturnal raptors in the last decade. Studies on urban raptors mainly focus on reproductive success, foraging pattern, and aggression in urban areas. Studies on top predatory bird species are essential while making management decisions, as these top predators might be responsible for keeping invasive faunal populations in check. For example, an increase in rodent populations on islands has led to the extinction of several island dwelling birds and other fauna [190]. Similar effects might affect bird populations in urban areas if proper management decisions are not taken at an appropriate time. Further, certain species are poorly represented in the research. Most of the studies that focus on species specific trait changes focus only on a few common species, such as great tits, Eurasian blackbirds, and house sparrows. Behavioral and physiological studies carried out on a couple of species might skew the literature in concluding effects of urbanization, either positive or negative and making generalizations. Caution should be taken that there is a lot of inter-species and intra-species variation and we need more studies looking at a variety of species and a combination of traits. For example, gulls seem to show extraordinary innovative ability around humans when compared to other urban birds. There is further scope to generate species specific databases in order to understand the effects of urbanization on diverse bird species and communities.

Urban areas provide a unique habitat in which to study community dynamics of birds. For example, in urban parks where there is a surge in activity during weekends, community composition of birds in these parks might be affected [78]. This could be very important, as, even though these are areas with good green cover, they might not be suitable nesting sites as park visitors might cause disturbance to nesting birds and might lead to abandoning the nest. On the other hand, these areas might be good areas to forage for fruits, seeds, and insects during the weekdays. An interesting hypothesis could be tested, looking at nesting success in the parks and outside the parks to actually see the effects of disturbance on bird communities and individual species.

5. Conclusions

The most significant trait changes that we identify have implications for urban avian diversity conservation, and they can be of use to park managers, citizen science groups, and urban planners. Omnivorous and cavity nesters have a better success in urban environments when compared to insectivores and ground nesting birds. Urban planners and park managers can maintain small patches of land that are enclosed to protect against common urban predators. such as feral dogs, cats, and rodents, in order to provide ground nesting birds that are especially endangered with refugia for roosting and nesting. Another important suggestion to bird enthusiasts and urban planners would be to include a variety of food items in bird feeders, as this might help birds with different diets to meet their nutrition requirements. Studies have shown that frugivores and granivores seem to do well in cities, as these birds feed on the seeds that are provided by people in bird feed. There has been evidence that, when artificial food is provided birds expand their ranges, thus having a positive effect on bird populations [78]. If dried insects and nuts are included in the food, it might help insectivorous birds in the city. If bigger cavity nests are provided for birds, like owls and other larger species, then these species might bounce back in good numbers in the cities. Additionally, maintaining patches of native vegetation in urban habitats might be a great solution for supporting not only the bird species, but also insects and small mammal communities that can serve as a prey base to the birds. Along with patches of native vegetation, it is also crucial to conserve the local water bodies in cities, as these act as refuges for several migratory birds and local waders [46,191]. In the Mediterranean, it was observed that the natural waterbodies were key habitat for several species of owls [192]. Hence, the health of the lakes, ponds, and rivers is crucial for the survival of many urban species.

Species specific studies and detailed knowledge of local bird populations can greatly help in effective management measures, as we find substantial, documented variation in how birds of specific species respond to the pressures of urbanization. We hope for the long- term monitoring of bird populations while using a combination of detailed scientific research and citizen science initiatives, as demonstrated in the papers reviewed here, can help to bridge gaps in knowledge and benefit the future survival of a diverse range of birds in urban environments. To conclude, using a trait-based approach will be useful for understanding the impacts of urbanization on bird species. Understanding the role of traits with an understanding of urban changes will be most useful. In particular, we stress the need for further research on the traits that influence bird survival in tropical cities, as well as on individual species. A meta-analyses kind of an approach encompassing multiple traits together was out of scope for this study, as enough information on each trait is still not available. Some traits have enough literature, others have hardly a couple of studies. However, this review could act as a baseline for further research on urban wildlife and its ecology.

Supplementary Materials: The following are available online at https://www.mdpi.com/2073-445X/10/2/92/s1, Figure S1: List of journals which published the (>1) articles reviewed in the current study; Table S1: All traits considered for selecting the literature were distributed under five broad trait groups. In total, 32 sub-traits were considered for this study. The broad description of these sub-traits is given below; Table S2: List of publications reviewed for this study.

Author Contributions: S.P.: Conceptualization, Methodology, Data Curation, Analysis, Writing—Original Draft Preparation, Reviewing and Editing. R.J.: Conceptualization, Methodology, Data Curation, Analysis, Writing—Original Draft Preparation, Reviewing and Editing. K.R.S.: Conceptualization, Supervision, Writing—Reviewing and Editing. H.N.: Conceptualization, Methodology, Supervision, Writing—Original Draft Preparation, Reviewing and Editing, Project administration. All authors have read and agreed to the published version of the manuscript.

Funding: This research received no external funding.

Acknowledgments: The authors thank Azim Premji University, Bangalore for supporting this research. This research did not receive any specific grant from funding agencies in the public, commercial, or not-for-profit sectors.

Conflicts of Interest: The authors declare no conflict of interest.

References

1. McKinney, M.L. Urbanization, biodiversity and conservation. *BioScience* **2002**, *52*, 883–890. [CrossRef]
2. Kowarik, I. Novel urban ecosystems, biodiversity, and conservation. *Environ. Pollut* **2011**, *159*, 1974–1983. [CrossRef] [PubMed]
3. UNDESA. *2018 The Sustainable Development Goals Report*; United Nations: New York, NY, USA, 2018; Available online: https://www.un.org/development/desa/publications/the-sustainable-development-goals-report-2018.html (accessed on 20 June 2020).
4. Lepczyk, C.A.; Aronson, M.F.; Evans, K.L.; Goddard, M.A.; Lerman, S.B.; MacIvor, J.S. Biodiversity in the city: Fundamental questions for understanding the ecology of urban green spaces for biodiversity conservation. *BioScience* **2017**, *67*, 799–807. [CrossRef]
5. Sol, D.; Duncan, R.P.; Blackburn, T.M.; Cassey, P.; Lefebvre, L. Big brains, enhanced cognition, and response of birds to novel environments. *Proc. Natl. Acad. Sci. USA* **2005**, *102*, 5460–5465. [CrossRef] [PubMed]
6. Liu, Z.; He, C.; Wu, J. The relationship between habitat loss and fragmentation during urbanization: An empirical evaluation from 16 world cities. *PLoS ONE* **2016**, *11*, e0154613. [CrossRef] [PubMed]
7. Belaire, J.A.; Westphal, L.M.; Whelan, C.J.; Minor, E.S. Urban residents' perceptions of birds in the neighborhood: Biodiversity, cultural ecosystem services, and disservices. *Condor* **2015**, *117*, 192–202. [CrossRef]
8. Aronson, M.F.; la Sorte, F.A.; Nilon, C.H.; Katti, M.; Goddard, M.A.; Lepczyk, C.A.; Warren, P.S.; Williams, N.S.; Cilliers, S.; Clarkson, B.; et al. A global analysis of the impacts of urbanization on bird and plant diversity reveals key anthropogenic drivers. *Proc. R. Soc. B Biol. Sci.* **2014**, *281*, 20133330. [CrossRef]
9. Hensley, C.B.; Trisos, C.H.; Warren, P.S.; MacFarland, J.; Blumenshine, S.; Reece, J.; Katti, M. Effects of urbanization on native bird species in three southwestern US Cities. *Front. Ecol. Evol.* **2019**, *7*, 71. [CrossRef]
10. Whelan, C.J.; Wenny, D.G.; Marquis, R.J. Ecosystem services provided by birds. *Ann. N. Y. Acad. Sci.* **2008**, *1134*, 25–60. [CrossRef]
11. Sandström, U.G.; Angelstam, P.; Mikusiński, G. Ecological diversity of birds in relation to the structure of urban green space. *Landsc. Urban Plan* **2006**, *77*, 39–53. [CrossRef]
12. Goodness, J.; Andersson, E.; Anderson, P.M.; Elmqvist, T. Exploring the links between functional traits and cultural ecosystem services to enhance urban ecosystem management. *Ecol. Indic.* **2016**, *70*, 597–605. [CrossRef]
13. Heyman, E.; Gunnarsson, B.; Dovydavicius, L. Management of urban nature and its impact on bird ecosystem services. In *Ecology and Conservation of Birds in Urban Environments*; Murgui, E., Hedblom, M., Eds.; Springer: Cham, Switzerland, 2017; pp. 465–488.
14. Gil, D.; Brumm, H. *Avian Urban Ecology*, 1st ed.; Oxford University Press: Oxford, UK, 2014; pp. xiii–xv.
15. Kumar, N.; Gupta, U.; Malhotra, H.; Jhala, Y.V.; Qureshi, Q.; Gosler, A.G.; Sergio, F. The population density of an urban raptor is inextricably tied to human cultural practices. *Proc. R. Soc. B Biol. Sci.* **2019**, *286*, 20182932. [CrossRef] [PubMed]
16. Chace, J.F.; Walsh, J.J. Urban effects on native avifauna: A review. *Landsc. Urban Plan* **2006**, *74*, 46–69. [CrossRef]
17. Palacio, F.X. Urban exploiters have broader dietary niches than urban avoiders. *Ibis* **2019**, *162*, 42–49. [CrossRef]
18. Leveau, L.M. Bird traits in urban–rural gradients: How many functional groups are there? *J. Ornithol.* **2013**, *154*, 655–662. [CrossRef]
19. Lizée, M.H.; Mauffrey, J.F.; Tatoni, T.; Deschamps-Cottin, M. Monitoring urban environments on the basis of biological traits. *Ecol. Indic.* **2011**, *11*, 353–361. [CrossRef]
20. Rooney, T.P.; Wiegmann, S.M.; Rogers, D.A.; Waller, D.M. Biotic impoverishment and homogenization in unfragmented forest understory communities. *Conserv. Biol.* **2004**, *18*, 787–798. [CrossRef]
21. Shochat, E.; Lerman, S.; Fernández-Juricic, E. Birds in urban ecosystems: Population dynamics, community structure, biodiversity, and conservation. In *Urban Ecosystem Ecology*, 1st ed.; Aitkenhead-Peterson, J.A., Volder, A., Eds.; American Agronomy Society: Madison, WI, USA, 2010; pp. 75–86.
22. Sol, D.; González-Lagos, C.; Moreira, D.; Maspons, J.; Lapiedra, O. Urbanization tolerance and the loss of avian diversity. *Ecol. Lett.* **2014**, *17*, 942–950. [CrossRef] [PubMed]

23. Blumstein, D.T. Attention, habituation, and antipredator behavior: Implications for urban birds. In *Avian Urban Ecology*; Diego, G., Henrik, B., Eds.; Oxford University Press: Oxford, UK, 2014; pp. 41–53.

24. Bonier, F. Hormones in the city: Endocrine ecology of urban birds. *Horm. Behav.* **2012**, *61*, 763–772. [CrossRef]

25. Rayner, L.; Ikin, K.; Evans, M.J.; Gibbons, P.; Lindenmayer, D.B.; Manning, A.D. Avifauna and urban encroachment in time and space. *Divers. Distrib.* **2015**, *21*, 428–440. [CrossRef]

26. Leveau, L.M.; Leveau, C.M.; Villegas, M.; Cursach, J.A.; Suazo, C.G. Bird communities along urbanization gradients: A comparative analysis among three neo-tropical cities. *Ornitol. Neotrop.* **2017**, *28*, 77–87.

27. la Sorte, F.A.; Lepczyk, C.A.; Aronson, M.F.J.; Goddard, M.A.; Hedblom, M.; Katti, M.; MacGregor-Fors, I. The phylogenetic and functional diversity of regional breeding bird assemblages is reduced and constricted through urbanization. *Divers. Distrib.* **2018**, *24*, 928–938. [CrossRef]

28. Saari, S.; Richter, S.; Higgins, M.; Oberhofer, M.; Jennings, A.; Faeth, S.H. Urbanization is not associated with increased abundance or decreased richness of terrestrial animals-dissecting the literature through meta-analysis. *Urban Ecosyst.* **2016**, *19*, 1251–1264. [CrossRef]

29. Ibáñez-Álamo, J.D.; Rubio, E.; Benedetti, Y.; Morelli, F. Global loss of avian evolutionary uniqueness in urban areas. *Glob. Chang. Biol.* **2017**, *23*, 2990–2998. [CrossRef] [PubMed]

30. Batáry, P.; Kurucz, K.; Suarez-Rubio, M.; Chamberlain, D.E. Non-linearities in bird responses across urbanization gradients: A meta-analysis. *Glob. Chang. Biol.* **2018**, *24*, 1046–1054. [CrossRef]

31. Morelli, F.; Benedetti, Y.; Ibáñez-Álamo, J.D.; Jokimäki, J.; Mänd, R.; Tryjanowski, P.; Møller, A.P. Evidence of evolutionary homogenization of bird communities in urban environments across Europe. *Glob. Ecol. Biogeogr.* **2016**, *25*, 1284–1293. [CrossRef]

32. Hagen, O.E.; Hagen, O.; Ibáñez-Álamo, J.D.; Petchey, O.L.; Evans, K.L. Impacts of urban areas and their characteristics on avian functional diversity. *Front. Ecol. Evol.* **2017**, *5*, 84. [CrossRef]

33. Shochat, E.; Warren, P.S.; Faeth, S.H.; McIntyre, N.E.; Hope, D. From patterns to emerging processes in mechanistic urban ecology. *Trends Ecol. Evol.* **2006**, *21*, 186–191. [CrossRef]

34. MacGregor-Fors, I.; Morales-Pérez, L.; Schondube, J.E. Migrating to the city: Responses of neo-tropical migrant bird communities to urbanization. *Condor* **2010**, *112*, 711–717. [CrossRef]

35. Leveau, L.M. Urbanization, environmental stabilization and temporal persistence of bird species: A view from Latin America. *PeerJ* **2018**, *6*, e6056. [CrossRef]

36. Faeth, S.H.; Bang, C.; Saari, S. Urban biodiversity: Patterns and mechanisms. *Ann. N. Y. Acad. Sci.* **2011**, *1223*, 69–81. [CrossRef] [PubMed]

37. Paton, G.D.; Shoffner, A.V.; Wilson, A.M.; Gagne, S.A. The traits that predict the magnitude and spatial scale of forest bird responses to urbanization intensity. *PLoS ONE* **2019**, *14*, e0220120. [CrossRef] [PubMed]

38. Julliard, R.; Jiguet, F.; Couvet, D. Common birds facing global changes: What makes a species at risk? *Glob. Chang. Biol.* **2004**, *10*, 148–154. [CrossRef]

39. Callaghan, C.T.; Major, R.E.; Wilshire, J.H.; Martin, J.M.; Kingsford, R.T.; Cornwell, W.K. Generalists are the most urban-tolerant of birds: A phylogenetically controlled analysis of ecological and life history traits using a novel continuous measure of bird responses to urbanization. *Oikos* **2019**, *128*, 845–858. [CrossRef]

40. Callaghan, C.T.; Benedetti, Y.; Wilshire, J.H.; Morelli, F. Avian trait specialization is negatively associated with urban tolerance. *Oikos* **2020**, *129*, 1541–1551. [CrossRef]

41. Devictor, V.; Julliard, R.; Couvet, D.; Lee, A.; Jiguet, F. Functional homogenization effect of urbanization on bird communities. *Conserv. Biol.* **2007**, *21*, 41–751. [CrossRef]

42. Futuyama, D.J.; Moreno, G. The evolution of ecological specialization. *Annu. Rev. Ecol. Evol. S* **1988**, *19*, 207–233. [CrossRef]

43. Kitahara, M.; Sei, K.; Fujii, K. Patterns in the structure of grassland butterfly communities along a gradient of human disturbance: Further analysis based on the generalist/specialist concept. *Popul. Ecol.* **2000**, *42*, 135–144. [CrossRef]

44. Kark, S.; Iwaniuk, A.; Schalimtzek, A.; Banker, E. Living in the city: Can anyone become an 'urban exploiter? *J. Biogeogr.* **2007**, *34*, 638–651. [CrossRef]

45. Evans, K.L.; Chamberlain, D.E.; Hatchwell, B.J.; Gregory, R.D.; Gaston, K.J. What makes an urban bird? *Glob. Chang. Biol.* **2011**, *17*, 1365–2486. [CrossRef]

46. Morelli, F.; Benedetti, Y.; Su, T.; Zhou, B.; Moravec, D.; Šímová, P.; Liang, W. Taxonomic diversity, functional diversity and evolutionary uniqueness in bird communities of Beijing's urban parks: Effects of land use and vegetation structure. *Urban Urban Green.* **2017**, *23*, 84–92. [CrossRef]

47. Gutiérrez-Tapia, P.; Azócar, M.I.; Castro, S.A. A citizen-based platform reveals the distribution of functional groups inside a large city from the Southern Hemisphere: E-Bird and the urban birds of Santiago (Central Chile). *Rev. Chil. Hist. Nat.* **2018**, *91*, 3. [CrossRef]

48. Lim, H.C.; Sodhi, N.S. Responses of avian guilds to urbanization in a tropical city. *Landsc. Urban Plan* **2004**, *66*, 199–215. [CrossRef]

49. Boal, C.W. Urban raptor communities: Why some raptors and not others occupy urban environments. In *Urban Raptors*; Boal, C.W., Dykstra, C.R., Eds.; Island Press: Washington, DC, USA, 2018; pp. 36–50.

50. Hindmarch, S.; Elliott, J.E. A specialist in the city: The diet of barn owls along a rural to urban gradient. *Urban Ecosyst.* **2015**, *18*, 477–488. [CrossRef]

51. Rullman, S.; Marzluff, J.M. Raptor presence along an urban–wildland gradient: Influences of prey abundance and land cover. *J. Raptor Res.* **2014**, *48*, 257–272. [CrossRef]

52. Thomson, V.K.; Stevens, T.; Jones, D.; Huijbers, C. Carrion preference in Australian coastal raptors: Effects of Urbanisation on scavenging. *Sunbird* **2016**, *46*, 16.

53. Jones, D.N.; Reynolds, J.S. Feeding birds in our towns and cities: A global research opportunity. *J. Avian Biol.* **2008**, *39*, 265–271. [CrossRef]

54. Hodgson, P.; French, K.; Major, R.E. Comparison of foraging behavior of small, urban-sensitive insectivores in continuous woodland and woodland remnants in a suburban landscape. *Wildl. Res.* **2006**, *33*, 591–603. [CrossRef]

55. Marzluff, J.M.; Bowman, R.; Donnelly, R. (Eds.) A historical perspective on urban bird research: Trends, terms, and approaches. In *Avian Ecology and Conservation in an Urbanizing*; World Springer: Boston, MA, USA, 2001; pp. 1–17.

56. Ikin, K.; Knight, E.; Lindenmayer, D.B.; Fischer, J.; Manning, A.D. Linking bird species traits to vegetation characteristics in a future urban development zone: Implications for urban planning. *Urban Ecosyst.* **2012**, *15*, 961–977. [CrossRef]

57. Croci, S.; Butet, A.; Clergeau, P. Does urbanization filter birds on the basis of their biological traits. *Condor* **2008**, *110*, 223–240. [CrossRef]

58. Reale, J.A.; Blair, R.B. Nesting success and life-history attributes of bird communities along an urbanization gradient. *Urban Habitats* **2005**, *3*, 1–24.

59. Máthé, O.; Batáry, P. Insectivorous and open-cup nester bird species suffer the most from urbanization. *Bird Study* **2015**, *62*, 78–86. [CrossRef]

60. Dykstra, C.R. City Lifestyles: Behavioral Ecology of Urban Raptors. In *Urban Raptors*; Boal, C.W., Dykstra, C.R., Eds.; Island Press: Washington, DC, USA, 2018; pp. 18–35.

61. Gahbauer, M.A.; Bird, D.M.; Clark, K.E.; French, T.; Brauning, D.W.; Mcmorris, F.A. Productivity, mortality, and management of urban peregrine falcons in northeastern North America. *J. Wildl. Manag.* **2015**, *79*, 10–19. [CrossRef]

62. Mason, C.F. Avian species richness and numbers in the built environment: Can new housing developments be good for birds? In *Human Exploitation and Biodiversity Conservation*; Hawksworth, D.L., Bull, A.T., Eds.; Springer: Dordrecht, The Netherlands, 2006; pp. 2365–2378.

63. Thornton, M.; Todd, I.; Roos, S. Breeding success and productivity of urban and rural Eurasian Sparrowhawks Accipiter nisus in Scotland. *Écoscience* **2017**, *24*, 115–126. [CrossRef]

64. Partecke, J.; Van't-Hof, T.; Gwinner, E. Differences in the timing of reproduction between urban and forest European blackbirds (Turdus merula): Result of phenotypic flexibility or genetic differences? *Proc. R. Soc. B Biol. Sci.* **2004**, *271*, 1995–2001. [CrossRef]

65. Atwell, J.W.; Cardoso, G.; Whittaker, D.J.; Price, T.D.; Ketterson, E.D. Hormonal, behavioral, and life-history traits exhibit correlated shifts in relation to population establishment in a novel environment. *Am. Nat.* **2014**, *184*, E147–E160. [CrossRef]

66. Zhang, S.; Chen, X.; Zhang, J.; Li, H. Differences in the reproductive hormone rhythm of Tree sparrows (Passer montanus) from urban and rural sites in Beijing: The effect of anthropogenic light sources. *Gen. Comp. Endocrinol.* **2014**, *206*, 24–29. [CrossRef]

67. Davies, S.; Behbahaninia, H.; Giraudeau, M.; Meddle, S.L.; Waites, K.; Deviche, P. Advanced seasonal reproductive development in a male urban bird is reflected in earlier plasma luteinizing hormone rise but not energetic status. *Gen. Comp. Endocrinol.* **2015**, *224*, 1–10. [CrossRef]

68. Kettel, E.F.; Gentle, L.K.; Quinn, J.L.; Yarnell, R.W. The breeding performance of raptors in urban landscapes: A review and meta-analysis. *J. Ornithol.* **2018**, *159*, 1–18. [CrossRef]

69. Chamberlain, D.E.; Cannon, A.R.; Toms, M.P.; Leech, D.I.; Hatchwell, B.J.; Gaston, K.J. Avian productivity in urban landscapes: A review and meta-analysis. *Ibis* **2009**, *151*, 1–18. [CrossRef]

70. Marini, K.L.D.; Otter, K.A.; LaZerte, S.E.; Reudink, M.W. Urban environments are associated with earlier clutches and faster nestling feather growth compared to natural habitats. *Urban Ecosyst.* **2017**, *20*, 1291–1300. [CrossRef]

71. Antonov, A.; Atanasova, D. Small-scale differences in the breeding ecology of urban and rural magpies (Pica pica). *Ornis Fennica* **2003**, *80*, 21–30.

72. Shochat, E.; Lerman, S.B.; Katti, M.; Lewis, D.B. Linking optimal foraging behavior to bird community structure in an urban-desert landscape: Field experiments with artificial food patches. *Am. Nat.* **2004**, *164*, 232–243. [CrossRef] [PubMed]

73. Caizergues, A.E.; Grégoire, A.; Charmantier, A. Urban versus forest ecotypes are not explained by divergent reproductive selection. *Proc. R. Soc. B Biol. Sci.* **2018**, *285*, 20180261. [CrossRef]

74. Mennechez, G.; Clergeau, P. Effect of urbanization on habitat generalists: Starlings not so flexible? *Acta Oecol.* **2006**, *30*, 182–191. [CrossRef]

75. Liker, A.; Papp, Z.; Bókony, V.; Lendvai, A.Z. Lean birds in the city: Body size and condition of house sparrows along the urbanization gradient. *J. Anim. Ecol.* **2008**, *77*, 789–795. [CrossRef]

76. Auman, H.J.; Meathrel, C.E.; Richardson, A. Supersize me: Does anthropogenic food change the body condition of Silver Gulls? A comparison between urbanized and remote, non-urbanized areas. *Waterbirds* **2008**, *31*, 122–126. [CrossRef]

77. Bókony, V.; Seress, G.; Nagy, S.; Lendvai, A.Z.; Liker, A. Multiple indices of body condition reveal no negative effect of urbanization in adult house sparrows. *Landsc. Urban Plan* **2012**, *104*, 75–84. [CrossRef]

78. Leveau, L.M. Urbanization induces bird color homogenization. *Landsc. Urban Plan* **2019**, *192*, 103645. [CrossRef]

79. Sol, D.; Bacher, S.; Reader, S.M.; Lefebvre, L. Brain size predicts the success of mammal species introduced into novel environments. *Am. Nat.* **2008**, *172*, 63–71. [CrossRef]

80. Maklakov, A.A.; Immler, S.; Gonzalez-Voyer, A.; Rönn, J.; Kolm, N. Brains and the city: Big-brained passerine birds succeed in urban environments. *Biol. Lett.* **2011**, *7*, 730–732. [CrossRef] [PubMed]

81. Carrete, M.; Tella, J.L. Inter-individual variability in fear of humans and relative brain size of the species are related to contemporary urban invasion in birds. *PLoS ONE* **2011**, *6*, e18859. [CrossRef] [PubMed]

82. Møller, A.P. Behavioural and ecological predictors of urbanization. In *Avian Urban Ecology*; Gil, D., Brumm, H., Eds.; Oxford University Press: Oxford, UK, 2014; pp. 54–68.

83. Møller, A.P.; Erritzøe, J. Brain size and urbanization in birds. *Avian Res.* **2015**, *6*, 8. [CrossRef]

84. Grunst, M.L.; Rotenberry, J.T.; Grunst, A.S. Variation in adrenocortical stress physiology and condition metrics within a heterogeneous urban environment in the Song sparrow (Melospiza melodia). *J. Avian Biol.* **2014**, *45*, 574–583. [CrossRef]

85. Davies, S.; Haddad, N.; Ouyang, J.Q. Stressful city sounds: Glucocorticoid responses to experimental traffic noise are environmentally dependent. *Biol. Lett.* **2017**, *13*, 20170276. [CrossRef] [PubMed]

86. Russ, A.; Reitemeier, S.; Weissmann, A.; Gottschalk, J.; Einspanier, A.; Klenke, R. Seasonal and urban effects on the endocrinology of a wild passerine. *Ecol. Evol.* **2015**, *5*, 5698–5710. [CrossRef]

87. Dominoni, D.M. The effects of light pollution on biological rhythms of birds: An integrated, mechanistic perspective. *J. Ornithol.* **2015**, *156*, 409–418. [CrossRef]

88. Partecke, J.; Schwabl, I.; Gwinner, E. Stress and the city: Urbanization and its effects on the stress physiology in European blackbirds. *Ecology* **2006**, *87*, 1945–1952. [CrossRef]

89. Fokidis, H.B.; Deviche, P. Plasma corticosterone of city and desert Curve-billed Thrashers, (Toxostoma curvirostre), in response to stress-related peptide administration. *Comp. Biochem. Physiol. A Mol. Integr. Physiol.* **2011**, *159*, 32–38. [CrossRef]

90. Zhang, S.; Lei, F.; Liu, S.; Li, D.; Chen, C.; Wang, P. Variation in baseline corticosterone levels of Tree sparrow (Passer montanus) populations along an urban gradient in Beijing, China. *J. Ornithol.* **2011**, *152*, 801–806. [CrossRef]

91. Davies, S.; Rodriguez, N.S.; Sweazea, K.L.; Deviche, P. The effect of acute stress and long-term corticosteroid administration on plasma metabolites in an urban and Desert songbird. *Physiol. Biochem. Zool.* **2013**, *86*, 47–60. [CrossRef] [PubMed]

92. Fokidis, H.B.; Deviche, P. Brain arginine vasotocin immunoreactivity differs between urban and desert curve-billed thrashers, (Toxostoma curvirostre): Relationships with territoriality and stress physiology. *Brain Behav. Evol.* **2012**, *79*, 84–97. [CrossRef] [PubMed]

93. Yeh, P.J.; Price, T.D. Adaptive phenotypic plasticity and the successful colonization of a novel environment. *Am. Nat.* **2004**, *164*, 531–542. [CrossRef] [PubMed]

94. Rodewald, A.D.; Shustack, D.P. Urban flight: Understanding individual and population-level responses of Nearctic–Neotropical migratory birds to urbanization. *J. Anim. Ecol.* **2008**, *77*, 83–91. [CrossRef] [PubMed]

95. Schoech, S.J.; Bowman, R.; Hahn, T.P.; Goymann, W.; Schwabl, I.; Bridge, E.S. The effects of low levels of light at night upon the endocrine physiology of Western scrub-jays (Aphelocoma californica). *J. Exp. Zool. A Ecol. Integr. Physiol.* **2013**, *319*, 527–538. [CrossRef] [PubMed]

96. Brown, L.M.; Graham, C.H. Demography, traits and vulnerability to urbanization: Can we make generalizations? *J. Appl. Ecol.* **2015**, *52*, 1455–1464. [CrossRef]

97. Iglesias-Carrasco, M.; Duchêne, D.A.; Head, M.L.; Møller, A.P.; Cain, K. Sex in the city: Sexual selection and urban colonization in passerines. *Biol. Lett.* **2019**, *15*, 20190257. [CrossRef]

98. Dominoni, D.M.; Quetting, M.; Partecke, J. Long-term effects of chronic light pollution on seasonal functions of European blackbirds (Turdus merula). *PLoS ONE* **2013**, *8*, e85069. [CrossRef]

99. Dominoni, D.M.; Quetting, M.; Partecke, J. Artificial light at night advances avian reproductive physiology. *Proc. R. Soc. B Biol. Sci.* **2013**, *280*, 20123017. [CrossRef]

100. Fudickar, A.M.; Greives, T.J.; Abolins-Abols, M.; Atwell, J.W.; Meddle, S.L.; Friis, G.; Stricker, C.A.; Ketterson, E.D. Mechanisms associated with an advance in the timing of seasonal reproduction in an urban songbird. *Front. Ecol. Evol.* **2017**, *5*, 85. [CrossRef]

101. Partecke, J.; Gwinner, E. Increased sedentariness in European blackbirds following urbanization: A consequence of local adaptation? *Ecology* **2007**, *88*, 882–890. [CrossRef] [PubMed]

102. Longcore, T. Sensory ecology: Night lights alter reproductive behavior of Blue tits. *Curr. Biol.* **2010**, *20*, R893–R895. [CrossRef] [PubMed]

103. Dominoni, D.M.; Goymann, W.; Helm, B.; Partecke, J. Urban-like night illumination reduces melatonin release in European blackbirds (Turdus merula): Implications of city life for biological time-keeping of songbirds. *Front. Zool.* **2013**, *10*, 60. [CrossRef] [PubMed]

104. Reynolds, S.J.; Ibáñez-Álamo, J.D.; Sumasgutner, P.; Mainwaring, M.C. Urbanization and nest building in birds: A review of threats and opportunities. *J. Ornithol.* **2019**, *160*, 841–860. [CrossRef]

105. Wawrzyniak, J.; Kaliński, A.; Glądalski, M.; Bańbura, M.; Markowski, M.; Skwarska, J.; Kaliński, P.; Cyżewska, I.; Bańbura, J. Long-term variation in laying date and clutch size of the Great tit (Parus major) in central Poland: A comparison between urban parkland and deciduous forest. *Ardeola* **2015**, *62*, 311–323. [CrossRef]

106. Møller, A.P. Successful city dwellers: A comparative study of the ecological characteristics of urban birds in the Western Palearctic. *Oecologia* **2009**, *159*, 849–858. [CrossRef] [PubMed]

107. Isaksson, C. Urbanization, oxidative stress and inflammation: A question of evolving, acclimatizing or coping with urban environmental stress. *Funct. Ecol.* **2015**, *29*, 913–923. [CrossRef]

108. Watson, H.; Videvall, E.; Andersson, M.N.; Isaksson, C. Transcriptome analysis of a wild bird reveals physiological responses to the urban environment. *Sci. Rep.* **2017**, *7*, 44180. [CrossRef]

109. Costantini, D.; Greives, T.J.; Hau, M.; Partecke, J. Does urban life change blood oxidative status in birds? *J. Exp. Biol.* **2014**, *217*, 2994–2997. [CrossRef]

110. Fuller, R.A.; Warren, P.H.; Gaston, K.J. Daytime noise predicts nocturnal singing in urban robins. *Biol. Lett.* **2007**, *3*, 368–370. [CrossRef]

111. Nemeth, E.; Brumm, H. Birds and anthropogenic noise: Are urban songs adaptive? *Am. Nat.* **2010**, *176*, 465–475. [CrossRef] [PubMed]

112. Slabbekoorn, H.; Ripmeester, E.A.P. Birdsong and anthropogenic noise: Implications and applications for conservation. *Mol. Ecol.* **2008**, *17*, 72–83. [CrossRef] [PubMed]

113. Slabbekoorn, H.; Boer-Visser, A. Cities change the songs of birds. *Curr. Biol.* **2006**, *16*, 2326–2331. [CrossRef] [PubMed]

114. Laiolo, P. The Rufous-Collared sparrow (Zonotrichia capensis) utters higher frequency songs in urban habitats. *Rev. Catalana d'Ornitologia* **2011**, *27*, 25–30.

115. Luther, D.A.; Derryberry, E.P. Birdsongs keep pace with city life: Changes in song over time in an urban songbird affects communication. *Anim. Behav.* **2012**, *83*, 1059–1066. [CrossRef]

116. Job, J.R.; Kohler, S.L.; Gill, S.A. Song adjustments by an open habitat bird to anthropogenic noise, urban structure, and vegetation. *Behav. Ecol.* **2016**, *27*, 1734–1744. [CrossRef]

117. LaZerte, S.E.; Otter, K.A.; Slabbekoorn, H. Mountain chickadees adjust songs, calls and chorus composition with increasing ambient and experimental anthropogenic noise. *Urban Ecosyst.* **2017**, *20*, 989–1000. [CrossRef]

118. Cardoso, G.C.; Hu, Y.; Francis, C.D. The comparative evidence for urban species sorting by anthropogenic noise. *R Soc. Open Sci.* **2018**, *5*, 172059. [CrossRef]

119. Moiron, M.; González-Lagos, C.; Slabbekoorn, H.; Sol, D. Singing in the city: High song frequencies are no guarantee for urban success in birds. *Behav. Ecol.* **2015**, *26*, 843–850. [CrossRef]

120. Luther, D.A.; Phillips, J.; Derryberry, E.P. Not so sexy in the city: Urban birds adjust songs to noise but compromise vocal performance. *Behav. Ecol.* **2016**, *27*, 332–340. [CrossRef]

121. Narango, D.L.; Rodewald, A.D. Urban-associated drivers of song variation along a rural–urban gradient. *Behav. Ecol.* **2016**, *27*, 608–616. [CrossRef]

122. Phillips, J.N.; Derryberry, E.P. Urban sparrows respond to a sexually selected trait with increased aggression in noise. *Sci. Rep.* **2018**, *8*, 7505. [CrossRef] [PubMed]

123. Ríos-Chelén, A.A.; Quirós-Guerrero, E.; Gil, D.; Macías-Garcia, C. Dealing with urban noise: Vermilion flycatchers sing longer songs in noisier territories. *Behav. Ecol. Sociobiol.* **2013**, *67*, 145–152. [CrossRef]

124. Lowry, H.; Lill, A.; Wong, B.B.M. How noisy Does a noisy miner have to be? Amplitude adjustments of alarm calls in an avian urban 'adapter'. *PLoS ONE* **2011**, *7*, e29960. [CrossRef] [PubMed]

125. Potvin, D.A.; Mulder, R.A. Immediate, independent adjustment of call pitch and amplitude in response to varying background noise by Silvereyes (Zosterops lateralis). *Behav. Ecol.* **2013**, *24*, 1363–1368. [CrossRef]

126. Knight, C.R.; Swaddle, J.P. Eastern Bluebirds alter their song in response to anthropogenic changes in the acoustic environment. *Integr. Comp. Biol.* **2015**, *55*, 418–431. [CrossRef]

127. Díaz, M.; Parra, A.; Gallardo, C. Serins respond to anthropogenic noise by increasing vocal activity. *Behav. Ecol.* **2011**, *22*, 332–336. [CrossRef]

128. Redondo, P.; Barrantes, G.; Sandoval, L. Urban noise influences vocalization structure in the House wren (Troglodytes aedon). *Ibis* **2013**, *155*, 621–625. [CrossRef]

129. Moseley, D.L.; Phillips, J.N.; Derryberry, E.P.; Luther, D.A. Evidence for differing trajectories of songs in urban and rural populations. *Behav. Ecol.* **2019**, *30*, 1734–1742. [CrossRef]

130. Potvin, D.A.; Mulder, R.A.; Parris, K.M. Silvereyes decrease acoustic frequency but increase efficacy of alarm calls in urban noise. *Anim. Behav.* **2014**, *98*, 27–33. [CrossRef]

131. Potvin, D.A.; Parris, K.M.; Mulder, R.A. Geographically pervasive effects of urban noise on frequency and syllable rate of songs and calls in Silvereyes (Zosterops lateralis). *Proc. R. Soc. B Biol. Sci.* **2011**, *278*, 2464–2469. [CrossRef] [PubMed]

132. Seger-Fullam, K.D.; Rodewald, A.D.; Soha, J.A. Urban noise predicts song frequency in Northern cardinals and American robins. *Bioacoustics* **2011**, *20*, 267–276. [CrossRef]

133. Shannon, G.; McKenna, M.F.; Angeloni, L.M.; Crooks, K.R.; Fristrup, K.M.; Brown, E.; Warner, K.A.; Nelson, M.D.; White, C.; Briggs, J.; et al. A synthesis of two decades of research documenting the effects of noise on wildlife. *Biol. Rev.* **2016**, *91*, 982–1005. [CrossRef] [PubMed]

134. Fernández-Juricic, E.; Poston, R.; Collibus, K.D.; Morgan, T.; Bastain, B.; Martin, C.; Jones, K.; Treminio, R. Microhabitat selection and singing behavior patterns of male house finches (Haemorhous mexicanus) in urban parks in a heavily urbanized landscape in the western U.S. *Urban Habitats* **2005**, *3*, 49–69.

135. Badyaev, A.V.; Young, R.L.; Oh, K.P.; Addison, C. Evolution on a local scale: Developmental, functional, and genetic bases of divergence in bill form and associated changes in song structure between adjacent habitats. *Evolution* **2008**, *62*, 1951–1964. [CrossRef] [PubMed]

136. Nordt, A.; Klenke, R. Sleepless in town—Drivers of the temporal shift in dawn song in urban European blackbirds. *PLoS ONE* **2013**, *8*, e71476. [CrossRef]

137. Dorado-Correa, A.M.; Rodríguez-Rocha, M.; Brumm, H. Anthropogenic noise, but not artificial light levels predict song behavior in an equatorial bird. *R Soc. Open Sci.* **2016**, *3*, 160231. [CrossRef]

138. Kempenaers, B.; Borgström, P.; Loës, P.; Schlicht, E.; Valcu, M. Artificial night lighting affects dawn song, extra-pair siring success, and lay date in songbirds. *Curr. Biol.* **2010**, *20*, 1735–1739. [CrossRef]

139. Randler, C. Disturbances by dog barking increase vigilance in coots Fulica atra. *Eur. J. Wildl. Res.* **2006**, *52*, 265–270. [CrossRef]

140. Bonnington, C.; Gaston, K.J.; Evans, K.L. Fearing the feline: Domestic cats reduce avian fecundity through trait-mediated indirect effects that increase nest predation by other species. *J. Appl. Ecol.* **2013**, *50*, 15–24. [CrossRef]

141. Rebolo-Ifrán, N.; Carrete, M.; Sanz-Aguilar, A.; Rodríguez-Martínez, S.; Cabezas, S.; Marchant, T.A.; Bortolotti, G.R.; Tella, J.L. Links between fear of humans, stress and survival support a non-random distribution of birds among urban and rural habitats. *Sci. Rep.* **2015**, *5*, 13723. [CrossRef] [PubMed]

142. Samia, D.S.M.; Nakagawa, S.; Nomura, F.; Rangel, T.F.; Blumstein, D.T. Increased tolerance to humans among disturbed wildlife. *Nat. Commun.* **2015**, *6*, 8877. [CrossRef] [PubMed]

143. Eötvösa, C.B.; Magura, T.; Lövei, G.L. A meta-analysis indicates reduced predation pressure with increasing urbanization. *Landsc. Urban Plan* **2018**, *180*, 54–59. [CrossRef]

144. Weaver, M.; Ligon, R.A.; Mousel, M.; McGraw, K.J. Avian anthrophobia? Behavioral and physiological responses of house finches (Haemorhous mexicanus) to human and predator threats across an urban gradient. *Landsc. Urban Plan* **2018**, *179*, 46–54. [CrossRef]

145. McGiffin, A.; Lill, A.; Beckman, J.; Johnstone, C.P. Tolerance of human approaches by Common mynas along an urban-rural gradient. *EMU Austral Ornithol.* **2013**, *113*, 154–160. [CrossRef]

146. Møller, A.P.; Grim, T.; Ibáñez-Álamo, J.D.; Markó, G.; Tryjanowski, P. Change in flight initiation distance between urban and rural habitats following a cold winter. *Behav. Ecol.* **2013**, *24*, 1211–1217. [CrossRef]

147. Møller, A.P.; Díaz, M.; Flensted-Jensen, E.; Grim, T.; Ibáñez-Álamo, J.D.; Jokimäki, J.; Mänd, R.; Markó, G.; Tryjanowski, P. Urbanized birds have superior establishment success in novel environments. *Oecologia* **2015**, *178*, 943–950. [CrossRef]

148. Vines, A.; Lill, A. Boldness and urban dwelling in little ravens. *Wildl. Res.* **2014**, *42*, 590–597. [CrossRef]

149. Gravolin, I.; Key, M.; Lill, A. Boldness of urban Australian magpies and local traffic volume. *Avian Biol. Res.* **2014**, *7*, 244–250. [CrossRef]

150. Samia, D.S.M.; Blumstein, D.T.; Díaz, M.; Grim, T.; Ibáñez-Álamo, J.D.; Jokimäki, J.; Tätte, K.; Markó, G.; Tryjanowski, P.; Møller, A.P. Rural-urban differences in escape behavior of European birds across a latitudinal gradient. *Front. Ecol. Evol.* **2017**, *5*, 66. [CrossRef]

151. Morelli, F.; Mikula, P.; Benedetti, Y.; Bussière, R.; Jerzak, L.; Tryjanowski, P. Escape behavior of birds in urban parks and cemeteries across Europe: Evidence of behavioral adaptation to human activity. *Sci. Total Environ.* **2018**, *631*, 803–810. [CrossRef] [PubMed]

152. Biondi, L.M.; Fuentes, G.M.; Córdoba, R.S.; Bó, M.S.; Cavalli, M.; Paterlini, C.A.; Castano, M.V.; García, G.O. Variation in boldness and novelty response between rural and urban predatory birds: The Chimango Caracara (Milvago chimango) as study case. *Behav. Process* **2020**, *173*, 104064. [CrossRef] [PubMed]

153. Fleming, P.A.; Bateman, P.W. Scavenging opportunities modulate escape responses over a small geographic scale. *Ethology* **2017**, *123*, 205–212. [CrossRef]

154. Feng, C.; Liang, W. Behavioral responses of black-headed gulls (Chroicocephalus ridibundus) to artificial provisioning in China. *Glob. Ecol. Conserv.* **2020**, *21*, e00873. [CrossRef]

155. Pavlova, O.; Wronski, T. City gulls and their rural neighbours: Changes in escape and agonistic behaviour along a rural-to-urban gradient. In *Advances in Environmental Research*; Nova Science Publishers, Inc.: Hauppauge, NY, USA, 2020.

156. Møller, A.P. Interspecific variation in fear responses predicts urbanization in birds. *Behav. Ecol.* **2010**, *21*, 365–371. [CrossRef]

157. Kitchen, K.; Lill, A.; Price, M. Tolerance of human disturbance by urban magpie-larks. *Aust. Field Ornithol.* **2010**, *27*, 1.

158. Gendall, J.; Lill, A.; Beckman, J. Tolerance of disturbance by humans in long-time resident and recent colonist urban doves. *Avian Res.* **2015**, *6*, 7. [CrossRef]

159. Sol, D.; Maspons, J.; Gonzalez-Voyer, A.; Morales-Castilla, I.; Garamszegi, L.Z.; Møller, A.P. Risk-taking behavior, urbanization and the pace of life in birds. *Behav. Ecol. Sociobiol.* **2018**, *72*, 59–66. [CrossRef]

160. Sol, D.; Griffin, A.S.; Bartomeus, I.; Boyce, H. Exploring or avoiding novel food resources? The novelty conflict in an invasive bird. *PLoS ONE* **2011**, *6*, e19535. [CrossRef]

161. Shanahan, D.F.; Strohbach, M.W.; Warren, P.S.; Fuller, R.A. The challenges of urban living. In *Avian Urban Ecology*; Diego, G., Henrik, B., Eds.; Oxford University Press: Oxford, UK, 2014; pp. 3–20.

162. Griffin, A.S.; Netto, K.; Peneaux, C. Neophilia, innovation and learning in an urbanized world: A critical evaluation of mixed findings. *Curr. Opin. Behav. Sci.* **2017**, *17*, 15–22. [CrossRef]

163. Seress, G.; Bókony, V.; Heszberger, J.; Liker, A. Response to predation risk in urban and rural house sparrows. *Ethology* **2011**, *117*, 896–907. [CrossRef]

164. Møller, A.P.; Ibáñez-Álamo, J.D. Escape behavior of birds provides evidence of predation being involved in urbanization. *Anim. Behav.* **2012**, *84*, 341–348. [CrossRef]

165. Tryjanowski, P.; Møller, A.P.; Morelli, F.; Biaduń, W.; Brauze, T.; Ciach, M.; Czechowski, P.; Czyż, S.; Dulisz, B.; Goławski, A.; et al. Urbanization affects neophilia and risk-taking at bird-feeders. *Sci. Rep.* **2016**, *6*, 28575. [CrossRef] [PubMed]

166. Audet, J.N.; Ducatez, S.; Lefebvre, L. The town bird and the country bird: Problem solving and immune-competence vary with urbanization. *Behav. Ecol.* **2016**, *27*, 637–644. [CrossRef]

167. Papp, S.; Vincze, E.; Preiszner, B.; Liker, A.; Bókony, V. A comparison of problem-solving success between urban and rural house sparrows. *Behav. Ecol. Sociobiol.* **2015**, *9*, 471–480. [CrossRef]

168. Ducatez, S.; Audet, J.N.; Ros, J.; Kayello, R.L.; Lefebvre, L. Innovativeness and the effects of urbanization on risk-taking behaviors in wild Barbados birds. *Anim. Cogn.* **2017**, *20*, 33–42. [CrossRef]

169. Spelt, A.; Soutar, O.; Williamson, C.; Memmott, J.; Shamoun-Baranes, J.; Rock, P.; Windsor, S. Urban gulls adapt foraging schedule to human-activity patterns. *Ibis* **2020**, *163*, 274–282. [CrossRef]

170. Yosef, R.; Zduniak, P.; Poliakov, Y.; Fingerman, A. Behavioural and reproductive flexibility of an invasive bird in an arid zone: A case of the Indian House Crow (Corvus splendens). *J. Arid Environ.* **2019**, *168*, 56–58. [CrossRef]

171. Bókony, V.; Kulcsár, A.; Tóth, Z.; Liker, A. Personality traits and behavioral syndromes in differently urbanized populations of house sparrows (Passer domesticus). *PLoS ONE* **2012**, *7*, e36639. [CrossRef]

172. Kozlovsky, D.Y.; Weissgerber, E.A.; Pravosudov, V.V. What makes specialized food-caching mountain chickadees successful city slickers? *Proc. R. Soc. B Biol. Sci.* **2017**, *284*, 20162613. [CrossRef]

173. Greggor, A.L.; Clayton, N.S.; Fulford, A.J.C.; Thornton, A. Street smart: Faster approach towards litter in urban areas by highly neophobic corvids and less fearful birds. *Anim. Behav.* **2016**, *117*, 123–133. [CrossRef]

174. Lill, A.; Hales, E. Behavioral and ecological keys to urban colonization by Little Ravens (Corvus mellori). *Open Ornithol. J.* **2015**, *8*, 22–31. [CrossRef]

175. Campbell, M. An animal geography of avian feeding habits in Peterborough, Ontario. *Area* **2008**, *40*, 472–480. [CrossRef]

176. Evans, J.; Boudreau, K.; Hyman, J. Behavioral syndromes in urban and rural populations of Song sparrows. *Ethology* **2010**, *116*, 588–595.

177. Scales, J.; Hyman, J.; Hughes, M. Behavioral syndromes break down in urban Song sparrow populations. *Ethology* **2011**, *117*, 887–895. [CrossRef]

178. Hasegawa, M.; Ligon, R.A.; Giraudeau, M.; Watanabe, M.; McGraw, K.J. Urban and colorful male house finches are less aggressive. *Behav. Ecol.* **2014**, *25*, 641–649. [CrossRef]

179. Hardman, S.I.; Dalesman, S. Repeatability and degree of territorial aggression differs among urban and rural Great tits (Parus major). *Sci. Rep.* **2018**, *8*, 5042. [CrossRef]

180. Galbreath, D.M.; Ichinose, T.; Furutani, T.; Yan, W.; Higuchi, H. Urbanization and its implications for avian aggression: A case study of urban black kites (Milvus migrans) along Sagami Bay in Japan. *Landsc. Ecol.* **2014**, *29*, 169–178. [CrossRef]

181. McClelland, S.C.; Ribeiro, R.D.; Mielke, H.W.; Finkelstein, M.E.; Gonzales, C.R.; Jones, J.A.; Komdeur, J.; Derryberry, E.; Saltzberg, E.B.; Karubian, J. Sub-lethal exposure to lead is associated with heightened aggression in an urban songbird. *Sci. Total Environ.* **2019**, *654*, 593–603. [CrossRef]

182. Kunca, T.; Yosef, R. Differential nest-defense to perceived danger in urban and rural areas by female Eurasian sparrowhawk (Accipiter nisus). *PeerJ* **2016**, *4*, e2070. [CrossRef]

183. Senar, J.C.; Garamszegi, L.Z.; Tilgar, V.; Biard, C.; Moreno-Rueda, G.; Salmón, P.; Rivas, J.; Sprau, P.; Dingemanse, N.J.; Charmantier, A.; et al. Urban Great tits (Parus major) show higher distress calling and pecking rates than rural birds across Europe. *Front. Ecol. Evol.* **2017**, *5*, 163. [CrossRef]

184. Minias, P.; Jedlikowski, J.; Włodarczyk, R. Development of urban behavior is associated with time since urbanization in a reed-nesting waterbird. *Urban Ecosyst.* **2018**, *21*, 1021–1028. [CrossRef]

185. Alberti, M.; Marzluff, J.; Hunt, V.M. Urban driven phenotypic changes: Empirical observations and theoretical implications for eco-evolutionary feedback. *Philos. Trans. R. Soc. B* **2017**, *372*, 20160029. [CrossRef]

186. Capilla-Lasheras, P.; Dominoni, D.M.; Babayan, S.A.; O'Shaughnessy, P.J.; Mladenova, M.; Woodford, L.; Pollock, C.J.; Barr, T.; Baldini, F.; Helm, B. Elevated immune gene expression is associated with poor reproductive success of urban Blue tits. *Front. Ecol. Evol.* **2017**, *5*, 64. [CrossRef]

187. Mueller, J.C.; Partecke, J.; Hatchwell, B.J.; Gaston, K.J.; Evans, K.L. Candidate gene polymorphisms for behavioral adaptations during urbanization in blackbirds. *Mol. Ecol.* **2013**, *22*, 3629–3637. [CrossRef]

188. Partecke, J.; Gwinner, E.; Bensch, S. Is urbanization of European blackbirds (Turdus merula) associated with genetic differentiation? *J. Ornithol.* **2006**, *147*, 549–552. [CrossRef]

189. Richards, D.R.; Passy, P.; Oh, R.R. Impacts of population density and wealth on the quantity and structure of urban green space in tropical Southeast Asia. *Landsc. Urban Plan* **2017**, *157*, 553–560. [CrossRef]

190. Harper, G.A.; Bunbury, N. Invasive rats on tropical islands: Their population biology and impacts on native species. *Glob. Ecol. Conserv.* **2015**, *3*, 607–627. [CrossRef]

191. Ferenc, M.; Sedláček, O.; Fuchs, R. How to improve urban greenspace for woodland birds: Site and local-scale determinants of bird species richness. *Urban Ecosyst.* **2014**, *17*, 625–640. [CrossRef]

192. Moreno-Mateos, D.; Rey Benayas, J.M.; Pérez-Camacho, L.; Montaña, E.D.L.; Rebollo, S.; Cayuela, L. Effects of land use on nocturnal birds in a Mediterranean agricultural landscape. *Acta Ornithol.* **2011**, *46*, 173–182. [CrossRef]

land

MDPI

Article

Integrating Diverse Perspectives for Managing Neighborhood Trees and Urban Ecosystem Services in Portland, OR (US)

Lorena Alves Carvalho Nascimento * and Vivek Shandas

Nohad A. Toulan School of Urban Studies and Planning, Portland State University, Portland, OR 97201, USA;
vshandas@pdx.edu
* Correspondence: lorena.nascimento@pcc.edu

Abstract: Municipalities worldwide are increasingly recognizing the importance of urban green spaces to mitigate climate change's extreme effects and improve residents' quality of life. Even with extensive earlier research examining the distribution of tree canopy in cities, we know little about human perceptions of urban forestry and related ecosystem services. This study aims to fill this gap by examining the variations in socioeconomic indicators and public perceptions by asking *how neighborhood trees and socioeconomic indicators mediate public perceptions of ecosystem services availability.* Using Portland, Oregon (USA) as our case study, we assessed socioeconomic indicators, land cover data, and survey responses about public perceptions of neighborhood trees. Based on over 2500 survey responses, the results indicated a significant correlation among tree canopy, resident income, and sense of ownership for urban forestry. We further identified the extent to which the absence of trees amplifies environmental injustices and challenges for engaging communities with landscape management. The results suggested that Portland residents are aware of tree maintenance challenges, and the inclusion of cultural ecosystem services can better address existing environmental injustices. Our assessment of open-ended statements suggested the importance of conducting public outreach to identify specific priorities for a community-based approach to urban forestry.

Keywords: urban forestry; cultural ecosystem services; public survey; tree maintenance

Citation: Alves Carvalho Nascimento, L.; Shandas, V. Integrating Diverse Perspectives for Managing Neighborhood Trees and Urban Ecosystem Services in Portland, OR (US). *Land* **2021**, *10*, 48. https://doi.org/10.3390/land1001 0048

Received: 30 November 2020
Accepted: 31 December 2020
Published: 7 January 2021

Publisher's Note: MDPI stays neutral with regard to jurisdictional claims in published maps and institutional affiliations.

1. Introduction

By 2050, the United Nations suggests that the world human population will near 10 billion, with most living in urban centers. In the United States (US), eighty percent of the population already live in urban areas, which corresponds to only 3% of national land [1]. The fast pace of urbanization and landscape change caused by humans are the major factors for transforming forests, urban and otherwise [2]. Some have argued that the transformation of urban landscapes brings an 'extinction of experience' with nature [3], which impacts the well-being, public health and empathy for natural features. The management of urban areas requires the consideration of multiple land-use possibilities for conservation of built or natural environments. Roads, buildings, urban renewal, green infrastructure and new developments compete for limited urban space. This fact requires municipalities to use strategic approaches to manage urban growth with civic services, including sewer, roadways and other gray infrastructures. Often with priories of gray infrastructure, the available locations for tree canopy is reduced and existing tree canopy is removed—a phenomenon that has been well documented across the US [4,5].

Several scholars have argued that urban tree canopy can be better managed by characterizing ecosystem services, which describes the benefits that trees and other natural landscape features provide to humans. For urban trees and tree canopy, these are often described as improvement of air quality, reduction of heat, filtering and infiltration of stormwater and a host of cultural attributes that improve the overall quality of life for residents [6,7]. Ecosystem services, in the form of tree canopy, come in three essential categories: (a) those in parks, schools, open spaces and other natural areas; (b) those on private lands, owned by somebody other than a public agency; (c) those on streets and other public rights-of-way.

Municipal governments generally manage the urban tree canopy through a distributed model that embeds tree specialists in different public agencies or through a central division that works with other agencies. Few municipalities, if any, explicitly examine the host of ecosystem services that constitute the urban forest. We argue that urban ecosystem services as a management approach can help situate the urban forest within broader and potentially more inclusive management of natural features. The rationale behind our approach for ecosystem services lies in these three points: (a) they offer a cost-effective and functionally-based solution to major challenges facing urban areas, including rising temperatures, air pollution and flooding; (b) when properly applied, ecosystem services can support a more equitable distribution of canopy, which is currently highly centered on wealthy and white communities; (c) the consideration of ecosystem services can improve human-nature relationships, create a sense of ownership of places and provide stewardship and community engagement opportunities. Together these functions of urban trees and forests compel their careful management, though the extent to which communities view or even understand these ecosystems services remains unclear.

This study examines the role of community perspectives concerning urban canopy management by assessing the relationship between the quantity of neighborhood tree canopy, public perceptions of ecosystem services and socioeconomic indicators to support urban environmental planning. Many studies have found a positive correlation between tree canopy and residential income [8–10]. However, few examine the extent to which public opinion about planting priorities, maintenance challenges and the expectations for urban ecosystem services can be central to decision-making processes. We posit that engaging the public in myriad and creative ways in urban forestry efforts is increasingly essential. Planting and maintaining trees can promote a connection between residents and urban environmental services according to each neighborhood's needs, regarding socioeconomic, cultural and historical aspects.

Background

The use of urban ecosystem services (UES) in describing trees and forests is a relatively new idea [6,7]. Four categories of UES classify the services provided by the maintenance, preservation and conservation of urban forestry. First, supporting UES contributes to nutrient cycling and soil formation through tree debris and habitat for decomposers. Second, provisioning UES for urban foraging of food and natural medicine [11]. Third, regulating UES, as local climate resilience that prevents urban heat islands [12], air quality that mitigates respiratory problems [13] and stormwater catchment that controls flood and water flow [14]. Fourth, cultural UES bring socio-ecological values through self-actualization, esteem and belonging [7], connection with biodiversity and personalized ecology [15]. The literature of ecosystem services [16] is gaining popularity, with research that includes public perception evaluations and how the ecosystem services adjust to reflect community identity.

UES relies on urban forestry and green infrastructure management, including public participation in strategic planning to recognize multiple ecological needs in diverse contexts. People's involvement in managing urban forests is often heralded as necessary for ensuring a just and equitable distribution of ecosystem services. However, communities' engagement may be mediated by intersectional factors that are often not considered in planning decisions [7]. For example, older adults and children are more vulnerable to respiratory diseases and may be highly sensitive to degraded air quality; black communities have been historically excluded from desirable green areas [17]; queer identities struggle to be accepted in heteronormative nature spaces [18]; low-income residents have less canopy access [8].

Urban forestry scholars have already acknowledged the link between socioeconomic factors and access to ecosystem services in the municipalities. McPhearson et al. [19] called attention to incorporating UES into urban planning since cities globally are rapidly increasing in population. The persistent need for environmental justice and climate resilience created a framework for using ecosystem services as planning metrics. Wilkerson et al. [7] developed an urban sociological framework to explain the intersectionality between UES planning and social demands because the variation of socioeconomic factors impacted the accessibility to green spaces. Empirical studies that measured urban accessibility based on socioeconomic indicators found geographic mismatches within vulnerable groups for the balance (demand/flow ratio) of ecosystem services of climate regulation [12], food supply and recreation [20].

While these earlier studies call out distributional inequities, we need to establish programs and policies to ensure that historically underserved communities are at the center of urban forestry programs. We argue that urban forest planning needs to acknowledge and incorporate voices from diverse communities when managing distributional equity. Moreover, we need to find effective and practical approaches for hearing voices from communities, especially about their perceptions of trees and the factors that mediate the level of saliency for expanding tree canopy in historically disinvested neighborhoods. Since the fields of forestry and more recently urban forestry, have been mainly heteronormative, white and male [18], advancing a call for expanding the participatory process to include opinions that have not traditionally heard is instrumental to ensuring canopy equity in cities.

To provide a basis for our argument, we evaluated the presence of trees and public perceptions of ecosystem services through a community survey in Portland, Oregon (OR), US. The study aimed to understand what people expect about greening strategies, tree maintenance and tree planting priorities. We specifically asked, *how do neighborhood trees and socioeconomic indicators mediate the public perceptions of ecosystem services availability?* We addressed this question through integrative analysis, involving a survey, socioeconomic indicators and spatial analysis of neighborhoods in the study area. Portland has several advantages for a study of this kind, including an inequitable distribution of trees [10]; historically unserved areas that reflect racist planning policies [21]; and late incorporation as a city in the United States, leaving with its existing tracts of large trees, even today [22]. At the same time, Portland's regional culture seems to have an explicit appreciation for urban forestry, as evidenced by establishing the first Parks Commission in 1900 and green corridors West of Willamette River planned by the Olmsted Brothers in the early development of the city services [23].

2. Materials and Methods

2.1. Study Area

Located in the Pacific Northwest region of the United States, Portland bears the earliest and most preserved forest formations in the country [24]. Once called Silicon Forest [25], over the past 20 years, Portland attracts people looking for jobs in the tech industry and individuals seeking outdoor recreation and lifestyles. The physical geography has forest fragments in hilly areas to reduce the chance of landslides, as illustrated in Figure 1. Portland's canopy cover follows geological points of interest, like rivers, wetlands and elevated formations.

Most of the city's canopy is west of the Willamette River in the Northwest (NW) and Southwest (SW) zip code sectors (Figure 1). Portland's western sectors also contain the most topographical relief for which trees protect from erosion, landslides and riverbanks [26,27]. Western sectors of the study region had extensive early conservation policies by the Parks Commission, with the Olmsted Brothers' support. This landscape architecture firm promoted urban ecology practices, such as green corridors, large urban parks, wildlife habitat and biodiversity in the early 20th Century [23].

Across the Willamette River, the eastern zip code sectors have flattened surfaces containing fewer trees than the western sectors. The eastern sectors are North (N), Northeast (NE), Southeast (SE) and East (E). The eastern sectors have a larger proportion of industrial and commercial areas and the largest percent of the city's population [22]. The Columbia River, the second largest river in the United States [28], surrounds North and East Portland. The high susceptibility to floods resulted in a 21-mile levee system created to allow urban development in areas along the Columbia River's riverbanks. Among the 654,741 people living in Portland [29], at least 25.6% live in the East sector [22], which is growing fast in terms of ethnic and income diversity. Still, Portland has a history of racist urban planning, redlining [30,31], gentrification [21,32,33] and late incorporation of eastern neighborhoods [22] that explains the separation between low-income communities and canopy abundance.

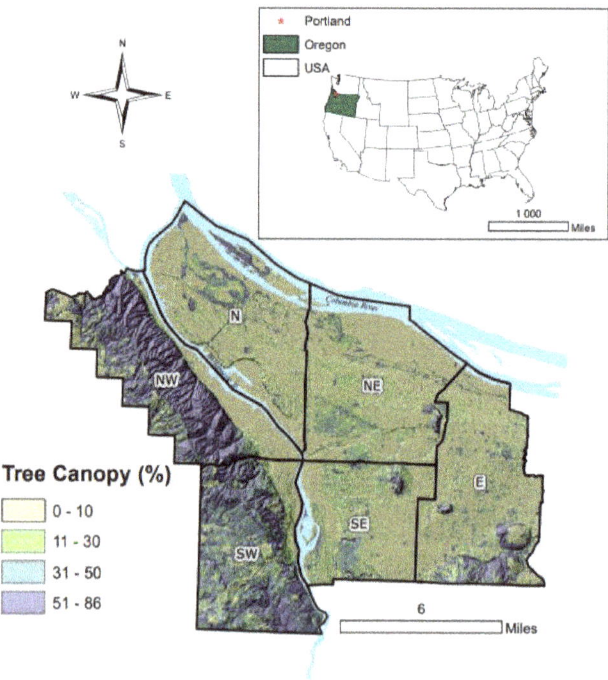

Figure 1. Map of Portland, Oregon, USA, with canopy distribution over the six zip code sectors: Southwest (SW), Northwest (NW), North (N), Northeast (NE), East (E) and Southeast (SE) City of Portland [34]. Oregon Spatial Data Library [35], USGS [36].

2.2. Research Design

We used a cross-sectional research design that applied satellite-derived measurements of the tree canopy cover, demographic analysis and assessments about the public perceptions of the urban forest. This study used three parameters that encompassed tree canopy measurements from the National Land Cover Database (NLCD), socioeconomic indicators from the US Census and a public survey about urban forestry perceptions applied in Portland. The aim was to assess the extent of the connection between people and UES following the level of satisfaction and accessibility to urban tree canopy [9]. By integrating the biophysical with survey responses, we argued that we were better able to describe how the presence or absence of neighborhood trees mediated differences in the perception of tree maintenance and tree ownership in the study area. Evaluation of public perceptions can offer insights, perhaps a first step, to understand the extent to which urban forest management can better support communities in the maintenance, ownership and accessibility to trees among different socioeconomic groups.

2.2.1. External Datasets for Tree Canopy Cover and Demographic Data

US Census data from 2017 estimates [29] regarding income, race and homeownership had two specific roles in this study. First, we wanted to note the fidelity of our survey sample. We compared the values of socioeconomic indicators from the survey and census. Second, we used census data as the parameters for socioeconomic indicators. The census data had higher representability and more complex data collection than the demographic questions from the survey. We aggregated the census data by zip code, following the delimitation of study areas from the survey.

The canopy cover percentage was obtained from the NLCD with 30 m $\times$ 30 m spatial resolution. The dataset informed the percentage of canopy per pixel [36,37]. On ArcGIS software, we used the Extract by Mask tool to mask Portland boundaries and the Tabulate Area tool to measure the canopy area per zip code. Equation (1) calculated the average canopy percentage by zip code. The shapefiles for zip code and neighborhood boundaries were from the Portland Maps web library [34]:

$$\text{Canopy cover } (\%) \; = \; \sum (\text{ZAi} \times \text{Ci}) / \text{area} \tag{1}$$

where ZA is the zip code area per canopy percentage, C is the percentage of tree canopy per pixel and area is the total zip code area.

2.2.2. Survey Data for Public Perceptions of Urban Forestry

To explore the public perceptions of UES, we used the results of an online public survey conducted between May and July of 2017. Portland is known for the neighborhood bonding and strong connection that residents have with their vicinities [25]. Therefore, we used zip codes as the determinant scale and the unit of analysis. A total of 26 zip codes were large enough to have representative samples to assess patterns of responses and small enough to provide variations in our sample. We excluded zip codes from neighborhoods with less than fifteen answers for a better variability of answers and representation.

The survey had 26 questions that explored the views of Portlanders about the quality of the local urban forest, strategies for planting programs, possibilities for tree maintenance and a socioeconomic questionnaire [38]. For this study, we combined 12 relevant questions that addressed: (a) the sense of ownership of trees; (b) the sense of maintenance for trees in public spaces; (c) the perception of UES on strategies to increase urban forestry; (d) and socioeconomic indicators (Appendix A).

The first six questions of our study encompassed the perceptions of tree ownership and maintenance [Appendix A—Questions 1–6]. Three questions about tree ownership asked the participants about: (1) the importance of trees; (2) the satisfaction with the number of trees and (3) the satisfaction with trees' health [Appendix A, Questions 1–3]. Three questions about tree maintenance inquired about (1) the maintenance of existing trees in the right-of-way; (2) maintenance of trees in the right-of-way in low-income communities and (3) planting new trees in the right-of-way [Appendix A, questions 4–6]. The participants used a Likert scale ranging between 1 and 5 to inform how much they agree or disagree with the six sentences related to tree ownership and tree maintenance. Table 1 shows the range of Likert scale values, tree ownership and maintenance questions and multi-metric evaluation.

Instead of analyzing the six questions separately, we created two multi-metric indexes: Tree Ownership Satisfaction Index (TOSI) and Tree Maintenance Satisfaction Index (TMSI). TOSI combined the three questions about tree ownership and TMSI the three questions about tree maintenance. TOSI and TMSI also used a Likert scale and we calculated them by finding the average values of the three questions of each index, as displayed in Table 1. TOSI and TMSI resemble multi-metric indexes used in the biological assessment of watersheds [39,40].

Table 1. Where Q is the question number, Tree Ownership Satisfaction Index (TOSI) is the tree ownership satisfaction index and Tree Maintenance Satisfaction Index (TMSI) is the tree maintenance satisfaction index.

Questions	Likert-Scale Values					Multi-Metric Index
	Strongly Agree	Agree	Don't Know	Disagree	Strongly Disagree	
Q1: "Portland's trees are important to me"						
Q2: "My neighborhood has enough trees"	5	4	3	2	1	$TOSI = \dfrac{(Q1 + Q2 + Q3)}{3}$
Q3: "The trees in my neighborhood are in good condition and healthy"						
Q4: "The city should maintain all trees along the street"						
Q5: "The city should prioritize maintenance of trees along the street in low-income communities"	5	4	3	2	1	$TMSI = \dfrac{(Q4 + Q5 + Q6)}{3}$
Q6: "The city should plant trees in all available spaces along the street"						

One open-ended question asked the participants about municipality strategies to increase tree canopy [Appendix A, question 7]. After observing the responses, we coded repetitive values from the answers and created a typology based on UES and management challenges. Table 2 shows the coded values that translated the public concerns about ecosystem services, tree maintenance and financial concerns.

The last survey question about urban forestry asked if participants had trees on their property [Appendix A, question 8]. We measured the answers per zip code, informing the percentage of participants that had trees on their yards.

Four questions explored the socioeconomic characteristics of the participants [Appendix A, Questions 9–12]. We asked questions about the participants' race, income and zip code for the socioeconomic indicators. These indicators were traditional demographic parameters in other urban forestry and ecosystem services studies [8,10,12]. We also asked about housing ownership, as house tenure is an indicator of membership and active participation in urban environmental planning [41]. We compared the survey data's socioeconomic indicators to the census data to reinforce the survey sample's validity.

Table 2. Coded values from the open-ended question regarding strategies to increase tree canopy.

Type	UES	Typology
Regulation and Provisioning services	Climate	Support microclimate, provide shade, mitigate urban heat island
	Air Quality	Promote clear air, mitigate pollution on transit corridors and industrial areas
	Water flow	Encourage tree planting in water facilities, acknowledge that trees consume water and support maintenance of groundwater
	Water purification	Improve watersheds and riparian areas, industries that pollute water should contribute to tree maintenance
	Erosion	Awareness that trees prevent landslides risks
	Natural Disaster Regulation	Flood prevention, stormwater mitigation in green infrastructures, reduction of impervious surfaces
	Pollination	Support bees and pollination
	Pest and disease	Physical and mental health, elders benefit from trees, industries that use pesticides should contribute to tree maintenance
	Waste	Pollution mitigation, energy savings, households with trees should have a discount on sewage bills, companies with more waste generation should contribute to tree maintenance
	Food	Encourage planting of fruit trees, urban agroforestry, mitigation of food deserts
Cultural services	Recreation and tourism	Recreation and relaxation, preference for trees over recreation facilities, canopy attracts tourists to the city
	Aesthetics and inspiration	Inspiration, beautification, common appreciation for trees
	Knowledge Systems	Educate people on how to plant and maintain trees, partnership with education institutions, job creation, encourage workshops about the importance of trees
	Religious and spiritual	Biophilia, spirituality, partnership with religious groups
	Cultural heritage	Cultural values of trees, community bonding, civic engagement, diversity of cultures and their interpretation of trees
	Natural heritage	Promote native trees, promote wildlife habitat, canopy preservation, encourage natural heritage stewardships
Management Challenges	Financial solutions	Tree giveaways, public/private/nonprofit partnerships, volunteers to reduce budget costs, use of taxes/donations/fundraisers resources
	Financial burden	Tree permit taxes, maintenance costs, other civic priorities over trees
	High maintenance	Right tree/right place, water and pruning care, public utilities, sidewalk maintenance, planning for climate change, regulations for trees in developed areas, tree maintenance strategies

2.2.3. Survey Data for Public Perceptions of Urban Forestry

To integrate the analysis from the NLCD tree canopy, census and survey datasets, we developed a conceptual model that described the analytical steps for addressing our research aims (Figure 2). The conceptual model contained, at its core, the research question: *How do neighborhood trees and socioeconomic indicators mediate the public perceptions of available ecosystem services?* The conceptual model also integrated the datasets vis-a-vie specific questions that reference each of the three datasets we employed:

1. Census data and socioeconomic survey questions: Does the variation of socioeconomic indicators in the survey provide a good representation of the census data?
2. Census data and tree canopy data: Is there a relationship between socioeconomic indicators and tree canopy?
3. Tree canopy data and urban forestry survey questions: Does the presence of trees influence the public perceptions of urban ecosystem services?

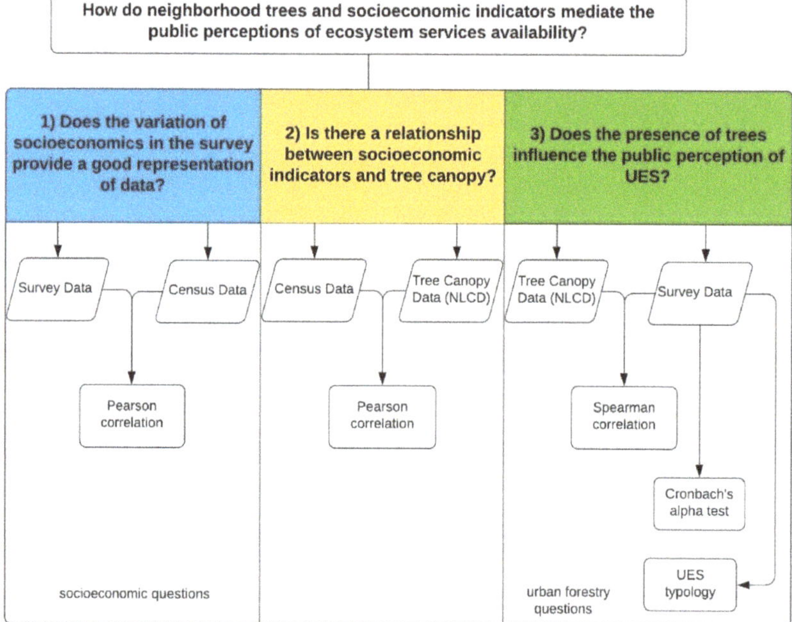

Figure 2. Flowchart with the research questions and research design.

To answer the first two questions, we performed the Pearson correlation test between the socioeconomic survey questions and Census data and between tree canopy data and Census data, as the variables were parametric. To answer the third question, we performed the Spearman correlation test between the tree canopy data and the survey urban forestry questions, as the variables were nonparametric. For both Pearson and Spearman correlation we considered the results that were significant with $p < 0.05$. For the urban forestry questions in the survey that used a Likert scale (Table 1), we performed Cronbach's alpha, which is a reliability test for the internal consistency of scaled questions and their variance. All statistical tests were performed with SPSS software v.26. Using the open-ended responses, we created a UES typology table (Table 2), which also served as additional data for evaluating and corroborating the statistical analysis.

3. Results

The survey for public perceptions of urban forestry and UES had 2548 valid answers from 26 zip codes within Portland. The survey responses ranged between 15 and 249 answers per zip code [Appendix B, Figure A1; Appendix B, Table A1]. As a voluntary online survey distributed in community engagement platforms (municipal listserv, Nextdoor, social media channels, focus groups, public meetings), there was a high chance that the participants had previous interests in urban forestry and city planning. We obtained completed answers and discarded those who did not complete the survey.

In the following subsections, we will answer the specific research questions regarding the correlation between the variables of census data, survey data and tree canopy data. In the last subsection, we will summarize the core question *"how do neighborhood trees and socioeconomic indicators mediate the public perceptions of ecosystem services availability?"* using the open-ended statements and their associations with the statistical findings.

3.1. Does the Variation of Socioeconomics in the Survey Provide a Good Representation of the Census Data?

To answer this question, we compared the census data and survey's socioeconomic indicators, which were both collected in 2017 (Table 3). The correlation between survey answers and the number of households per zip code was moderately strong (R = 0.554) and significant ($p < 0.01$). This result suggested that the number of respondents reflects the total population size within the zip codes.

We found strong Pearson correlation values between the census and the survey for the variables of house ownership (R = 0.796; $p < 0.01$) and income (R = 0.922; $p < 0.01$). The percentage of house ownership was higher among the survey respondents (82.09%) than the values indicated by the census data (51.54%). We believed that our survey targeted participants aware of the local public budget [39], as property owners have more responsibility with taxes that support tree maintenance.

For the race variable, both data from the census and the survey showed that Portland has a majority white population in all zip codes. Due to this fact, we labeled all non-white races as people of color. People of color (POC) is a term commonly used in the US to describe a population that is not white. The correlation values between the census and survey data for the POC variable was moderate (R = 0.402, $p < 0.05$). The percentage between POC in the census (22.22%) and survey (23.29%) had similar values.

Overall, the results suggest that for the specific characteristics the survey contained a representative sample for the city as a whole, which provides support to address the remaining questions. Our survey had a consistent representation with the Census data, with significant values for population size, race, income and house ownership.

Table 3. Pearson correlation values between socioeconomic indicators from the survey and census.

Variable	Mean Value Per Zip Code	Pearson Correlation
Surveys (N)	98 ± 66	0.554 **
Households (N)	11318 ± 5076	
Housing ownership census (%)	51.54 ± 15.20	0.796 **
Housing ownership survey (%)	82.09 ± 12.87	
People of color census (%)	22.22 ± 8.56	0.402 *
People of color survey (%)	23.19 ± 7.26	
Mean income census (US $)	87,813.04 ± 28,397.40	0.922 **
Mean income survey (US $)	93,319.65 ± 20,275.61	

* Level of significance = 0.05; ** level of significance = 0.01.

3.2. Is There a Relationship between Tree Canopy and Socioeconomic Indicators?

Earlier research from several studies across the United States [6–8] and Portland [36] suggested that historically underserved communities are less likely to have immediate access to tree canopy. In contrast, white and wealthier populations have greater access, partly due to historical policies that segregated neighborhoods [38]. This study was no exception and corroborated previous findings. Using Equation (1), we observed that zip codes in NW and SW had higher tree canopy than zip codes in the eastern sectors (E, N, NE, SE) of the Willamette River. NW and SW had respectively 42.5% and 37% of canopy cover and \$125,739 and \$100,696 of household incomes (Table 4). East had the lowest income (\$57,104) and 12.4% of average canopy. The values of household income were obtained from the census data.

Table 4. Average tree canopy and income within the zip codes sectors in the study area.

Variables	E	N	NE	NW	SE	SW
Canopy	12.39%	7.80%	7.30%	42.45%	10.24%	37.02%
Income	\$57,104	\$86,824	\$95,714	\$125,739	\$93,200	\$100,696

Pearson correlation values between tree canopy and census socioeconomic indicators for income, race and house ownership (Table 5). The strongest correlation across all the sociodemographic was between tree canopy and income ($R = 0.625$, $p < 0.01$). This result supported the findings observed in Figure 1 and Table 4, with a high percentage of canopy cover in affluent neighborhoods of West Portland, suggesting that people with higher income in Portland have more access to the urban tree canopy. The results for the correlation between tree canopy with house ownership ($R = 0.206$; $p > 0.3$) and race ($R = -0.186$; $p > 0.3$) did not have significant values. As such, we conclude that income is the only significant (and positively correlated) variable in relation to the amount of tree canopy for the study area, as found in other urban forestry studies that used Portland as a case study [10,38,42].

Table 5. Pearson correlation values between tree canopy and census socioeconomic indicators.

Correlation Variables	Pearson Correlation
Tree Canopy and Income	0.625 **
Tree Canopy and Race	−0.186
Tree Canopy and House ownership	0.206

** Level of significance = 0.01.

3.3. Does the Presence of Trees Influence the Public Perception of UES?

3.3.1. TOSI and TMSI Indicators

We combined three questions to build the TOSI, regarding: (1) the satisfaction with the number of trees; (2) the good condition of trees and (3) the importance of trees. The TMSI combined three questions about: (4) the maintenance of street trees; (5) the maintenance of trees in low-income communities and (6) the planting of new trees. Table 6 shows Cronbach's alpha results for the questions that encompassed the indexes.

While the Cronbach's alpha value of 0.490 can increase to 0.612 by removing Question 2 of the TOSI and TMSI multi-metric, we maintained the question because only 78.63% of respondents addressed all six questions, while 8.70% respondents answered five questions, 6.98% answered four questions, 5.61% answered three questions and 0.08% answered two questions. While multi-metric methods (ecosystem services coding, Likert scale, TOSI, TMSI) are reduced by including additional questions, some of which may not be addressed, doing so also increases the diversity and reliability of responses. In addition, surveys about public perceptions of urban forestry are relatively limited and the development of such metrics and observations, we expect, can contribute to further comparative studies.

Table 6. Cronbach's alpha for the Likert scale questions of survey.

Total Statistics Cronbach's Alpha = 0.490	Item Statistics		
Questions	Mean	Std. Deviation	Cronbach's Alpha If Item Deleted
Q1: "Portland's trees are important to me"	4.81	0.49	0.502
Q2: "My neighborhood has enough trees"	3.07	1.26	0.612
Q3: "The trees in my neighborhood are in good condition and healthy"	3.39	0.91	0.505
Q4: "The city should maintain all trees along the street"	3.70	1.23	0.287
Q5: "The city should prioritize maintenance of trees along the street in low-income communities"	3.98	1.2	0.287
Q6: "The city should plant trees in all available spaces along the street"	3.61	1.32	0.340

The variation of TOSI and TMSI answers are presented in a Likert scale across a series of maps (Figure 3). The Likert scale ranged from 1 to 5, where 1 represents strongly disagree and 5 represented strongly agree. The TOSI and TMSI were the three questions' [Appendix A, Questions 1–6] average values combined on the respective indexes.

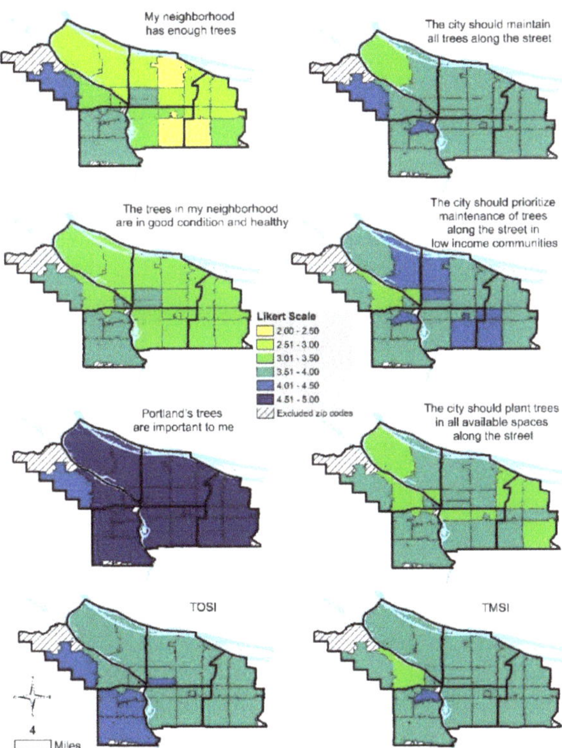

Figure 3. Range of answers to individual Likert scale questions that are used to generate the multimeric indices of ownership [TOSI] and maintenance [TMSI] for the study area.

In the questions associated with TOSI and TMSI, most of the answers per zip code were higher than 3 on the Likert scale, except for satisfaction with neighborhood trees. The eastern zip codes had lower satisfaction with the number of trees than the western zip codes. In the question about tree health, the western zip codes had a higher rate on the

Likert scale. A zip code in NW, a high-income and high canopy part of the study area, scored the lowest value for trees' importance.

3.3.2. Public Perceptions of Urban Ecosystem Services

The results indicated that most of the participants had trees on their private property. The average percentage of participants that informed having trees on their yard was 94.2%. The lowest rate of private property trees was 75.8%, in the zip code 97209, an urban renewal area (redevelopment of industrial and low-income areas in inner-city) in Northwest Portland [31].

The last question about urban forestry perceptions was open-ended and we coded the answers using UES typology (Table 7). According to the coded answers, the strategies recommended a focus on urban forestry management (56.1%), cultural ecosystem services (31.5%) and regulating and provision services (12.4%).

Table 7. Public perceptions of relevant strategies to increase urban tree canopy.

Type	UES	Survey Answers (%)	Survey Answers Aggregated (%)
Regulating and provisioning ecosystem services	Climate	1.67	12.36
	Air Quality	2.67	
	Water flow	2.02	
	Water purification	0.67	
	Erosion	0.19	
	Natural hazard	1.14	
	Pollination	0.16	
	Pest and disease	2.00	
	Waste	1.02	
	Food	0.81	
Cultural ecosystem services	Recreation and tourism	0.70	31.54
	Aesthetics and Inspiration	3.18	
	Knowledge	11.01	
	Religious and spiritual	0.33	
	Cultural heritage	9.55	
	Natural heritage	6.78	
Management Challenges	Financial Solution	33.59	56.10
	Financial Burden	4.53	
	High Maintenance	17.98	

Within the management challenges, 33.6% of the answers suggested financial solutions. The answers recommended the municipality to seek partnerships with volunteer associations, donation of seedlings and financial support for homeowners and renters, such as tax and water/sewage bill discounts. The second most mentioned typology was the maintenance of trees, with 18% of answers showing that respondents were aware of tree health and upkeep's essential requirements. The participants informed that proper tree pruning, tree species selection, tree debris removal and regulation for trees in developed areas were their top priorities.

Among the UES, knowledge systems and cultural heritage were the most significant concerns, holding 11% and 9.6% of the answers, respectively. Knowledge systems associated with urban forestry solutions that educate the population on how to take care of trees. Connection with teaching opportunities through schools and training programs can prepare present and future generations to maintain trees and understand urban ecology interactions. Awareness of cultural heritage indicated the respondents' related urban forestry to community development, community engagement and neighborhood pride. The open-ended answers recommended tree planting events to promote activities that bring interaction among neighbors to praise trees' values for multiple cultures and ethnicities.

As most of the answers indicated strategies using financial solutions and tree maintenance, we conducted a separate analysis only with the UES typology. Excluding the management challenges (Table 7), knowledge systems and cultural heritages lead the answers with 25.1% and 21.8% of the responses, respectively. Natural heritage was the third most important, with 15.5% of answers. Natural heritage responses concerned the loss of mature trees, biodiversity and wildlife habitat. The respondents commented that small trees could take longer to provide the ecosystem services promoted by centenary trees susceptible to removal for new developments or infrastructure challenges.

The final analysis described the Spearman correlation between survey answers and the percentage of tree canopy per zip code (Table 8). The variables that had the strongest positive correlation with tree canopy were satisfaction with neighborhood trees (R = 0.767), satisfaction with tree health (R = 0.704), TOSI (R = 0.758) in a 0.01 significance level and aesthetics and inspiration (R = 0.453) in a 0.05 significance level. Though these are general findings, we note that these levels of significance and strength of the relationship varied by zip code.

Table 8. Correlation between tree canopy and survey answers about public perceptions of urban forestry.

Public Perceptions	Spearman Correlation (R)	Public Perceptions	Spearman Correlation (R)
Climate	−0.136	Natural heritage	−0.191
Air quality	0.226	Financial Solution	0.039
Water flow	0.009	Financial Burden	−0.235
Water purification	0.194	High Maintenance	−0.162
Erosion	0.289	Food	−0.231
Natural hazard	0.034	Trees on property	0.048
Pollination	−0.233	Neighborhood trees	0.767 **
Pest and disease	0.078	Tree health	0.704 **
Waste	0.158	Importance of trees	−0.289
Recreation and tourism	−0.037	TOSI	0.758 **
Aesthetics and inspiration	0.453 *	Street trees	0.241
Knowledge	0.227	Trees in low income areas	−0.27
Religious and spiritual values	−0.149	Plant street trees	−0.104
Cultural heritage	−0.009	TMSI	0.005

* Level of significance = 0.05; ** Level of significance = 0.01.

3.4. How Do Neighborhood Trees and Socioeconomic Indicators Mediate the Public Perceptions of Ecosystem Services Availability?

Seven out of eight zip codes in Portland's western sectors had a canopy rate higher than the eastern sectors. The justification for the uneven distribution is associated with the early conservation practices of urban forestry in the western zip code sectors [23], the hilly geological formation [26,27] and the higher income [8]. Tree canopy had a significant positive relationship to TOSI indicators and cultural UES of aesthetics and inspiration (Table 8). However, the excessive canopy does not please everyone, as expressed in the following statements from participants living in high canopy areas:

> Too many trees already. While they have benefits, the trees need to be healthy and co-exist safely with residents. This requires regular, vigilant maintenance,

which a lot of people (...) fail to do. We've repeatedly witnessed the tragedy of human deaths and property destructions, especially this last winter. Even one death is too many! We need to take better care of the existing trees before we consider adding more. (97221—Southwest)

I don't necessarily think the city should plant more trees. While the trees here certainly help relieve heat we need to be mindful of how little sun we get here (in Portland rainy weather). I think the city needs to maintain a balance between densely wooded areas (e.g., Forest Park) and highly exposed areas (e.g., central eastside). I think students and volunteers could plant lots of trees. (97221—Southwest)

(...) determine the most aesthetic and functional places to plant trees and then only plant trees that make logical sense for the conditions present in the chosen locations. (97210—Northwest)

I think the city should plant fruit and nut trees when they plant trees. They grow just as easy as any other tree. Most have beautiful flowers and foliage. And better yet they make healthy snacks especially in low income neighborhoods and food deserts. (97236—East)

(...) I can't see cars coming at intersections because there are too many trees already in my neighborhood. There is a near miss almost every day at my house because people can't see the cars coming. (97212—Northeast)

East of the Willamette River, five zip codes stood out with more than 10% of canopy cover. The zip code 97236 had 23.5% canopy cover in the Pleasant Valley area, an early incorporated neighborhood near an affluent suburb, Happy Valley. This zip code area also bears Powell Butte Natural Area, a remaining forest fragment. 97212 had 17% of tree coverage and the fourth largest average income citywide. 97202 had 13.8% of canopy cover and a protected riparian zone in the eastbound of Willamette River. 97215 had 13% of canopy cover and a preserved forest fragment on Mount Tabor Park. 97266 had 12.4% of tree coverage and was the third-lowest income zip code. However, it bears a forest fragment on Kelly Butte Natural Area. These findings showed that tree canopy follows income and geological formation features, such as riverbanks, forest fragments and hilly areas.

In low-income communities, trees' maintenance is observed as a financial burden, which can be classified as an ecosystem disservice [43]. Two zip codes from the East sector answered that 8.3% and 7.7% of the increasing tree canopy strategies have financial burden as a management challenge. The average answer mentions for financial burden was 4.5% per zip code. Seven out of eleven neighborhoods with answers above average are in the East, the sector with the lowest income and low canopy (Table 4). The following statements extracted from the open-ended question about strategies for increase tree canopy reflect the concerns for financial burden within residents of low-income neighborhoods:

"Offer to plant them (trees), offer low income solutions to families" (97233—East)

(...) lots of trees (are) in bad places and they die so better planning would do just fine and offering classes for ppl (people) who want to learn how to maintain the trees better and if they have it why does low income not have access to the classes and knowledge of them? (97233—East)

Don't charge for leaf cleanup. Offer a small tax credit for trees planted and maintained to property owner(s). Education regarding the importance of trees for everyone. Offer education to grow trees in a pot. Then everyone can grow a tree. (97220—East)

(...) more financial and volunteer support to groups like that (street tree planting nonprofit). Also when Portland had the ice storm this past winter, many residential trees came down. (...) people could bring downed trees and branches, maybe even for a donation. Free mulch for the city and donations! Tree culture need(s) to be supported in more ways than just plantings (...) to make owning trees easy. (97220—East)

I'm sure a lot of people are scared away from that program (street tree planting program) due to the need to care for the tree and the possible damage to sidewalks that they will eventually be forced to repair at their expense later on down the line. (97220—East)

The answers for increasing canopy strategies also indicated concerns for other UES, such as climate change, knowledge systems, natural heritage and cultural heritage. Excluding the answers about management challenges, responses about regulating and provisioning UES represented 28% and cultural UES accounted for 72% of answers. We believed that people highly value cultural benefits from trees in urban areas due to the "extinction of experience" [3]. Within cultural ecosystem services, we distinguished patterns in responses that seek environmental education, multiple ethnic values for forest biodiversity and conservation of heritage trees. The answers associated with knowledge systems indicated that besides incentives for tree planting, people also need to know how to care for trees and their importance regarding ecosystem services. Public surveys assessing urban forestry and management of UES have suggested the enforcement of knowledge systems [44]. As observed in the previous statements, the survey participants repetitively suggested partnerships with education institutions, urban forestry jobs, internships and free workshops. The responses indicated that the residents expect more personal accountability for ownership if they have access to knowledge, tools and technical support from municipality and nonprofits.

In answers that mentioned cultural heritage, people requested more planting events to bond with neighbors and create a sense of community. The participants asked for multilingual tree support, public participation in urban forestry planning and access to trees with ethnic values regarding inclusion and diversity measures. Natural heritage answers indicate the population's willingness to perpetuate biodiversity, urban ecology and centenary trees. Together, cultural and natural heritage are UES that reflect landscape interpretations, which are individual perspectives of the environment based on personal background, memories, experiences and expectations. In general, environmental planning bears management tools that can perpetuate systemic racism by restricting access to ecosystem services based on socioeconomic values, reducing maintenance costs in areas with low-income and people of color and not acknowledging the diversity of behaviors in public space [45,46]. Surveys, interviews and focus groups can collect ideas, perspectives and expectations from historically unheard voices and open a path for public participation in urban forestry planning.

Portland had a complex history of gentrification that burdened the black community with displacement, loss of sense of spatial identity and identity representation [21]. The increase in population promoted real estate development for housing, business and other civic infrastructures. In Portland, there is an inverse relationship between canopy cover and urban development indicators, as water pipers [10]. The survey answers indicated that people are aware that new developments threaten trees, impacting their maintenance and natural heritage. The following statements are from the zip codes with lower housing ownership [Appendix B, Table A1] and most gentrified areas [32]:

"Yes, we need many more trees but (...) focus on protecting the most mature trees as they have been shown to provide the greatest benefits." (97227—North)

Demolition review to ensure maintenance of entire tree canopy as development can remove existing trees. The accelerated development in Portland has not been counterbalanced with a comprehensive plan to prevent tree removal and plant new trees. It has greatly reduced potential green spaces which could offset somewhat the unbridled concrete development. (97232—Northeast)

Trees are natural green infrastructures that support stormwater catchment, avoid erosion and improve air quality. Other forms of green infrastructure such as rain gardens, green roofs, artificial wetlands and parks can bring green gentrification—a gentrification process caused by the implementation of green infrastructures. The following Discussion section will explore how the results are associated with environmental justice and landscape management.

4. Discussion and Conclusions

This study aimed to address questions about the relationship between the existing amounts of neighborhood tree canopy with sociodemographic data and community perspectives. One of the explicit goals of the present study was to understand the relationship between tree ownership, maintenance and the amount of tree canopy. We found that zip codes with higher tree canopy were consistent with greater sense of ownership and quality of trees, as measured by the TOSI. Specifically, the two TOSI questions about the number of trees and the good condition of trees had a strong correlation and high significance values with tree canopy. This finding is consistent with a low Cronbach's alpha, suggesting that this metric can be explored further, perhaps in a different setting.

Our findings also indicate that a sense of ownership comprises the importance and satisfaction with the quantity and the quality of trees in the neighborhood, including trees in private property, public spaces and the right-of-way. Affluent zip codes had higher canopy cover had a higher correlation with public maintenance of street trees, as measured by the TMSI. While earlier research suggests that tree canopy and income are correlated in the U.S. [8–10], the perception of tree ownership is a new concept that brings accountability of ownership and maintenance in relation to urban ecosystem services. We observed a lower correlation between the tree canopy and TMSI than with TOSI (Table 8). TMSI also presented a lower range of mean values on the Likert scale response (Table 6), suggesting a common concern about tree care citywide (Figure 3). In the question about increasing tree canopy strategies, the responses about tree maintenance represented about 18% of the answers (Table 7).

Perhaps one of the most germane findings in our study is the fact that while a canon of literature describes the importance of trees in providing UES (e.g., pollination, air quality, climate regulation, etc.), our survey findings indicate that issues about management and cultural ecosystems services feature prominently among the respondents. This finding, while perhaps mundane, is significant for several reasons, including the fact that respondents seem to recognize the financial burden and maintenance when considering trees. If this finding is consistent across the city, then municipal goals for expanding tree canopy will face formidable obstacles if they present trees an important for traditional regulating ecosystem services. Rather, recognizing that communities are considering, perhaps less these regulatory services, than those surrounding maintenance and financing, may provide more effective.

Suitable messaging may not be the only implication of this finding. If aesthetics, inspiration and level of ownership are correlated with the amount of neighborhood tree canopy (Table 8), then attempts to create distributional equity will require considerable recognition of the maintenance costs involved. Maintenance often includes the planting appropriate species, pruning of trees, watering and a host of other monitoring to ensure healthy growth. Responses indicated the importance of maintenance and also suggested that municipalities provide support to those communities how may not have the financial resources to address maintenance concerns. Currently the City of Portland requires adjacent property owners to maintain all public trees, which increases the level of inequity already experienced by lower income communities. Perhaps the positive and significant correlation between income and presence of tree canopy is because lower income community may not prefer trees due to the cost of maintenance, which the open-ended responses indicated. The development of trees in the right-of-way interacts with other infrastructure, such as sidewalks, residences and transit features. Studies that observe the growth and health of street trees [47], suggest regular monitoring and maintenance such as root pruning and interaction with underground infrastructures [48–50] which can help to ensure a healthy and robust tree canopy.

While these findings offer a first step towards integrating community perspectives into urban forest management, the findings suggest the importance of engaging communities in the management of tree canopy. The open-ended results suggested that respondents genuinely understand the challenges facing urban forest management and the importance of finding systematic ways to maintain canopy and provide equitable access to all residents. Our findings indicate, for example, that promoting financial solutions that optimize public and private budgets toward urban forestry and cultural heritage practices that engage the community in participatory planning and empowerment are the priority strategies for increasing tree canopy. With the multiple goals for achieving environmental equity through urban forestry, these strategies must also include raising awareness about the inequitable distribution of existing tree canopy, planting more trees in vulnerable communities, exploring diverse perspectives about climate resilience and exploring the role of trees among historically marginalized communities [46].

This study offers a means for understanding the importance of ownership and maintenance in addressing urban ecosystem services. While our survey can help to underscore some of these priorities, we recognize that engaging communities about urban forestry may pose several challenges. If employed effectively, other data collection methods, including listening sessions, focus groups and interviews, can help to contextualize urban forestry within the broader set of community needs that may be priorities. The COVID-19 pandemic has made clear that priorities such has housing, food and medical care are often front-and-center among POC and lower income communities, which may pose several challenges for discussions about urban forestry. Since POC and low-income neighborhoods have been excluded from planning for a healthy and abundant urban forest [8], a pattern that may be associated to redlining practices in the U.S. [51] perhaps the built and natural environment in neighborhoods can be a direct means for understanding other pressing priorities. By engaging historically disinvested communities and address distributional injustices that have created current inequities in the distribution of tree canopy cover, municipal agencies may find creative solutions that 'multi-solve' the myriad pressing challenges.

Our study found that survey respondents seeks more measures and strategies to address cultural UES, such as cultural heritage, natural heritage, aesthetics and inspiration. The gap of systematic descriptions for these cultural UES within municipal plans may require greater levels of public involvement, which would build on diverse perspectives [46]. While government agencies are often responsible for the management of public spaces, the same agencies may not be trusted allies with communities that have been historically disinvested. As such, management options that engage community-based organization (CBOs) may be a more trusted and effective approach for soliciting plausible solutions. Such CBOs can work with community members to explore their expectations for land use, tree canopy, species selection, planting events, tree giveaways and volunteer workforce.

Examples of such cross-sectoral urban forestry management are emerging. In Toronto, CBOs had a more diverse species list than municipal agencies, landscape architects and nurseries [52]. In Detroit, interviews with CBO staff and recipients of giveaway trees informed that the ability to choose the tree species is a fact that impacts the willingness of residents to care for private trees, as well as live in areas with lower canopy. In both cases, studies have found that the major challenges are concerns with maintenance practices and costs, such as pruning, sidewalk damages and tree debris removal [53]. Both studies suggested stewardships for a functioning and healthy urban forestry, where CBOs would have the goal to promote understanding, while supporting cultural ecosystem services.

Author Contributions: L.A.C.N. have performed the analysis and wrote the manuscript. V.S. have collected the survey data, coordinated the researched, and reviewed the manuscript. Both authors have read and agreed to the published version of the manuscript.

Funding: This research was funded by Conselho Nacional de Desenvolvimento Científico e Tecnológico (CNPq), grant number: 221077/2014-6—LASPAU—Brasil/CNPq—GDE—EUA; and by the US Forest Service's National Urban and Community Forestry Challenge Grant (17-DG-11132544-014) and the Robert Wood Johnson Climate and Health project.

Conflicts of Interest: The authors declare no conflict of interest. The founding sponsors had no role in the design of the study; in the collection, analyses, or interpretation of data; in the writing of the manuscript, and in the decision to publish the results.

Appendix A

Survey:

Tree Ownership Satisfaction Index (TOSI):

Q1 Portland's trees are important to me. Strongly Agree/Agree/Don't Know/Disagree/Strongly Disagree.

Q2 My neighborhood has enough trees: Strongly Agree/Agree/Don't Know/Disagree/Strongly Disagree.

Q3 The trees in my neighborhood are in good condition and healthy. Strongly Agree/Agree/Don't Know/Disagree/Strongly Disagree.

Tree maintenance Satisfaction Index (TMSI).

Q4 The city should maintain all trees along the street (in the public right-of-way, next to the sidewalk area): Strongly Agree/Agree/Don't Know/Disagree/Strongly Disagree.

Q5 The city should prioritize maintenance of trees along the street (in the public right-of-way, next to the sidewalk area) in low-income communities: Strongly Agree/Agree/Don't Know/Disagree/Strongly Disagree.

Q6 The city should plant trees in all available spaces along the street (in the public right-of-way, next to the sidewalk area): Strongly Agree/Agree/Don't Know/Disagree/Strongly Disagree.

Strategies for increase tree canopy:

Q7 How do you think the city should get more trees planted?

Presence of trees in private properties:

Q8 Do you have trees at the property where you live? Yes/No.

Demographic Questions:

Q9 What is your household income? Less than $10,000/$10,000–$19,999/$20,000–$29,999/$30,000–$39,999/$40,000–$49,999/$50,000–$59,999/$60,000–$69,999/$70,000–$79,999/$80,000–$89,999/$90,000–$99,999/$100,000–$149,999/$150,000–$199,999/$200,000 or more/I don't know.

Q10 Which best describes your race or ethnicity? Choose as many as apply:

❑ Alaska Native ❑ American Indian/Native American ❑ East Asian ❑ South Asian ❑ Southeast Asian ❑ West Asian ❑ Middle Eastern ❑ Black or African American ❑ African ❑ Hispanic or Latino ❑ Native Hawaiian or Pacific Islander ❑ Slavic or Eastern European ❑ White ❑ Other (please specify).

Q11 What is your home zip code?

Q12 Do you rent or own the place where you live? Rent/Own.

Appendix B

Table A1. Socio indicator data from survey and Census.

Zip Code	Area	Valid Survey Answers (N)	Population Census (N)	Households Census (N)	Housing Ownership Census (%)	Housing Ownership Survey (%)	Housing Ownership Census (N)	Housing Ownership Survey (N)	People of Color Census (%)	People of Color Survey (%)	People of Color Census (N)	People of Color Survey (N)	Mean Income Census (US$)	Mean Income Survey (US$)
97201	SW	50	17,566	9009	31.58	T52.00	2845	26	23.17	26.53	4070	13	92,276	88,666
97202	SE	211	42,189	18,135	49.68	81.52	9010	172	14.47	13.74	6103	29	92,391	96,865
97203	North	107	34,089	12,091	55.43	85.98	6702	92	24.2	21.5	8250	23	67,963	80,663
97205	SW	31	7122	4881	17.62	58.06	860	18	20.33	10	1448	3	61,537	81,087
97206	SE	249	50,655	20,100	62.83	84.34	12,628	210	19.94	34.94	10,102	87	71,658	83,602
97209	NW	33	16,507	11,376	25.58	60.61	2910	20	15.68	15.15	2589	5	82,582	100,645
97210	NW	35	11,676	6253	40.76	74.29	2549	26	9.76	25.71	1140	9	140,398	132,407
97211	NE	181	34,856	13,081	65.81	85.08	8609	154	27.21	19.89	9484	36	91,179	98,866
97212	NE	143	26,601	10,890	64.05	89.51	6975	128	15.81	11.89	4206	17	125,138	112,073
97213	NE	165	32,284	13,783	61.63	85.45	8495	141	15.98	17.58	5160	29	87,715	89,455
97214	SE	144	25,398	12,190	35.64	81.25	4344	117	12.91	22.22	3280	32	84,012	94,801
97215	SE	91	17,939	7802	63.86	92.31	4982	84	10.84	25.27	1945	23	101,953	97,530
97216	East	51	17,112	6530	46.66	90.2	3047	46	29.93	33.33	5122	17	53,870	70,306
97217	North	170	34,327	14,520	62.41	82.94	9062	141	22.34	25.29	7669	43	84,004	88,975
97218	NE	76	15,556	5543	59.1	90.79	3276	69	28.43	27.63	4422	21	68,539	73,450
97219	SW	202	41,534	16,561	70.32	90.1	11,646	182	13.05	17.82	5419	36	119,199	104,608
97220	East	93	30,374	11,488	56.27	91.4	6464	85	34.62	24.73	10,514	23	60513	80,632
97221	SW	41	12,363	5170	75.92	95.12	3925	39	11.79	34.15	1457	14	146,948	121,333
97227	North	15	4648	2259	31.03	86.67	701	13	24.94	26.67	1159	4	77,831	90,833
97229	NW	21	65,285	23,913	70.72	100	16,912	21	31.83	23.81	20,779	5	137,990	144,166
97230	East	77	39,884	14,967	56.97	93.51	8527	72	32.96	23.38	13,147	18	63,112	87,847
97232	NE	57	12,865	6641	31.86	56.14	2116	32	14.42	15.79	1855	9	86,325	104,727
97233	East	54	42,001	13,496	45.82	66.67	6184	36	35.16	37.04	14,767	20	49,769	51,875
97236	East	53	40,274	12,996	54.59	90.57	7094	48	36.56	30.19	14,726	16	60,022	69,200
97239	SW	67	16,682	8322	49.54	85.07	4123	57	17.89	14.93	2985	10	120,876	110,350
97266	East	131	34,757	12,280	54.24	84.73	6661	111	33.46	23.66	11,628	31	55,339	71,349

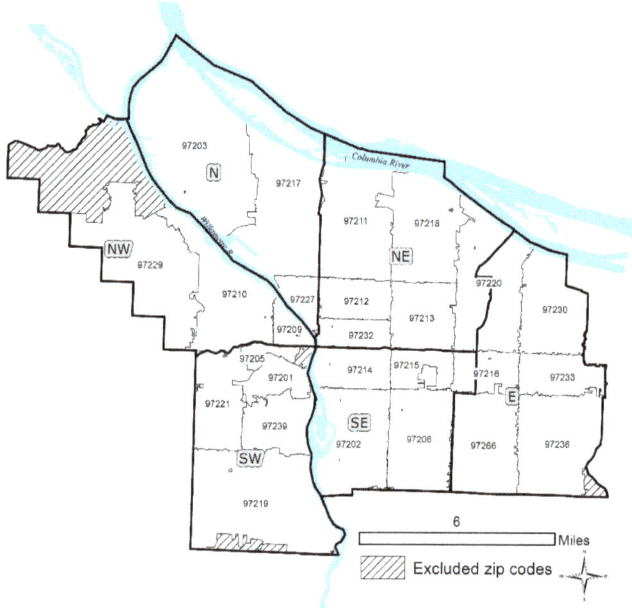

Figure A1. Zip codes in the study area.

References

1. United States Census. New Census Data Show Differences between Urban and Rural Populations. Available online: https://www.census.gov/newsroom/press-releases/2016/cb16-210.html (accessed on 6 July 2020).
2. Hall, J.R. Social Futures of Global Climate Change: A Structural Phenomenology. *Am. J. Cult. Sociol.* **2016**, *4*, 1–45. [CrossRef]
3. Soga, M.; Gaston, K.J. Extinction of experience: The loss of human-nature interactions. *Front. Ecol. Environ.* **2016**, *14*, 94–101. [CrossRef]
4. Nowak, D.J.; Greenfield, E.J. US Urban Forest Statistics, Values, and Projections. *J. For.* **2018**, *116*, 164–177. [CrossRef]
5. Nowak, D.J.; Greenfield, E.J. Declining Urban and Community Tree Cover in the United States. *Urban For. Urban Green.* **2018**, *32*, 32–55. [CrossRef]
6. Salmond, J.A.; Tadaki, M.; Vardoulakis, S.; Arbuthnott, K.; Coutts, A.; Demuzere, M.; Dirks, K.N.; Heaviside, C.; Lim, S.; MacIntyre, H.; et al. Health and Climate Related Ecosystem Services Provided by Street Trees in the Urban Environment. *Environ. Health A Glob. Access Sci. Source* **2016**, *15* (Suppl. 1). [CrossRef] [PubMed]
7. Wilkerson, M.L.; Mitchell, M.G.E.; Shanahan, D.; Wilson, K.A.; Ives, C.D.; Lovelock, C.E.; Rhodes, J.R. The Role of Socio-Economic Factors in Planning and Managing Urban Ecosystem Services. *Ecosyst. Serv.* **2018**, *31*, 102–110. [CrossRef]
8. Schwarz, K.; Fragkias, M.; Boone, C.G.; Zhou, W.; McHale, M.; Grove, J.M.; O'Neil-Dunne, J.; McFadden, J.P.; Buckley, G.L.; Childers, D.; et al. Trees Grow on Money: Urban Tree Canopy Cover and Environmental Justice. *PLoS ONE* **2015**, *10*. [CrossRef] [PubMed]
9. Locke, D.H.; Landry, S.M.; Grove, J.M.; Roy Chowdhury, R. What's Scale Got to Do with It? Models for Urban Tree Canopy. *J. Urban Ecol.* **2016**, *2*, juw006. [CrossRef]
10. Ramsey, J. Tree Canopy Cover and Potential in Portland, OR: A Spatial Analysis of the Urban Forest and Capacity for Growth. Master's Thesis, Portland State University, Portland, OR, USA, 2019. [CrossRef]
11. McLain, R.J.; Hurley, P.T.; Emery, M.R.; Poe, M.R. Gathering "Wild" Food in the City: Rethinking the Role of Foraging in Urban Ecosystem Planning and Management. *Local Environ.* **2014**, *19*, 220–240. [CrossRef]
12. Voelkel, J.; Hellman, D.; Sakuma, R.; Shandas, V. Assessing Vulnerability to Urban Heat: A Study of Disproportionate Heat Exposure and Access to Refuge by Socio-Demographic Status in Portland, Oregon. *Int. J. Environ. Res. Public Health* **2018**, *15*, 640. [CrossRef] [PubMed]
13. Rao, M. Investigating the Potential of Land Use Modifications to Mitigate the Respiratory Health Impacts of NO_2: A Case Study in the Portland-Vancouver Metropolitan Area. Ph.D. Thesis, Portland State University, Portland, OR, USA, 2016.
14. Kondo, M.C.; Low, S.C.; Henning, J.; Branas, C.C. The Impact of Green Stormwater Infrastructure Installation on Surrounding Health and Safety. *Am. J. Public Health* **2015**, *105*, e114–e121. [CrossRef] [PubMed]

15. Gaston, K.J.; Soga, M.; Duffy, J.P.; Garrett, J.K.; Gaston, S.; Cox, D.T.C. Personalised Ecology. *Trends Ecol. Evol.* **2018**, *33*, 916–925. [CrossRef] [PubMed]
16. Millennium Ecosystem Assessment. *Ecosystems and Human Well-Being: Synthesis (Millennium Ecosystem Assessment Series)*; Island Press: Washington, DC, USA, 2005.
17. Kuthy, D. Redlining and Greenlining: Olivia Robinson Investigates Root Causes of Racial Inequity. *Art Educ.* **2017**, *70*, 50–57. [CrossRef]
18. Mortimer-Sandilands, C.; Erickson, B. *Queer Ecologies: Sex, Nature, Politics, Desire (Book Collections on Project MUSE)*; Indiana University Press: Bloomington, IN, USA, 2010.
19. McPhearson, T.; Andersson, E.; Elmqvist, T.; Frantzeskaki, N. Resilience of and through Urban Ecosystem Services. *Ecosyst. Serv.* **2015**, *12*, 152–156. [CrossRef]
20. Ortiz, M.S.O.; Geneletti, D. Assessing Mismatches in the Provision of Urban Ecosystem Services to Support Spatial Planning: A Case Study on Recreation and Food Supply in Havana, Cuba. *Sustainability* **2018**, *10*, 2165. [CrossRef]
21. Hern, M. *What a City is for: Remaking the Politics of Displacement*; MIT Press: Cambridge, CA, USA, 2016.
22. Griffin-Valade, L.; Kahn, F.; Scott, J. East Portland: History of City Services Examined. Available online: https://www.portlandoregon.gov/auditservices/article/488003 (accessed on 6 July 2020).
23. Hawkins, W. *The Legacy of Olmsted Brothers in Portland, Oregon*; Priv Printed: Portland, OR, USA, 2014.
24. Robertson, G.; Mason, A. Assessing the sustainability of agricultural and urban forests in the United States. 2016. Available online: https://www.fs.usda.gov/treesearch/pubs/52278 (accessed on 6 July 2020).
25. Gibson, K.; Abbott, C. Portland, Oregon. *Cities* **2002**, *19*, 425–436. [CrossRef]
26. Alberti, M. *Advances in Urban Ecology: Integrating Humans and Ecological Processes in Urban*; Springer: New York, NY, USA, 2008.
27. Turner, M.; Gardner, R.H.; Golley, F.B. Oak Ridge National Laboratory. In *Landscape Ecology in Theory and Practice: Pattern and Process*, 2nd ed.; Springer: New York, NY, USA, 2015.
28. Wherry, S.; Wood, T.; Moritz, H.; Duffy, K. Assessment of Columbia and Willamette River Flood Stage on the Columbia Corridor Levee System at Portland, Oregon, in a Future Climate. Scientific Investigations Report, 1, 2018. Available online: https://pubs.usgs.gov/sir/2018/5161/sir20185161.pdf (accessed on 6 July 2020).
29. United States Census Data. Available online: https://data.census.gov/cedsci/ (accessed on 6 July 2020).
30. Ause, C.W. Black and Green: How Disinvestment, Displacement and Segregation Created the Conditions for Eco-Gentrification in Portland's Albina District, 1940–2015. Bachelor's Thesis, Portland State University, Portland, OR, USA, 2016.
31. Hughes, J. Historical Content of Racist Planning: A history of how planning segregated Portland. In *City of Portland*; 2019. Available online: https://www.portland.gov/sites/default/files/2019-12/portlandracistplanninghistoryreport.pdf (accessed on 6 July 2020).
32. Bates, L.K. *Gentrification and Displacement Study: Implementing an Equitable Inclusive Development Strategy in the Context of Gentrification*; Portland State University Urban Studies and Planning Faculty Publications and Presentations: Portland, OR, USA, 2013.
33. Goodling, E.; Green, J.; McClintock, N. Uneven Development of the Sustainable City: Shifting Capital in Portland, Oregon. *Urban Geogr.* **2015**, *36*, 504–527. [CrossRef]
34. City of Portland. Portland Maps-Open Data. Available online: https://gis-pdx.opendata.arcgis.com/ (accessed on 6 July 2020).
35. Oregon State University, State of Oregon. Oregon Spatial Data Library. Available online: https://spatialdata.oregonexplorer.info/geoportal/ (accessed on 6 July 2020).
36. United States Geological Survey. Multi-Resolution Land Characteristics Consortium. Available online: https://www.mrlc.gov/data (accessed on 6 July 2020).
37. Coulston, J.W.; Moisen, G.G.; Wilson, B.T.; Finco, M.V.; Cohen, W.B.; Brewer, C.K. Modeling percent tree canopy cover: A pilot study. *Photogr. Eng. Remote Sens.* **2012**, *78*, 715–727. [CrossRef]
38. Shandas, V.; Boden, K.; Voelkel, J. Citywide Tree Planting Report. 2017. Available online: https://www.portlandoregon.gov/parks/article/705823 (accessed on 6 July 2020).
39. Atique, U.; An, K.G. Stream Health Evaluation Using a Combined Approach of Multi-Metric Chemical Pollution and Biological Integrity Models. *Water (Switzerland)* **2018**, *10*, 661. [CrossRef]
40. Yirigui, Y.; Lee, S.W.; Pouyan Nejadhashemi, A. Multi-Scale Assessment of Relationships between Fragmentation of Riparian Forests and Biological Conditions in Streams. *Sustainability* **2019**, *11*, 5060. [CrossRef]
41. Arifwidodo, S.D.; Chandrasiri, O. The Relationship between Housing Tenure, Sense of Place and Environmental Management Practices: A Case Study of Two Private Land Rental Communities in Bangkok, Thailand. *Sustain. Cities Soc.* **2013**, *8*, 16–23. [CrossRef]
42. Donovan, G.H.; Prestemon, J.P. The Effect of Trees on Crime in Portland, Oregon. *Environ. Behav.* **2012**, *44*, 3–30. [CrossRef]
43. Speak, A.; Escobedo, F.J.; Russo, A.; Zerbe, S. An Ecosystem Service-Disservice Ratio: Using Composite Indicators to Assess the Net Benefits of Urban Trees. *Ecol. Indic.* **2018**, *95*, 544–553. [CrossRef]
44. Young, R.F. Mainstreaming Urban Ecosystem Services: A National Survey of Municipal Foresters. *Urban Ecosyst.* **2013**, *16*, 703–722. [CrossRef]
45. Brownlow, A. An Archaeology of Fear and Environmental Change in Philadelphia. *Geoforum* **2006**, *37*, 227–245. [CrossRef]

46. Schell, C.J.; Dyson, K.; Fuentes, T.L.; Des Roches, S.; Harris, N.C.; Miller, D.S.; Woelfle-Erskine, C.A.; Lambert, M.R. The Ecological and Evolutionary Consequences of Systemic Racism in Urban Environments. *Science* **2020**, *4497*, eaay4497. [CrossRef]

47. Batala, E.; Tsitsoni, T. Street Tree Health Assessment System: A Tool for Study of Urban Greenery. *Int. J. Sustain. Dev. Plan.* **2009**, *4*, 345–356. [CrossRef]

48. Benson, A.R.; Morgenroth, J.; Koeser, A.K. The Effects of Root Pruning on Growth and Physiology of Two Acer Species in New Zealand. *Urban For. Urban Green.* **2019**, *38*, 64–73. [CrossRef]

49. Kuliczkowska, E.; Parka, A. Management of Risk of Environmental Failure Caused by Tree and Shrub Root Intrusion into Sewers. *Urban For. Urban Green.* **2017**, *21*, 1–10. [CrossRef]

50. Ordóñez Barona, C. Adopting Public Values and Climate Change Adaptation Strategies in Urban Forest Management: A Review and Analysis of the Relevant Literature. *J. Environ. Manag.* **2015**, *164*, 215–221. [CrossRef]

51. Hoffman, J.S.; Shandas, V.; Pendleton, N. The Effects of Historical Housing Policies on Resident Exposure to Intra-Urban Heat: A Study of 108 US Urban Areas. *Climate* **2020**, *8*, 12. [CrossRef]

52. Conway, T.M.; Vander Vecht, J. Growing a Diverse Urban Forest: Species Selection Decisions by Practitioners Planting and Supplying Trees. *Landsc. Urban Plan.* **2015**, *138*, 1–10. [CrossRef]

53. Carmichael, C.E.; McDonough, M.H. Community Stories: Explaining Resistance to Street Tree-Planting Programs in Detroit, Michigan, USA. *Soc. Nat. Resour.* **2019**, *32*, 588–605. [CrossRef]

 *land*

Article

Land-Based Financing Elements in Infrastructure Policy Formulation: A Case of India

Raghu Dharmapuri Tirumala * and **Piyush Tiwari**

Faculty of Architecture, Building and Planning, Melbourne School of Design, The University of Melbourne, Melbourne 3010, Australia; Piyush.tiwari@unimelb.edu.au
* Correspondence: dtvraghu@unimelb.edu.au

Abstract: A rapid increase in land and property values has been one of the driving forces of urban ecosystem development in many countries. This phenomenon has presented project proponents/policymakers with multiple options and associated challenges, nudging them to configure or incorporate elements of land-based financing in their policies and legislations. Specifically, the Government of India and various state governments have sought to monetize land through diverse instruments, for augmenting the financial viability of infrastructure and area development projects. This paper compares Indian central and state infrastructure policies/acts with regard to land monetization strategies. The analysis indicates that policies and legislations are taking a turn towards promoting land monetization mechanisms as a financing tool for cities and project implementation agencies. However, the approach is cautiously used and implementation is often seen to fall behind actual project timelines. Based on the findings, key determinants of a successful policy that captures an increase in land values, are identified. The learnings provide useful inputs for states to strengthen their policy documents and legislative/institutional frameworks, for ensuring the effectiveness of land-based financing tools.

Keywords: land-based financing; land monetisation; policy; infrastructure; Sustainable Development Goals

Citation: Tirumala, R.D.; Tiwari, P. Land-Based Financing Elements in Infrastructure Policy Formulation: A Case of India. *Land* **2021**, *10*, 133. https://doi.org/10.3390/land10020133

Academic Editor: Alessio Russo
Received: 31 December 2020
Accepted: 27 January 2021
Published: 29 January 2021

Publisher's Note: MDPI stays neutral with regard to jurisdictional claims in published maps and institutional affiliations.

1. Introduction

The infrastructure and urban development landscape across the world witnessed a substantial transformation in the last two decades, attributable to economic growth, internationalization, and greater expectations of citizens [1]. The developing world has struggled with wide-ranging systemic constraints that include technical, institutional, and financial aspects in managing this burgeoning change. The growing fiscal stress of cities ("urban local bodies" (ULBs)) is forcing nations and subnational entities across the world to explore newer sources of finance—broadening the tax base and introducing different charges for utilities [2]. Land-based financing is increasingly seen as a tool for developing infrastructure and providing ecosystem services, across the world. The success of Singapore and Hong Kong in using these tools has spurred the interests of various governments the world over, to explore various tools and instruments that capture the increased value of land, and to finance infrastructure or area-based developments. Driven by speculation, these approaches have led to increased land prices and created more inequality in the developing world. In developed markets such as Hong Kong, London, Paris, and Tokyo, it is estimated that a person earning a national average income would need to work for more than 60 years to buy a residential property. Similar assessment for India ranges between 100 years for property across Indian metros, and nearly 580 years for a property at the highest rate in Mumbai [3].

Land is an important backbone for infrastructure development, which led to numerous statutory interventions by governments, to ensure its availability for economic purposes. Land is conventionally seen as one of the factors of production, and the emphasis is on making this increasingly scarce resource available for economic growth. The traditional land acquisition acts in many countries were all premised on procuring land, in some instances under the ambit of eminent domain, to propel economic and ecosystem growth. Further, to prevent runaway misuse of increasing land and property values, some countries put in place, urban land ceiling acts, to cap the rentals that could be charged [4].

The prominence attached to land is very explicit in the Sustainable Development Goals when compared to the Millennium Development Goals (MDGs). In the MDGs, land (and property) is viewed as a resource under Goal 7 (Ensuring Environmental Sustainability) for improving the lives of slum dwellers. In contrast, the importance of land as a contributing parameter has been featured in many SDGs. In SDG 1 ("Removing Poverty in all its forms everywhere"), land features as a factor in Target 1.4 that describes rights to economic resources for all (particularly the poor and vulnerable) including control over land and other forms of property. Access to land in a secure and equal manner is considered an important determinant for improving the productivity and incomes of small-scale food producers in Target 2.3 of SDG 2 ("End hunger, achieve food security and improved nutrition, and promote sustainable agriculture"). Having rights to ownership and control of land is also seen as an important feature for achieving SDG 5 ("Achieve gender equality and empower all women and girls"—Target 5a). Access to land is also an important factor for achieving SDG 11 ("Make cities inclusive, safe, resilient and sustainable"), which addresses access to the infrastructure [5]. Land has always been considered as a fundamental factor that influences poverty, hunger, and sustainability [6]. The development path of the urban ecosystem will be influenced in some sense, by how land develops [7].

This innate relationship between land and development, coupled with the poor resources of the governments to raise finances for providing urban infrastructure facilities to an increasing urban population, gave rise to the concept of land-based financing tools. Land-based financing is modeled on the premise that land as a resource adjacent to an infrastructure facility/service has a substantial increase in value, a part of which could be used for financing development projects in that area. Many countries including India and global forums such as Global Land Tool Network, Royal Institution of Chartered Surveyors have adopted tools for this value capture in the form of area linked development charges, impact fees, transfer of development rights, urban land value tax, surcharges on stamp duty, etc. [8–10].

The idea of levying a tax on increased land value due to the efforts of the government or the community was first stated by Henry George in his seminal book "Progress and Poverty" [11]. Since then the concept of land-based financing or variants of land value capture has been propagated through the years [2,12]. The instruments span across a transfer of development rights, impact fees, land leases and sale proceeds, property, and land tax variants, betterment charges, etc. [13]. The very steep increase in land values, particularly in the developing world, has presented project proponents and policymakers with a range of opportunities to fund projects using land as one of the resources. [14].

While the benefits from land value capture are evident, land-based financing does have its share of criticisms. Leveraging the increased value of land distorts the land rights regime and equity across various sections of the society. In India, for instance, many informal and slum settlements are situated on government-owned lands. The dwellers in these settlements do not own a legal title to the land but have perceived rights to reside. Attempts to gain from monetizing such land parcels might negatively impact these human settlements [15]. The equitable benefits of land-based financing across all sections of the society are not conclusive, as there have been instances of the system bypassing the poor [13] where the expected revenues do not subsidize the needy.

One of the major criticisms of land-based financing models is the intrinsic nature to front end the investment that is recouped over a period, during which time the affordability of the land to the general public is constrained. Thus, lawmakers and policy actors need to assess policies against an equity framework that addresses aspects relating to where the economic value is being created and recovered from, to who is going to benefit from the initiative. The linkage between the taxation and budgeting process and distributing the financial risks equitably needs to be established. Projects such as Hudson Yards and Atlanta Beltline have been assessed under such criteria [16–18]. The evaluation of the effectiveness of the policy or the project being implemented is based on (i) whether there is value creation (ii) how do policies enable this value creation, (iii) how is the value shared between various stakeholders, and (iv) whether the re-deployment of value generated further increases value at other places in the region [16].

Cities, traditionally, have better control and flexibility in managing their non-tax revenues or own assets, due to the relatively limited need for approvals and adherence to the fiscal frameworks [19]. The discourse on land-based financing is largely limited to land value capture instruments and how cities can utilize them to fund their operations. There is limited research on how the potential increase in land values can be used for financing direct infrastructure projects, and how the policies at national or sub-national levels have incorporated the elements that promote land-based financing aspects. This paper discusses the extent of land-based financing being formally incorporated in policy documents, using the policies and legislations in India as a case example.

The rest of the article is organized as follows. The approach adopted for identifying the various policies that have elements related to land-based financing and the comparison framework of these policies is described in Section 2. A few Indian project experiences that used land-based financing are presented in Section 3 to provide an overview of the diversity of regions and project structures that are being developed across the country. Various national and subnational policies, schemes, and acts that incorporate or relate to land monetization are set out in Section 4. Contours of a framework that could be used for incorporating land-based financing aspects in policy are outlined in Section 5. The findings of this analysis and lessons for its wider adoption in land value capture policies are summarized in Section 6.

2. Method

The intent of the governments to encourage land-based financing tools and instruments is typically communicated through various policy documents (and in some cases specific programmes and schemes). The land-based financing mechanisms that are in practice internationally and captured in many forums such as the Global Land Tool Network [8] emphasize the need for such mechanisms to be contextually tailored. The effectiveness of land-based mechanisms and the equity principles promoted by the policies could be gauged by five questions [16] as set out in Table 1 below.

Table 1. Framework for Comparative Analysis.

Parameters	Description
Contributors	Who are the stakeholders from whom the increased value created by public sector interventions/public infrastructure is recovered?
Beneficiaries	Who are the stakeholders to whom the benefits of the value created will accrue?
Process	How is the value–recovery mechanism linked to budgets of the policy proponents and land monetization benefits related to taxation?
Financial Risks	Who are the stakeholders likely to bear the financial risks associated with future cash flows with current investments?
Governance Actors	Who are the stakeholders (actors) involved in the governance of value recovery, distribution, and allocation?

To identify the programs/policies for a comparative study, the initiatives by the Government of India, and by three frontrunner states to undertake PPP projects over the last two decades, were screened for the following criteria (i) policies/programs that specifically address urban area development (ii) programs/policies that mention land-based financing and (iii) the programs/policies that allow for private sector participation to achieve the respective objectives. For this research, land-based financing included all the forms of land value capture, land monetization instruments generally adopted in global networks like Global Land Tool Network [8] and mentioned by the Government of India in its land value capture policy. The sections of the policies that refer to the land-based financing mechanisms were listed out. The land-based financing elements of these policies have been compared with the five parameters (as set out in Table 1) to understand if these elements address the fundamental questions of effectiveness and equity.

3. Indian Project Experiences of Land-Based Financing

Historically, urban land has been used to develop cities in India. Mumbai (known earlier as Bombay) was developed as a port city by selling the leasehold rights by the then English administration. A similar model was adopted to develop Kolkata (earlier known as Calcutta). Many surrounding areas of the capital city, Delhi, were also developed using the sale of urban land (through Development Authorities) [20]. When the Indian economy started to blossom in early 2000, many sectors attracted international investments which subsequently led to a sharp increase in land and real estate prices. For example, allowing foreign direct investments in privately built townships and special economic zones. The domestic investments followed suit, rapidly changing the sector into an attractive investment class [21]. The Government of India and various state governments took advantage of this increase and used the land as a means to fund infrastructure projects. A recent example is the greenfield development of Amaravati as the capital city of Andhra Pradesh (explained in Table 2).

Over the last two decades, India attracted a variety of domestic and international investors, developers, and other stakeholders to participate in its growth story, who have sought progressively to move up the risk curve by exploring newer sectors and implementation arrangements [22]. The launch of major infrastructure programmes such as the Golden Quadrilateral, East West North South corridors, redevelopment of airports, ports, telecom, and energy sector improvements gave a fillip to public and private investment in infrastructure. The subnational entities, state-owned organizations, city administrations, and parastatals followed suit with a range of projects spread across infrastructure subsectors [23]. The buoyancy and aggressiveness of the investment programme coupled with the increasing risk appetite of the private sector resulted in many projects being deemed financially unsustainable solely based on user fees and charges [23]. The project proponents then began structuring projects, primarily those implemented under public–private partnership arrangements, with land-based revenue as an additional source of financing. The trend continued across all the subsectors including transport, urban, tourism, and agriculture marketing projects. The project sponsors were also spread across the national, state, and city levels. While a few of the projects were large and had attracted substantial national and regional media attention, there were also a plethora of smaller projects spread across different states that benefitted. An indicative list of projects that used land-based financing mechanisms and have attracted substantial research attention are listed in Table 2.

The greenfield development project of Amaravati city that was based on land pooling, was designed to provide the development authority with considerable land to use for development and also to raise finances for infrastructure creation, through capturing the increase in land value. The challenge, however, will be from competing developments in fringe areas that could affect the land value appreciation in the development areas of the city. The development authority would need to identify measures to counter the impact of competing developments, for instance, by permitting higher built-up areas and capturing the value from this increase. Another challenge for the authority to balance is the mismatch in the timing of initial infrastructure investments required and the value capture that would be realized over a longer time horizon [15].

Table 2. Illustrative projects that incorporated land-based financing elements.

Project	Key Features
Amaravati City	Amaravati, city assumed importance in 2014 when it was designated as the administrative capital of the newly carved state of Andhra Pradesh. Anticipating the challenges that would come from rapid urbanization, population increase, and the associated increase in the value of land in the urban area, the Government of Andhra Pradesh introduced the "Land Pooling Scheme (LPS)" as an innovative land-use planning instrument to capture the land value increases in a manner that is equitable and balances development with urbanization. The unique feature of the LPS was to ensure that landowners benefit directly from the increase in land value when the capital city would be developed, by making them stakeholders in the development process. The scheme encouraged landowners to contribute their plot of land in return for a smaller plot of urban serviced land (returnable plot), expected to be higher in value and was also promised annuity payments (to support livelihoods) and skill upgrading programs for setting up self-owned enterprises [15].
The Bangalore-Mysore Infrastructure Corridor	The project was conceived in the 1980s as an expressway of about 111 km connecting the heritage city of Mysore to the capital city Bangalore. The project proposed by Nandi Infrastructure Corridor Enterprises Ltd. was also to include residential, industrial, and commercial facilities. The objective was to de-congest Bangalore city and attract people to shift to the upcoming townships along the corridor. The government of Karnataka entered into a Framework Agreement with the developer in the year 1997 agreeing to provide land for the project. However, several rounds of litigation ensued, challenging the land acquisition process for the project and refusal to grant permission by the Planning Authority for the construction of group housing [24].
Hyderabad Metro Rail Project	The rail project comprising 66 stations in 72 km estimated at Rs. 14,132 crores are being developed by L&T Hyderabad Metro Rail Private Limited, Hyderabad. Nearly, 40% of the revenues are estimated to come from real estate development and lease revenues therefrom [25].
Modernization of Delhi International Airport	Rajiv Gandhi International Airport was a greenfield airport built on approximately 5000 acres of land. Nearly 45 acres were identified as phase 1 development for commercial and hospitality facilities comprising about 14 assets. Out of 5000 rooms, 3000 rooms were budget rooms and the remaining were categorized as luxurious rooms. The cost of construction was estimated to be Rs. 75 lakhs to Rs. 1 crore per room. The concession agreement was signed on 4.4.2006 and construction completed on 31.3.2010. The other major components of the project included the renovation of Terminals 1A, 1B, 1C, and Terminal 2 of the existing airport [26].
Dhamra Port, Odisha	Dhamra is a port close to the mineral-rich industrial states of Odisha, Jharkhand and Chattisgarh developed on a Build Own Operate Share Transfer framework. The 25.0 MTPA project with a concession period of 408 months was bid out on a revenue share basis. The concession agreement was signed on 2.4.1998. Excess land allotted to the project was met with challenges [27].
Modern Central Bus Terminus cum Commercial Complex, Haldia, West Bengal	Eleven acres of land near HPL Township was identified for the project, out of which the private partner was to develop a modern Central Bus Terminus on 7 acres of land and operate it for 20 years. The remaining 4 acres of land was to be developed commercially by the PPP partner who was also expected to operate and maintain the facilities for 50 years [28].
Construction of central bus terminal, Makarpura, Vadodara	Constructed on 6.3 acres as a BOT project for a concession period of 378 months. The concession agreement was signed in 2010 with Hubtown (Vadodara) Bus Terminal Ltd. at a project cost of Rs. 60.28 crores [28].

Source: Authors compilation based on a review of the literature.

The Bangalore Mysore Infrastructure Corridor, a road project that linked two major cities Bangalore and Mysore in the southern state of Karnataka, had attracted significant public attention as a large quantum of land was promised to the project developer. The business case of the project hinged on making land available to the developer along the corridor, for developing and maintaining townships for a defined period, which would offset the costs for developing the road and associated infrastructure. While there was opposition to the excessive land sought to be acquired for the project, the contours of this model served as a template for infrastructure projects in the country to use land-based financing approaches. Therefore, whenever the business case of a project resulted in suboptimal revenues (from the user charges and associated cash flow streams), the project sponsors, typically the public sector entities offered additional land or rights to use the land for shoring up the sources. For instance, in the case of Delhi International Airport (and other airports at Hyderabad, Bangalore, and Mumbai), the project sponsor offered land surrounding the main airport for commercial development. The income arising from such additional land development was expected to offset the capital and operations and maintenance (O&M) expenditure being incurred on the project.

Similar models have also been adopted in port projects (Dhamra port) and metro train projects (Hyderabad metro). The viability of revenue streams was a risk that was sought to be addressed using land and real estate as a sweetener in many instances. The approach, though, has not been a runaway success due to the limitations in monetizing the land value, especially after the global financial crisis and its aftermath [23]. Political controversies have also stalled the progress of land acquisition and the implementation of projects in many cases as there were attempts to convert agricultural land to industrial and infrastructure purposes [14]. However, the belief in such a model had nevertheless been seeded in many project sponsors.

4. Elements of Land Monetization in Key Legislation and Policies in India

There have been initiatives (though on a limited scale and fragmented) to reform the institutional, governance, and financial systems for encouraging private sector participation and coexisting with the social contexts surrounding land [14,29]. The policy and legislative framework that enables land-based financing in India include Municipal Corporations Acts, Municipality Acts, Development Authority Acts, Town Planning Acts, Stamp Duty, and Registration Acts, various schemes of the central and state government, national policies, and state policies. The powers to legislate different types of instruments varies by the type of agency.

The proponents of infrastructure projects are mostly city administrations and the state level parastatals, whose access to different sources of finances are limited by the powers to influence the prevailing legislations. In most cases, they have very limited say in introducing a new element or changing the base or the rate of a tax system, though they have better control over the administration of any tool once the same has been enacted. Moreover, the revisions to policies and legislations are not carried out regularly by central and state governments, leading to a potential delay in accessing the upside in the land value increases. In the meantime, the expenditure for the infrastructure service provision keeps escalating. Another area of criticism is the lack of transparency that exists in the distribution of land and the direction of transfer (typically from public ownership to private control). Therefore, an effective institutional and governance framework is vital for conducting transparent and equitable land-based financing transactions.

Various government constituted expert committees and interest groups have suggested mechanisms and instruments for utilizing unused land to finance infrastructure in India. One of the earlier initiatives was the report of the Committee on Roadmap for Fiscal Consolidation that recommended the monetization of land resources for financing civic infrastructure [30]. Instruments such as land pooling, monetization of underutilized/unutilized urban lands, land readjustment, land value capture were suggested as part of land-based financing systems [31,32]. The levy of area-based development charges was in vogue till 2010 in Mumbai, where the revenues were collected and retained by the city municipal corporation. Ahmedabad city has been adopting the premium floor space index tool on a stand-alone basis, and to supplement the transit-oriented development initiatives in the city.

The design of recent government of India's schemes and programmes reflect this intent with sustainability as the centerpiece. Various urban development schemes were launched by the Government of India in mid-2000 to facilitate better infrastructure facilities and ecosystem services for citizens. The schemes were customized for different categories of towns and cities (based on population, characteristics, etc.). The prominent amongst them being the Jawaharlal Nehru National Urban Renewal Mission (JnNURM) Scheme, Atal Mission for Rejuvenation and Urban Transformation (AMRUT), and the Smart Cities Mission. These schemes (and similar ones for different cities) aim to accelerate investments in city infrastructure while proposing to reform the governance, institutional, and financing systems. These schemes seek to improve land registration and cadastral systems and make the real estate available more freely for development projects through repealing land ceiling regulations.

Though direct methods of land monetization are not suggested in the JnNURM guidelines, it provided for legal and policy changes to be made by the States and simplify rules related to the conversion of land from agriculture to non-agriculture purposes. This implied that more land needs to be made available for the creation of assets meant for ecosystem services. As a consequence of this mission, the State needed to earmark land also for commercial use to serve the economically weaker sections (EWS) and low-income groups (LIG) allottees. This would include public markets, parks, schools, etc., that may generate continued revenues to the urban local bodies (ULB). As a result of the development of EWS and LIG housing projects, the land value in adjacent areas may increase due to additional habitation. The Smart City Mission guidelines place the onus on State level public agencies to create frameworks and policies for monetizing land and land-based assets in a Smart City. Land is also recognized as a key resource for implementing the AMRUT scheme, even though the scheme stops short of suggesting that excess land can be monetized for improving the sustainability of projects.

The primary legislation that enables the acquisition of land for infrastructure and development projects is the Land Acquisition Act, 2013 (which was promulgated replacing the archaic act of 1894). The Act sets out compensations that are much more aligned to market values than its predecessor. Section 26(3)(c) of this Act envisages that a Company can acquire land as equity and make landowners as its shareholders. It is possible, therefore, for a Government company to monetize land this way by acquiring it, developing necessary public infrastructure, and benefitting from the revenues that accrue.

The intention of various tiers of governments to provide direction for funding infrastructure developments is set out in the respective infrastructure policies and acts. The three states that are at the forefront are Gujarat, Andhra Pradesh, and Karnataka. While land monetization is not specified in the Andhra Pradesh law, it may be construed that in the process of development of a project for carrying out the objects of the Act, the Infrastructure Authority can suggest ways and means monetize land acquired for a project. Though the powers to acquire land is not vested in the Infrastructure

Authority directly, however, in Schedule V of the law [under Section 2(rr)], it is envisaged that the State Government will extend support to acquire land necessary for a project. It also provides for asset-based support by the Government, whereby the project proponent provides land at a subsidized lease for a predefined period (not exceeding 33 years).

The Gujarat Infrastructure Development Act does not empower the project authority to acquire land. However, in Chapter II—Infrastructure Projects (Section 6), the law envisages assistance by Government agencies for conferment of the right to develop any land. It also provides for Government agencies to participate in the equity of a project (not exceeding 49% of total equity), extends senior or subordinated loans, and similar such conditions to provide assistance to any person developing an infrastructure project. Thus, the authority has a role of a facilitator in the development of infrastructure in the State. The Infrastructure Policy of the Government of Karnataka does not explicitly mention that land-based financing can be used, or land can be monetized, though there were numerous examples of land being used as a financing source across the state in the last two decades.

Traditionally, states and cities have been raising funds by the sale of lands which is a less efficient form of resource mobilization. Typically, land value has been captured by a levy of the impact fee, betterment charges, etc. For example, infrastructure project agencies in Maharashtra, (Mumbai Metropolitan Region Development Authority (MMRDA) and City and Industrial Development Corporation (CIDCO)) have used value capture methods to finance infrastructure development. Haryana and Gujarat have used land pooling schemes, but there is no systematic approach outlined for land value capture. The government of India's latest policy on land value capture (LVC) seeks to capture value from increases in private land valuation from public investments and public policy actions. However, it does not address direct monetization (sale/leasing) of public land. The policy document also lists out various value capture methods used in India. These include land value tax, fees for change in land use, betterment levy, TDRs, relations of rules or additional FSI, vacant land tax, etc. The guidance note recognizes that presently there is no tool to assess the increased value of the land as a result of development. However, the impact fee tool is discussed at length in the note. The other types of LVC are not dealt with in detail. While success stories of LVC implementation in India and abroad have been dealt with in Section III (page 20 onwards) examples of tried, tested and failed initiatives by Government on value capture are not provided in the guidance note. Annexure 1 contains details of the type of value capture fund sources (Land Tax, Conversion tax, Betterment Levy, Impact Fee, etc.) as has been practiced in different states in the country.

The list of policies, schemes, and legislations that have included land-based financing sources are summarized in Table 3.

Table 3. Legislations/policies with land monetization elements.

Title	Salient Features Related to Land Monetization
JNNURM guidelines(Mission period starts—2005-06)	Amongst others, the guidelines propose that ULBs will develop and manage municipal assets to ensure sustainable public service delivery to the citizens especially the urban poor and people living in peri-urban areas. It is expected that the State Level Nodal Agency/ULBs would leverage financial resources in addition to financial assistance from the Central and State Governments. To ensure bankability, it is envisaged that liquidity support mechanism, up-front debt service reserve facility, deep discount bonds, contingent liability support, and equity support are to be established. The other reform proposed is to earmark 20–25% of developed land for EWS and LIG housing projects. This was suggested to enhance the supply of land for affordable houses for the urban poor and to provide them access to basic services. It was envisaged that with the increased housing supply, the ULBs would be able to garner higher amounts towards property tax. The other reform envisaged in the guidelines is the repealing of the Urban Land Ceiling Act, 1976. Its objective was to facilitate the availability of urban land at affordable prices by increasing its supply in the market. The objective of the reform in the rent control act is to bring out amendments in existing provisions for balancing the interests of landlords and tenants. Additionally, the increased investment in housing would lead to increased housing stock and increased revenue from property tax. So far twelve states have yet to make desired amendments in the rent control act [33].

Table 3. *Cont.*

Title	Salient Features Related to Land Monetization
Smart City guidelines	**"Para 3.1 (i)** Promoting mixed land use in area-based developments—planning for 'unplanned areas' containing a range of compatible activities and land uses close to one another to make land use more efficient. The States will enable some flexibility in land use and building bye-laws to adapt to change." **"Para 5.1.2** Redevelopment will effect a replacement of the existing built-up environment and enable co-creation of a new layout with enhanced infrastructure using mixed land use and increased density. Redevelopment envisages an area of more than 50 acres, identified by Urban Local Bodies (ULBs) in consultation with citizens." **"Para 5.1.3** Greenfield development will introduce most of the Smart Solutions in a previously vacant area (more than 250 acres) using innovative planning, plan financing and plan implementation tools (e.g., land pooling/land reconstitution) with provision for affordable housing, especially for the poor. Unlike retrofitting and redevelopment, greenfield developments could be located either within the limits of the ULB or within the limits of the local Urban Development Authority (UDA)." **"Para 11.3 (i)** The GOI funds and the matching contribution by the States/ULB will meet only a part of the project cost. Balance funds are expected to be mobilized from i. States/ULBs own resources from a collection of user fees, beneficiary charges and impact fees, land monetization, debt, loans, etc." **"Para 13.1.3** The National Mission Director will have the responsibility to … … (iii) Oversee capacity building and assisting in handholding of SPVs, State, and ULBs. This includes developing and retaining a best practice repository (Model RFP documents, Draft DPRs, Financial models, land monetization ideas, best practices in SPV formation, use of financial instruments and risk mitigation techniques) and mechanism for knowledge sharing across States and ULBs (through publications, workshops, seminars)" [34,35]
AMRUT guidelines	**"Para 6.10 Conditionalities**: … … , in the AMRUT no projects should be included which do not have land available and no project work order should be issued if all clearances from all the departments have not been received by that time. Moreover, the cost of land purchase will be borne by the States/ULBs." "..Explore innovative ways for resource mobilization, private financing, and land leveraging for funding of projects." It also provides that in the appraisal of the DPR, the National Mission Director may decide to "Mobilize external resources and improve internal resource generation of the ULBs. For instance, facilitate access to municipal bonds by credit rating ULBs, providing assistance to ULBs to monetize land and prepare Tax Increment Financing Proposals (TIF), obtain private funding, etc." [36]
Land Acquisition Act, 2013	The objective of the Act (amongst other things) is to ensure the development of infrastructural facilities and urbanization. The Act envisages that the affected persons become partners in development leading to an improvement in their social and economic status. Land can be acquired for a public purpose and it includes agro-processing, industrial corridors, water harvesting, education, sports, etc. (Section 2(1)). The law further provides on the method to be followed in the process of land acquisition for the determination of the social impact and public purpose (Chapter II—Sections 4 to 9) **Section 26 (3) (c)**—"Provided that in a case where the Requiring Body offers its shares to the owners of the lands (whose lands have been acquired) as a part compensation, for the acquisition of land, such shares in no case shall exceed twenty-five percent. of the value so calculated under sub-section (I) or sub-section (2) or sub-section (3) as the case may be": [4]

Table 3. *Cont.*

Title	Salient Features Related to Land Monetization
The Andhra Pradesh Infrastructure Development Enabling Act, 2001	To provide for the rapid development of physical and social infrastructure in the State and attract private sector participation An 'Infrastructure Authority' is established to conceptualize and identify projects and ensuring their conformance to the objectives of the State. The Authority also has the responsibility to categorize projects, prepare a project shelf, road map for project development, decide financial support, etc. It also has the responsibility to receive and consider projects from Government agencies and also advise them in the development of infrastructure. The prioritization of projects is to be carried out based on 'demand and supply gap, inter-linkages and any other relevant parameters'.
GIDB Act	The objective of the Act is to allow persons from the non-Government sector to 'participate in financing, construction, maintenance, and operation of projects. GIDB is the nodal agency for PPP projects in the State. The Board also acts as a policy advisor to the State Government and is vested with appropriate functions to prioritize various projects of the Government, to consider proposals received from private parties, to undertake technical and financial studies, to coordinate with concerned agencies.
Infrastructure Policy, Government of Karnataka	"Facilitating private participation in developing infrastructure facilities in the state by providing opportunities to private parties for participating in new infrastructure facilities development as well maintaining the existing infrastructure facilities. Infrastructure Development Department of the GoK is the nodal agency for appraisal and approval of infrastructure projects which is supported by a PPP Cell within the department. A Single Window Agency under the Chairmanship of Chief Secretary is set up for appraisal and approval of the projects. The State High-Level Committee chaired by the Chief Minister will approve Projects above Rs.50 Crores. GoK intends to put in mechanisms for expediting the land acquisition process and if necessary specific legislation would be passed in this regard. To enhance the commercial viability of projects, GoK may allow, wherever necessary, the Concessionaire/SPV to develop utilitarian services or other socially acceptable commercial activities, on the infrastructure project site."
Land Value Capture Policy, MoUD, GoI	The policy provides for four steps for project-based VCF. (1) Initiation (2) Planning (3) Design and strategy and (4) Execution and Operation. The policy document also provides a Guidance Note for the inclusion of VCF in projects. The Guidance Note envisages that VCF should be an integral part of the DPR for Central Government projects as has been stated by the Ministry of Finance in its OM dated 7 March 2017.

Source: Authors compilation based on a review of the policies/legislations.

The primary objective of these policies and legislation is to promote infrastructure development, and land monetization is suggested as a tool to achieve this goal.

5. Comparison of Policies

Over the years, project proponents have attempted to leverage the potential upside from increased land value due to infrastructure developments and have structured these increases as a source of revenue for the project. The project structures have attempted to balance concerns of the community and the political class pertaining to displacement and inequity. The history of policies, schemes, and laws also indicates an implicit real estate turn in the development landscape, as was witnessed in other Asian countries [14]. The following Figure 1 presents a timeline of the key projects and the policies, legislations discussed in the preceding sections.

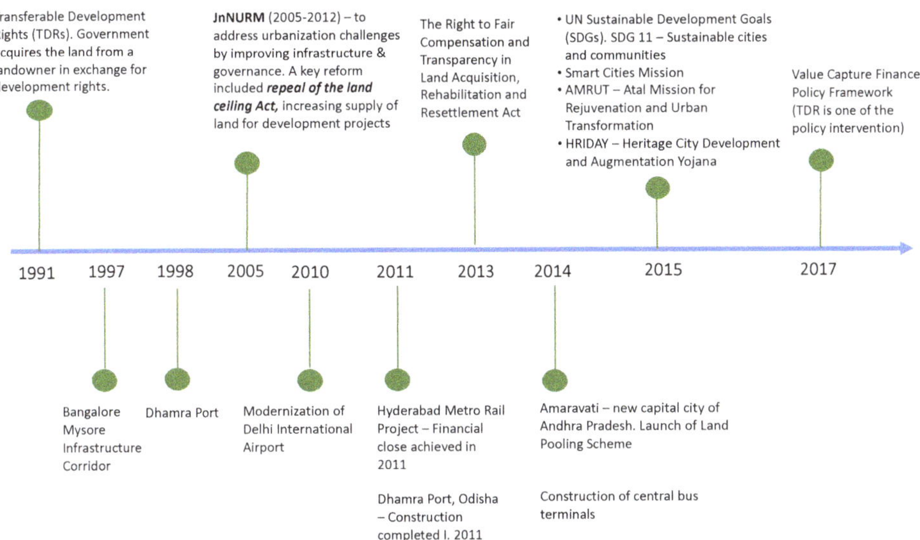

Figure 1. Key projects and legislations/policies.

Most of the projects were structured and implemented before the respective policies for land-based financing were in place. The public sector project proponents have attempted to use land-based financing techniques in the projects, though with mixed results. The capture of the potential upside of land value is sought to alleviate the relatively higher capital and O&M expenditure, in relation to the revenues that are likely to accrue. The use of such implementation arrangements has become mainstream options, though they differ in the mechanics of the application. Only the Smart Cities Mission and the Land Value capture Policy set out more elaborate provisions for monetizing land. However, the inclusion of land monetization elements in various policies and legislations have begun to appear only in the later schemes (post-2015). The permissibility of land-based financing tools under the earlier schemes was implicit, as the same were not directly prohibited. This points to a substantial lag between the inclination to use land monetization instruments for promoting investment activity (or to develop infrastructure projects), and the hesitancy in formalizing the options through well-defined policies and legislations. The approach was noticeably cautious given the challenges and controversies involved with the acquisition, distribution of the land, and its ever-changing value to the stakeholders concerned.

Table 4 sets out the comparison of the various policies, programmes, and schemes on the five parameters—from whom the increased value is being recovered, who benefits from the distribution of the land monetization benefits, how is the process linked to the overall budgets of the project proponents, the stakeholders bearing the financial risks (of future revenues and investments) and the stakeholders involved in governing the process.

Table 4. Comparative Analysis of Policies.

Title	Contributors	Beneficiaries	Process	Financial Risks	Governance Actors
JNNURM	Landowners, property developers	ULBs (cities), development authorities, parastatal agencies	Not explicitly stated. Assumed to be part of the consolidated fund	ULBs, development authorities, parastatal agencies	State government and ULBs
Smart City	Landowners, property developers	ULBs, smart city SPVs	Income to consolidated fund; revenue to SPVs	ULBs, SPVs	State Government, ULB
AMRUT	Landowners, property developers	ULBs, development authorities, parastatal agencies	Not explicitly stated. Assumed to be part of the consolidated fund	ULBs, development authorities, and parastatal agencies	State government and ULBs
Land Acquisition Act, 2013	Landowners, property developers	State Government	Proceeds go to the consolidated fund	State Government	State Government
The Andhra Pradesh Infrastructure Development Enabling Act, 2001	Landowners, property developers	ULBs, development authorities, parastatal agencies	Proceeds go to consolidated fund; PPP projects can appropriate value	Project proponents (public/private)	State Government
GIDB Act	Landowners, property developers	ULBs, development authorities, parastatal agencies	Proceeds go to consolidated fund; PPP projects can appropriate value	Project proponents (public/nongovernment)	State Government
Infrastructure Policy, Government of Karnataka	Landowners, property developers	ULBs, development authorities, parastatal agencies	Proceeds go to consolidated fund; PPP projects can appropriate value	Project proponents (public/nongovernment)	State Government
Land Value Capture Policy, MoUD, GoI	Landowners, property developers	ULBs, development authorities, parastatal agencies	Proceeds go to consolidated fund; PPP projects can appropriate value	Project proponents (public/nongovernment)	State Government

All the land-based financing aspects of the policies at the national and state level appear to be similar in content and process. The land value creation is proposed to be captured primarily from the landowners and the private developers who are in the region. There is no explicit mention/provision for widening the base to include other financial investors, who could create additional value (for example through the issuance of bonds as is the international practice [18], or to attract philanthropic investors [17]).

The economic value that is recovered is proposed to accrue to the city administrations, development authorities, and other public sector parastatal agencies. There are no formal statements on how the additional value will be distributed, and how the general public is benefited from most policies, except in relation to the land pooling system. The benefit to the contributors of the land value is made possible under the "land pooling" scheme wherein the contributors have access to additional value for the portion of the land they continue to own post-implementation of the scheme [15]. There is no specific mechanism that has been stated in any of the policies for tax abatement to any of the contributors.

The linkage of the value that is captured through land-based financing mechanism to the general budget or the taxation regimes is not touched upon in any of the policies. While there is an assumption that all the additional cash flows due to these activities will go to the consolidated fund of the various governments, there are no stated commitments to ensure that these additional sources are not spent on unrelated activities. The process appears to be at a preliminary stage of evolution. The policymakers are yet to set out the mechanism for investigating the relative advantages of land monetization options and to assess the impact of these choices on the wider society. Accordingly, the financial risks remain with the project proponents.

The governance of land-based financing structures has been retained by the government across all the policies. While there is no specific mention of region-specific urban development authorities to be created, such institutions exist across the country and have been used in the case of Amaravati city development [15]. Such institutional structures have added flexibility in raising capital, managing the project more efficiently, and insulating financial expenditure from the influence of political exigencies.

There have been substantial gaps in policy and planning with respect to incorporating land as a revenue source in the policy and legislative frameworks. Since urban development is a state subject, much of the federal schemes and guidelines highlight the need for adopting innovative land monetization principles but do not set out the finer implementation details. It is left to the states, to incorporate these principles based on their local context, relevance, and suitability. Many states have appropriately modified the relevant tax codes and other legislations to enable land value capture, though some are yet to initiate action. Therefore, at the state level, there is a wide disparity among states in their readiness to adopt or integrate land monetization approaches in their planning and project designs. While the value capture policy guidelines mention that land monetization principles should be an integral part of an assessment for all projects of the central government, the individual schemes under which projects are submitted (by individual states/departments) to the central government, do not insist on the same being an integral part. Therefore, there is some dissonance in the translation of a policy of the central government with policies/schemes of other departments. The dissonance only increases at the state level, where each state has different enabling mechanisms for promoting land-based revenue instruments.

In summary, the policies in India have lagged behind the projects in terms of incorporating land-based financing aspects. Driven by fiscal constraints, most states are now exploring various elements of land value capture to be included in their policies, though the Government of India policy on land value capture is dedicated to this aspect. The policy landscape appears to be in the initial stages of evolution and is yet to reach a higher-order process supported by appropriate legislations for mainstreaming land-based financing mechanisms in the design and planning processes. [8,16–18].

6. Policy Implications

For governments to use land-based financing tools effectively, it needs to be supported with appropriate laws/legislations or executive orders permitting value capture methods (through tax and non-tax revenues) and earmarking/distributing funds for specific projects/developments. Further, policies need to be formulated to provide a clear roadmap for (a) capturing the value (b) collecting the fees or charges (deposited to the consolidated funds of the state), (c) earmarking funds for specific projects/developments, and (d) ensuring timely disbursement of funds to project implementing agencies (through budgetary allocations or establishing project-specific funds/accounts) [8,16–18]. To enact the policies requires a strong institutional framework and collaboration between different stakeholders [2].

A reflection into the project structures and the elements of the national and subnational policies in India, and in international markets brings to the fore the conflicts between incorporating the private sector motivation of profit maximization against the public sector responsibilities for equitable access to ecosystem services and weaving these into state planning and policy statements [13]. In the Atlanta Beltline project, the project proponent has not made any provisions for reducing the negative effect of increases in taxes of low-income households or capping of any rents along the project influence area [17]. The extent of the control that the project proponent has over the land markets and the autonomy of these proponents in making appropriate changes is limited in most countries given the distribution of powers across various tiers of governance. [14] This aspect of control can provide a layout of how the policies can span out in addressing issues relating to the core principles of SDGs (equity, access, efficiency, sustainability, and delivering public value) [31]. These aspects need better articulation in most instances. All of these guiding principles have a continuum with low to high ends of spectrums, providing a framework for understanding when and to what extent the public sector planning and policy framework can embed the land-based financing elements. An indicative framework to analyze the incorporation of land monetization elements in policy is set out in Figure 2.

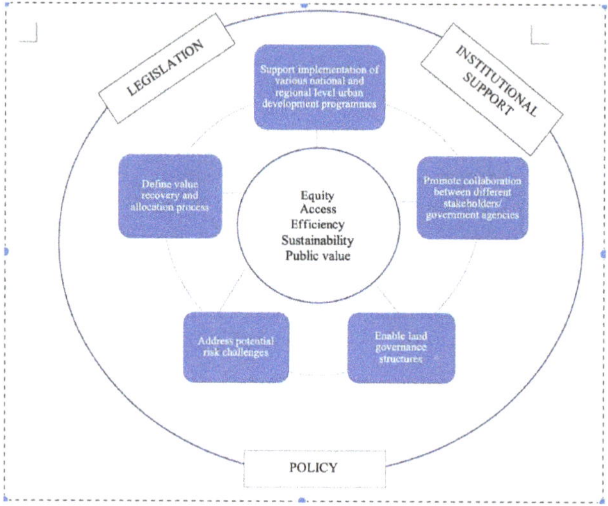

Figure 2. Framework for analyzing land-based financing elements in policies.

At the core lie the principles that define the expectations of and responsibilities towards various stakeholders concerned. These principles are derived from the philosophy of SDGs [5] and also reflect the intents of many governments. The land-based financing aspects would need to provide equal access to all, be sustainable and follow an effective process that continuously delivers public value.

The operational elements of the framework relate to the evaluation of enabling land governance structures, addressing potential risks and challenges, defining value recovery and allocation process, supporting the implementation of various national and regional level urban development programmes, and promoting collaboration between different stakeholders. The land governance structures define the extent of control or ownership of the land and real estate resources, and the flexibility of the policymakers in adapting the resources for use of non-state stakeholders. The ownership patterns of land (widely diffused through a diverse cross-section of public, private, and community owners, with substantial informal claimants in India [9]) renders the policy development more constrained. The fragmented nature of the ownership and more often than not, the unequal impact of land acquisition and land value increase is visible across most developing nations [14] An evaluation of the potential risks and challenges through the lens of guiding principles would keep the frame of evaluation grounded to the desired outcomes. In the instances where the public sector does not have substantial direct control of land, the policy initiatives have been focused on the greater role of private lands, as witnessed in the contents of the land value capture policy of the Government of India. The ability of the non-government stakeholders to influence the policy dialogue has been very minimal, even though the practice of contesting the implementation has been substantial [13]. A quantitative evaluation of the value recovery and the mechanics of the allocation across various stakeholders is essential to understand and configure the project structures and to promote sustainability of the land-based financing mechanisms. A broader perspective of how these elements support collaboration with prevailing or anticipated government rejuvenation programmes indicates the inclusiveness of the policy. A proactive approach to governing the land monetization process, with supporting institutional and regulatory aspects would provide a feedback overlay for the assessment.

7. Summary and Conclusions

The purpose of this research article is to sketch out the broad direction of how the various policies and legislations incorporate the land-based financing elements, given the projects being implemented are actively adopting such mechanisms. The increase in attention to land-based financing models and the adoption of land monetization instruments generally coincided with the development of infrastructure projects under the public–private partnership arrangements. The stress on finances of the project proponents coupled with the anticipation of upside in the land values post-implementation of infrastructure projects has given way to consider land as a revenue source, rather than just a factor of production [9,13]. The viability assessment of the projects, particularly those in the transport sector, improved substantially with the addition of a real estate component.

The institutional, governance, and policy formulation practices present in India are reflective of similar structures prevalent in Asia and other developing regions. Development agendas of many countries are moving in a similar direction, so are the challenges that they encounter in accelerating political and administrative actions. The use of land-based financing elements without appropriate structuring could lead to sub-optimal distributive justice to all the stakeholders concerned. The policies need to strike a balance between increasing the burden on the adversely affected landowners and users, while not exaggerating the benefits derived by those who are advantageously placed to the intervention. A framework that encompasses the generally accepted guiding principles for analyzing the extent to which the land-based financing elements could be incorporated in their respective policies and legislations could provide the policymakers and public sector project proponents a tool to comprehensively assess their needs, opportunities, and constraints. The project proponents also need to be conscious of how the financial risks are identified and borne, which ecosystem services are receiving lower funding due to the land value increase interventions [16]. A holistic assessment needs to be undertaken before incorporating land-based financing elements in policy to balance the realities of the contributors to the land value increments, and broad basing the beneficiaries pool (to include wider society where appropriate through lower taxes or better ecosystem services).

A reasoned discussion with stakeholders while framing the policies, supported by political advocacy, will lead to a more efficient public investment process. With the increasing attractiveness of land monetization options for funding economic growth, many developing societies will need to balance value capture with the expectations of affected stakeholders. This research contributes to the ongoing discourse of systematically understanding the elements for formulating public policies on land management.

Author Contributions: Conceptualization, methodology, formal analysis, resources, writing–original draft preparataion and writing–review and editing–equally contributed by R.D.T. and P.T. All authors have read and agreed to the published version of the manuscript.

Funding: This research received no external funding.

Institutional Review Board Statement: Not Applicable.

Informed Consent Statement: Not Applicable.

Data Availability Statement: The data presented in this study is contained within this article.

Conflicts of Interest: The authors declare no conflict of interest.

References

1. Suzuki, H.; Cervero, R.; Iuchi, K. *Transforming Cities with Transit. Transit and Land-Use Integration for Sustainable Urban Development*, 1st ed.; World Bank: Washington, DC, USA, 2013.
2. Bahl, R.W.; Linn, J.F.; Wetzel, D.L. *Financing Metropolitan Governments in Developing Countries*, 1st ed.; Lincoln Institute of Land Policy: Washington, DC, USA, 2013.
3. Chakravorty, S. A new price regime: Land markets in Urban and Rural India. *Econ. Polit. Wkly.* **2013**, *48*, 45–54.
4. Singh, S. Land Acquisition in India: An Examination of the 2013 Act and Options. *J. L. Rural Stud.* **2016**, *4*, 1. [CrossRef]
5. United Nations. SDG Indicators—SDG Indicators. 2017. Available online: https://unstats.un.org/sdgs/indicators/indicators-list/ (accessed on 24 July 2020).
6. Besley, T.; Burgess, R. Land reform, poverty reduction, and growth: Evidence from India. *Q. J. Econ.* **2000**, *2*, 389–430. [CrossRef]
7. UN-Habitat. *Affordable Land and Housing in Latin America and the Caribbean*, 1st ed.; UN-Habitat: Nairobi, Kenya, 2011; Available online: https://unhabitat.org/affordable-land-and-housing-in-latin-america-and-the-caribbean-2 (accessed on 20 January 2021).
8. Munoz-Gielen, D. Improving Public-Value Capturing in Urban Development. *Innov. L. Prop. Tax.* **2011**, 150–170. Available online: https://www.researchgate.net/publication/285574748_Improving_public-value_capture_in_urban_development (accessed on 15 December 2020).
9. Gandhi, S.; Phatak, V.K. Land-based Financing in Metropolitan Cities in India: The Case of Hyderabad and Mumbai. *Urbanisation* **2016**, *1*, 31–52. [CrossRef]
10. Rics. *RICS Valuation—Professional Standards (Red Book). Basis of Value*; Royal Institution of Chartered Surveyors: London, UK, 2012. Available online: https://www.rics.org/globalassets/rics-website/media/upholding-professional-standards/sector-standards/valuation/rics-valuation--global-standards-jan.pdf (accessed on 15 December 2020).
11. George, H. *Progress and Poverty (Edited and Abridged for Modern Readers by Bob Drake, 2006)*; Aziloth Books: London, UK, 1879.
12. Smolka, M.O. *Implementing Value Capture in Latin America: Policies and Tools for Urban Development Policy Focus Report Series*; Lincoln Institute of Land Policy: Cambridge, MA, USA, 2013.
13. Berrisford, S.; Cirolia, L.R.; Palmer, I. Land-based financing in sub-Saharan African cities. *Environ. Urban* **2018**, *1*, 35–52. [CrossRef]
14. Shatkin, G. The real estate turn in policy and planning: Land monetization and the political economy of peri-urbanization in Asia. *Cities* **2016**, *53*, 141–149. [CrossRef]
15. Ramachandraiah, C. Making of Amaravati. *Econ. Polit. Wkly.* **2016**, *51*, 68–75.
16. Wolf-Powers, L. Reclaim Value Capture for Equitable Urban Development. *Metropolitics* **2019**. Available online: https://metropolitics.org/Reclaim-Value-Capture-for-Equitable-Urban-Development.html (accessed on 20 January 2021).
17. Immergluck, D. Large redevelopment initiatives, housing values, and gentrification: The case of the Atlanta beltline. *Urban Stud.* **2009**, *46*, 1723–1745. [CrossRef]
18. Van der Veen, M.; Altes, W.K.K. Urban development agreements: Do they meet guiding principles for a better deal? *Cities* **2011**, *28*, 310–319. [CrossRef]
19. Peterson, G.E. *Unlocking Land Values to Finance Urban Infrastructure*; The World Bank: Washington, DC, USA, 2008.
20. Nallathiga, R. Urban infrastructure development in India: Resource requirements and mobilization methods. *IUP J. Infrastruct.* **2010**, *8*, 26–37.
21. RICS Research. *Real Estate and Construction Professionals in India by 2020*; RICS Research: London, UK, 2011.
22. RICS Research. *Bridging the Gap*; RICS Research: London, UK, 2020.
23. Kelkar, D.V. *Report of the Committee on Revisiting and Revitalising Public Private Partnership model of Infrastructure*; Government of India: New Delhi, India, 2015.
24. Balakrishnan, S. Highway urbanization and land conflicts: The challenges to decentralization in India. *Pac. Aff.* **2013**, *86*, 785–811. [CrossRef]
25. Kulshreshtha, R.; Kumar, A.; Tripathi, A.; Likhi, D.K. Critical Success Factors in Implementation of Urban Metro System on PPP: A Case Study of Hyderabad Metro. *Glob. J. Flex. Syst. Manag.* **2017**, *18*, 303–320. [CrossRef]
26. Chaudhuri, S. Impact of privatisation on performance of airport infrastructure projects in India: A preliminary study. *Int. J. Aviat. Manag.* **2011**, *1*, 40. [CrossRef]
27. Sahoo, K. Deregulation in development project: A case of Dhamra port project in Odisha. *Ocean Coast. Manag.* **2014**, *100*, 151–158. [CrossRef]
28. Sharma, K.K.; Misra, S.K.; Singla, A.K. Role of Public Private Partnership in Bus Terminals: A Case Study of Punjab. *Think India* **2019**, *22*, 116–128. [CrossRef]
29. Germán, L.; Bernstein, A.E. Land value capture. *Policy Br.* **2018**, 2016–2019. Available online: https://www.lincolninst.edu/sites/default/files/pubfiles/land-value-capture-policy-brief.pdf (accessed on 15 December 2020).
30. Kelkar, V.L.; Rajaraman, I.; Misra, S. *Report of the Committee on Roadmap for Fiscal Consolidation*; Government of India: New Delhi, India, 2012.
31. Kanuri, C.; Revi, A.; Espey, J.; Kuhle, H. *Getting Started With the SDGs in Cities*; Sustainable Development Solutions Network: New York, NY, USA, 2016.
32. Tiwari, P. India Habitat III, National Report, 2016. *Minist. Hous. Urban Poverty Alleviation* **2016**, *32*, 978–998.
33. Kundu, D. Urban Development Programmes in India: A Critique of JnNURM. *Soc. Chang.* **2014**, *44*, 615–632. [CrossRef]

34. Government of India MHUA. Smart Cities Mission, Government of India. Resource Website. 2015. Available online: http://smartcities.gov.in/content/ (accessed on 15 December 2020).
35. Roy, S. The Smart City Paradigm in India: Issues and Challenges of Sustainability and Inclusiveness. *Soc. Sci.* **2016**, *44*, 29.
36. Kundu, D.; Sietchiping, R.; Kinyanjui, M. *Developing National Urban Policies*; Springer: Singapore, 2020.

 *land*

Article

Who Pays the Bill? Assessing Ecosystem Services Losses in an Urban Planning Context

Harald Zepp [1,*] and Luis Inostroza [1,2]

1   Institute of Geography, The Ruhr-University Bochum, 44801 Bochum, Germany;
     luis.inostroza@ruhr-uni-bochum.de
2   Universidad Autonoma de Chile, 4810101 Temuco, Chile
*   Correspondence: Harald.Zepp@ruhr-uni-bochum.de

Abstract: While Ecosystem Services (ES) are crucial for sustaining human wellbeing, urban development can threaten their sustainable supply. Following recent EU directives, many countries in Europe are implementing laws and regulations to protect and improve ES at local and regional levels. However, urban planning regulations already consider mandatory compensation for the loss of nature, and this compensation is often restricted to replacing green with green in other locations. This situation might lead to the loss of ES in areas subject to urban development, a loss that would eventually be replaced elsewhere. Therefore, ES assessments should be included in urban planning to improve the environmental conditions of urban landscapes where development takes place. Using an actual planning and development example that involves a proposed road to a restructured former industrial area in Bochum, Germany, we developed an ad-hoc assessment to compare a standard environmental compensation approach applying ES. We evaluated the impact of the planned construction alternatives with both approaches. In a second step, we selected the alternative with a lower impact and estimated the ES losses from the compensation measures. Our findings show that an ES assessment provides a solid basis for the selection of development alternatives, the identification of compensation areas, and the estimation of compensation amounts, with the benefit of improving the environmental quality of the affected areas. Our method was effective in strengthening urban planning, using ES science in the assessment and evaluation of urban development alternatives.

Keywords: compensation measures; urban resilience; urban development; impact assessment

Citation: Zepp, H.; Inostroza, L. Who Pays the Bill? Assessing Ecosystem Services Losses in an Urban Planning Context. *Land* **2021**, *10*, 369. https://doi.org/10.3390/land10040369

Academic Editor: Alessio Russo

Received: 8 March 2021
Accepted: 31 March 2021
Published: 2 April 2021

Publisher's Note: MDPI stays neutral with regard to jurisdictional claims in published maps and institutional affiliations.

1. Ecosystem Services for Cities

Ecosystem Services (ES) science has provided a framework and empirical evidence for analysing, discussing and communicating environmental trade-offs arising from alternative development options in several planning contexts [1–3]. Resilience, sustainability and quality of life can be greatly improved in urban areas by including ES assessments [4,5]. Urban planning can benefit from the adequate use of the ES framework. Urban areas need to improve the application of laws and regulations following EU-directives to protect and improve the natural environment at local and regional levels [6,7].

This paper uses the ES concept in a practical way to illustrate its potential to support decision-making. The aim is to present a clear way to apply the ES framework in an actual urban development case to show the benefits of such an approach. We apply the ES concept in a real planning case study using a procedure that is ready to implement and easy to understand, avoiding unnecessary conceptual and operational complexity. Our approach can be used by a broader community of scholars and decision-makers who might not necessarily be familiar with the ES concept. We present the ES concept, focusing on how it can be implemented in practice for urban planning and the necessary steps to follow.

1.1. Ecosystem Services Assessments

ES are the benefits that society obtains due to the functioning of healthy ecosystems [8]. ES can be classified to assist in assessing them. The Economics of Ecosystems and Biodiversity (TEEB) and the Common International Classification of Ecosystem Services (CICES) [9] are two broadly accepted classification systems. We used the CICES, which classifies ES into three groups: provisioning Ecosystem Services (P-ES), regulating Ecosystem Services (R-ES), and cultural Ecosystem Services (C-ES) [10,11]. CICES version v5.1 offers a detailed and extensive list of ES that can be applied to identify relevant services in several geographical settings.

In the context of urban planning, the assessment of ES should always start with screening, identifiing and selecting relevant ES. This is a fundamental starting point, because ES are always context-specific. This means that the presence, intensity, distribution, and relevance of ES change from location to location [12]. ES also change due to land management practices or urbanisation intensity [12,13]. A good practice for sound identification is to refer to an established classification system, such as TEEB or CICES. Using a validated classification of ES can help ensure a systematic screening process that does not leave out any important service, and that help avoid the inclusion of benefits that are not ES. Having a consolidated list of relevant ES a follow-up good practice is to perform an exploratory assessment to provide a sound evaluation of ES intensities that can help in further prioritizing and mapping ES. Many studies have applied these two steps using expert assessments and the matrix approach [3,14,15]. A pool of experts can select relevant ES in a study area and score the intensity of the ES. The matrix approach can link such scores to specific land use/land cover (LULC) to map the spatial distribution of ES. Preliminary scoring and mapping of ES using the matrix approach have been shown to have high concordance with biophysical estimations [16]. These steps can be a robust guide in further evaluations and biophysical quantifications of ES in a more detailed analysis.

1.2. Ecosystem Services for Spatial Planning

The capacity of the ES framework to assist in decision-making in several fundamental aspects of urban development, such as green infrastructure, climate change adaptation and sustainable urban development, has been largely confirmed [4,17–20]. ES can greatly help planners understand the dynamics of decision-making in complex eco-sociotechnical systems [21]. In concrete terms, in the context of planning, ES knowledge can have both conceptual and instrumental uses. The conceptual use of ES is aimed at broadening understanding to shape decision-makers' and stakeholders' thinking. The instrumental use of ES is focused on supporting the decisions between policy options regarding gains and losses, and involving concrete decision options [22]. To have practical instrumental value, ES information must be presented in a meaningful manner [23] and be ready to be applied in real-world situations. However, there is a need for feasible methods, models and applications that can assist planners in the practical implementation of ES science [21]. Empirical evidence shows that there are several problems, such as data availability, uncertainties, and, most importantly, difficulties in translating abstract scientific knowledge into practical applications and linking assessments to the characteristics of a specific local context, that planners face when attempting to use ES knowledge [24]. Furthermore, the ES concept cannot easily be translated into a legal framework and technical guidelines established as routine workflows in cities and regions [18].

The ES framework can link environmental aspects under an urban development perspective to better understand their effects on human wellbeing [25]. Integrating ES into urban planning can provide important benefits, such as (1) supporting the implementation and design of adequate measures to address current urban challenges, such as climate adaptation; (2) enhancing the transparency of trade-offs and cobenefits arising from urban development, while increasing awareness about the hidden or underrepresented values of nature that could eventually be lost; and (3) directly addressing issues of environmental justice in land-use change decisions through the identification of ES demand and supply [19].

The need to incorporate ES into urban planning is not only related to the desire to improve practice with new knowledge. The inclusion of ES in policymaking has already resulted in important policy recommendations in the EU. The EU Biodiversity Strategy for 2030 explicitly addresses the need to incorporate ES mapping, monitoring and assessing into policy making; action 7 aims to ensure no net loss of biodiversity and ES [26]. The Territorial Agenda, a strategic policy document for guiding the spatial planning in regions and communities in Europe, explicitly highlights the relevance of ES to ensure their provision and public awareness of them [27]. Finally, the EU guidance on integrating ecosystems and their services into decision-making outlines concrete actions for the integration of ES into a range of decisions at different levels and areas, including spatial planning. This report emphasizes the inclusion of ES within existing planning frameworks to avoid the generation of parallel processes and assessments [28]. There is a set of criteria for addressing the potential negative impacts on ES. This mitigation hierarchy includes: *"(1) Avoidance: measures to identify and completely avoid detrimental impacts from the outset, such as careful spatial placement of infrastructure; (2) minimisation: measures to reduce the duration, intensity and/or extent of detrimental impacts (including direct, indirect and cumulative impacts) that cannot be completely avoided; (3) rehabilitation/restoration: measures to rehabilitate degraded ecosystems or restore cleared ecosystems following impacts that could not be completely avoided and/or minimised; (4) offsetting: measures to compensate for residual, significant, adverse impacts that could not be avoided, minimised or restored. Measures to overcompensate for losses can also lead to net societal gains by their contribution to well-being and prosperity"* [29]:13. This mitigation strategy is aimed at ensuring an increased delivery of multiple ES. On the other hand, according to this report, only a *"few cities have prioritised access to nature as a central objective of urban planning."* [28].

This paper addresses three research questions: (1) how can the ES framework be methodologically and operationally incorporated into urban planning? (2) How can the results of ES assessments be translated into urban planning tools for the public, stakeholders, and decision-makers? (3) Can ES help avoid the environmental deterioration occurring due to urban development?

2. Materials and Methods

The methods were designed for the instrumental use of ES, which supports decisions between policy options regarding gains and losses, and involves concrete decision options [22]. Our approach aimed to provide a sound assessment of environmental compensation accounting for the eventual loss of ES due to urban development. The approach was based on the analysis and selection of the best planning alternative by weighing the impacts of each development option in a comparative approach. The positive and negative environmental effects of the project to be implemented were investigated and compared to choose, modify, or reject the planning ideas. We used a double method to assess the impacts of the planning ideas. In the first assessment, we used a standard approach to calculate the environmental compensations; in the second assessment, we used the ES framework. We illustrated our method by using an example from Germany's Ruhr region. Our method is unique in that it articulates ES knowledge with a practical application based on a real planning situation, showing how the ES framework can support decision-making.

2.1. Case Study

The city of Bochum has 371,000 inhabitants, and it is part of Germany's Ruhr metropolis, which is one of the largest metropolitan areas in Europe with 5.1 million inhabitants. During the 19th and 20th-centuries, coal mining and steel production were fundamental economic activities. For the last 50 years, structural economic change has driven the closure of all coal mines and resulted in a decrease in steel production. This situation has transformed the economic base of the city to electronic devices manufacturing and car production, with a more diversified sectoral mix, including service industries and universities, as a part of the knowledge-based economy.

The study area is located in the eastern part of Bochum. It can be considered a typical example of "glocalisation" [30], depicting the local effects of economic globalisation. The multinational GM/Opel car factory is located at two sites within the city of Bochum, covering an area of approximately 100 hectares. After the company decided to end car production at the end of 2015, one site was given up, and the other site was developed to serve as the European logistics centre for the distribution of spare car parts and a new industrial area. The existing access road connecting the site by the freeway crosses a residential area, and will become overloaded by increasing traffic. To ameliorate the environmental impacts, four alternative corridors have been discussed. The four planning alternatives for the access roads to the former factories are generally presented in Figure 1.

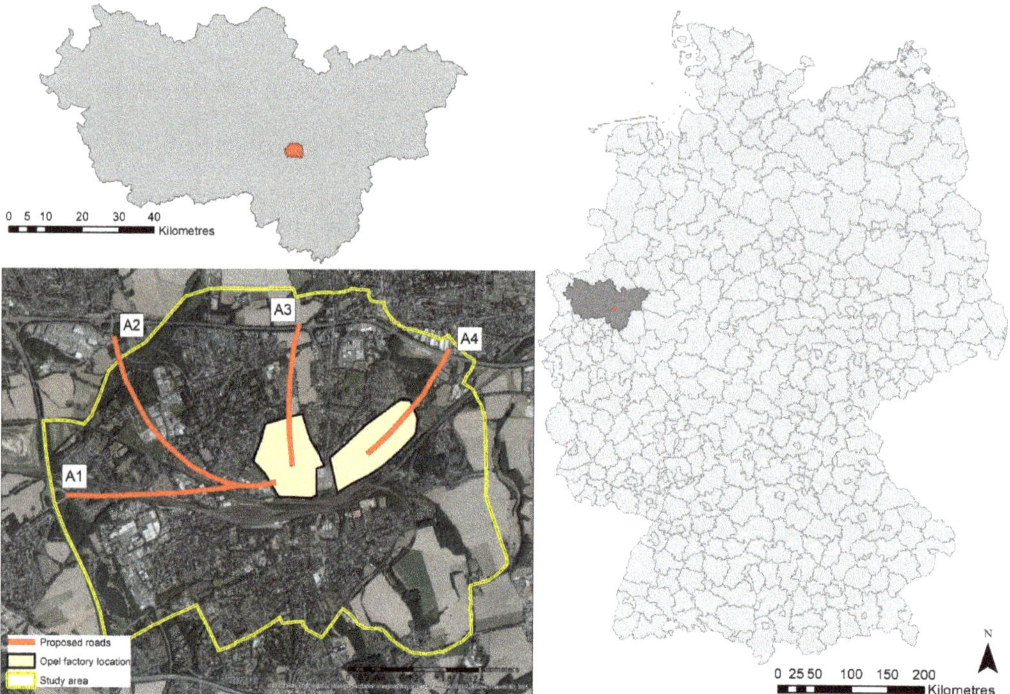

Figure 1. Location of the study area in Germany and the Ruhr area, and main features. Four alternative access roads to the GM/Opel European logistics centre are under discussion (A1, A2, A3 and A4). Satellite imagery: ESRI, DigitalGlobe, GeoEye, Earthstar Geographics, CNES/Airbus DS, USDA, USGS, AeroGRID, IGN, and the GIS User Community.

2.2. Mapping Urban Structural Types

We delineated the area named Lagendreer-Werne with a total surface area of 1632 hectares. We mapped the entire area in several field campaigns. This detailed mapping effort yielded several LULC classes. We grouped the LULC according to 22 urban structural subtypes (USSs), representing urban morphological units that embody the characteristics of the urban structure. The characteristics of urban structural types were related to factors such as the surface materials, the internal configuration of diverse open and sealed patches, the height of vegetation, and height [31]. We further differentiated the USSs listed in Table 1. A visual field survey was conducted after preparatory mapping based on aerial photos. Among the USSs, we identified open space, which normally contains the highest environmental values (Table 1).

Table 1. List of USSs and ES values used in the assessment. P-ES: provisioning ES; R-ES: regulating ES; C-ES: cultural ES. Column *p* indicates USSs that are terrestrial open space on natural soils. Values for P-ES, R-ES and C-ES are the mean of the single ES in each of the P-ES, R-ES and C-ES groups. The range is 0 to 5. High values indicate high ES supply provided by the USSs.

i	Urban Structural Subtypes	P-ES	R-ES	C-ES	*p*
1	Allotment gardens	2.1	3.4	3.6	1
2	Arable field	2.0	2.8	2.3	1
3	Cemeteries	1.1	3.3	3.5	1
4	Commercial and industrial uses with a high degree of surface sealing	0.8	1.0	0.7	
5	Detached and semi-detached houses	1.4	2.0	1.2	
6	Housing complexes with green areas, e.g. row houses single multi-story houses	1.2	1.6	0.8	
7	Lakes ponds	2.1	2.5	3.8	
8	Linear groves (including industrial & green buffers of industrial areas)	0.9	2.8	2.2	
9	Mixed commercial and residential uses with a low degree of surface sealing	0.9	1.3	0.8	
10	Parking lots and parking areas	0.2	0.4	0.3	
11	Parks and green belts	1.4	3.8	4.1	1
12	Pasture and meadow	2.0	3.7	3.6	1
13	Places and squares	0.3	0.4	0.5	
14	Playgrounds	0.6	1.5	2.3	
15	Public buildings and public institutions	1.0	1.4	1.5	
16	Railroad tracks	0.3	1.0	0.9	
17	Roads	0.2	0.4	0.3	
18	Sports grounds and leisure infrastructure	0.7	1.4	1.8	
19	Technical infrastructure (e.g., power transformation areas)	0.4	0.5	0.6	
20	Terrace houses	1.1	1.7	1.0	
21	Urban forest of considerable size and compactness	2.0	4.5	4.7	1
22	Villas	1.5	2.1	1.4	

2.3. Impact Assessment Using a Standard Approach

We analyzed the potential impacts of each of the four access roads using a parallel assessment. First, the impact assessment focused on three fundamental environmental aspects to estimate the necessary compensation, based on the protected environmental goods in German legislation: soil, biotope and recreation. We defined a buffer area of 50 m for each of the proposed access roads. We measured the high-quality soil, biotope and recreation values that would eventually be lost within each of these buffer areas.

2.3.1. Biotope

A biotope is evaluated based on vegetation cover from the perspective of nature conservation (biotope value). The value of biotopes refers to the scheme to manage compensation used in landscape planning [32]. Biotope value depends on the degree of naturalness, rareness, recoverability, and integrity. Biotope value is expressed on an ordinal scale from 0 (the lowest) to 10 (the highest). Our assessment focused only on the highest occurring biotope values (6–7 high and 8–9 very high), for which we calculated the respective hectares. We used the detailed biotope map provided by the cities of Bochum and Dortmund. The maps were prepared at a scale of 1:5000 and are regularly updated. We manually determined the values of empty areas by identifying equivalent biotopes.

2.3.2. Soil Values

Soil map units include soil quality based on a long-established assessment scheme [33], which assigns numbers according to a soil's relative capacity to bear and sustain crop production. Soil quality is assessed on the basis of the soil texture, which reflects a soil's capacity to store plant available water and nutrients; soil parent material and soil development also reflect natural nutrient provisioning. Climate was also considered. All aspects were combined in a dimensionless, ordinally scaled indicator to express the relative differences in net agricultural yield from 1 to 100. Generally, a soil map assigns one of five land value classes to each soil unit (very low, low, medium, high, and very high). The class, very low, did not occur in our area. The analysis concentrated on measuring the hectares lost in areas with medium and high soil values in open spaces, as there were no very high values areas in the impacted areas. We evaluated soil only in open spaces because these areas have not been sealed, and, in the case of constructing the road, such soil would eventually be lost.

2.3.3. Recreation Values

To estimate the recreational value of a particular USS, we used the assessment of cultural ES, as recreation is a cultural ES. We used the values obtained in an expert workshop with 11 scientists and professionals, as described in Section 2.4.1. We selected the five C-ES directly connected to recreation (column R in Table 2); therefore, this recreational value was slightly smaller than the estimated cultural ES. To estimate the recreational value, the analysis calculated the area loss only in terms of the USSs considered open space (column p in Table 1).

2.4. Impact Assessment Using ES

In the second step, we applied the ES land cover matrix and expert assessment approach [14,34,35] to evaluate the impacts of alternatives and to identify potential areas for compensation measures. The spatial scope of the compensation measures was restricted to the analyzed area for which the identification and assessment of ES were performed.

2.4.1. Mapping ES

For the identification of relevant ES, we relied on a workshop with 11 experts in planning and science working in the Ruhr area. Using CICES v5.1 [11], the experts identified the 25 most relevant ES for the study area (Table 2). Then, the expert panel assessed the potential ES supply of each USS (Table 1) using a scale from 0 to 5 (null to very high). As the evaluations of the experts slightly differed, we averaged the experts' scores. To calculate the respective bundle, we average the single ES values of the ten provisioning (P-ES), nine regulating (R-ES), and six cultural ES. Using these bundle values, we mapped the bundles of ES following the matrix approach [3,14,15,36]. The full list of identified ES, with their respective CICES codes, is presented in Table 2.

We calculated the total ES supply within the 50 m buffer area for the four analyzed access roads using Equation (1):

$$E_k = \sum_{i=1}^{n}(a_i e_i) \tag{1}$$

where:

E_k = total ES supply in buffer k (P-ES, R-ES, or C-ES)
a_i = Surface of USS i present in the buffer area k
e_i = Supply of ES of USS i (according to Table 1).

2.4.2. Identifying Hot and Cold Spots of Supply

To identify the areas with the highest and lowest supplies of ES, we mapped the hot and cold spots for P-ES, R-ES and C-ES at 90%, 95% and 99% confidence levels using the Getis-Ord Gi* tool in ArcGIS 10.1 ©. Using this method, we ensured a spatially explicit and meaningful identification of areas containing clusters of high and low ES supplies. These areas were also used in the design of compensation. The analysis of hot–cold spots was performed over a hexagonal grid of 1 ha.

Table 2. Selected ES used in the assessment. Column R indicates the ES used to estimate the recreation value.

	Bundle	Class	CICES Code	R
1		Cultivated terrestrial plants (including fungi and algae) grown for nutritional purposes	1.1.1.1	
2		Animals reared for nutritional purposes	1.1.3.1	
3	Provisioning (Biotic)	Animals reared by in situ aquaculture for nutritional purposes	1.1.4.1	
4		Wild plants (terrestrial and aquatic, including fungi and algae) used for nutrition	1.1.5.1	
5		Wild animals (terrestrial and aquatic) used for nutritional purposes	1.1.6.1	
6		Surface water used as a material (non-drinking purposes)	4.2.1.2	
7		Freshwater surface water used as an energy source	4.2.1.3	
8	Provisioning (Abiotic)	Ground (and subsurface) water for drinking	4.2.2.1	
9		Ground water (and subsurface) used as a material (non-drinking purposes)	4.2.2.2	
10		Ground water (and subsurface) used as an energy source	4.2.2.3	
11		Filtration/sequestration/storage/accumulation by micro-organisms, algae, plants, and animals	2.1.1.2	
12		Noise attenuation	2.1.2.2	
13		Visual screening	2.1.2.3	
14		Hydrological cycle and water flow regulation (Including flood control)	2.2.1.3	
15	Regulation & Maintenance (Biotic)	Pollination	2.2.2.1	
16		Maintaining nursery populations and habitats (Including gene pool protection)	2.2.2.3	
17		Decomposition and fixing processes and their effect on soil quality	2.2.4.2	
18		Regulation of the chemical condition of freshwater by living processes	2.2.5.1	
19		Regulation of temperature and humidity, including ventilation and transpiration	2.2.6.2	
20		Characteristics of living systems that that enable activities promoting health, recuperation or enjoyment through active or immersive interactions	3.1.1.1	R
21		Characteristics of living systems that enable activities promoting health, recuperation or enjoyment through passive or observational interactions	3.1.1.2	R
22	Cultural (Biotic)	Characteristics of living systems that enable scientific investigation or the creation of traditional ecological knowledge	3.1.2.1	
23		Characteristics of living systems that enable education and training	3.1.2.2	R
24		Characteristics of living systems that enable aesthetic experiences	3.1.2.4	R
25		Elements of living systems that have symbolic meaning	3.2.1.1	R

2.5. Evaluation of Compensation

We used the direct loss per buffer calculated with Equation (1) (E_k) as a value for ES compensation. To evaluate the possible compensation for ES loss, our criteria were twofold: (1) to maintain the same amount of ES supply existing in the selected buffer and that redistributed in a sector located near the buffer area, and (2) to maintain the same amount of open space that will eventually be lost; this surface will be relocated in the nearby impacted area. To identify the area for compensation, we used the analysis of hot–cold spots to select the cold spot close to the selected access road. We performed the calculations for compensation using a hexagonal grid of 1 ha. Using a grid approach helped to understand the analyzed impacts that were more in line with aspects of urban form, as, normally, LULC units are hierarchically arranged in space and time following six fundamental aspects [37,38]. The hexagonal grid is a powerful tool to depict the spatial structure of urban environments [39], because it can summarise the high heterogeneity the land uses that occur over short distances in a comparable manner.

For the estimation of ES supply for each hexagonal cell, we used the same Equation (1) used for the calculation within buffers. The total amount of ES to be compensated in the new area corresponded to the total amount of ES in the respective buffer (E_k), an assumption that only maintains the current situation—no quantitative ES improvement. This ES total amount was the sum product of the ES (P-ES, R-ES and C-ES) and the respective surface in hectares covered by each of the USSs within the buffer. To estimate the amount of ES to be compensated per cell, we used Equation (2):

$$E_c = \frac{E_k - \left(\sum_{i=1}^{p}(e_i - e_h)\right)}{(m + |p|)} \tag{2}$$

where:

E_c = ES to be compensated in cell h (P-ES, R-ES, or C-ES)
e_h = existing ES supply in cell h (calculated with Equation (1))
E_k = total supply of ES in buffer k (P-ES, R-ES and C-ES respectively
e_i = ES supply of open space USS (Table 2, P-ES, R-ES, C-ES where $p = 1$)
m = number of cells in the cluster to be compensated
p = number of cells to be replaced with open space, as shown in the following:

$$p = |p_k| \tag{3}$$

p_k = number of open space USS (Table 1), and
$k \in Z: 1 \leq i \leq 5$.

Then, the new ES value per cell to be mapped was:

$$E_t = E_c + e_h \tag{4}$$

3. Results

3.1. Land-Use Change Impact

The study area contained a core of residential and industrial uses surrounded by open spaces. Massive railroad infrastructure running from East to West separates the area into two parts. The three predominant USSs were "housing complexes with green areas", e.g., row houses and single multi-story houses, "arable fields", and "urban forest", covering 17.8%, 13.8% and 12% of the total area, respectively. Open space covered 36% of the total area at approximately 591 ha (see Figure 2).

Due to the particular spatial distribution of the USS and the differences in the surfaces covered by each of the proposed access roads, the potential impacts in terms of land-use change varied. Access A1 will affect 25.2 ha, of which 3.8 ha (15%) corresponds to open space. In the case of A2, the area impacted will be 25.3 ha, containing 7.7 ha (30%) of open space. Access A3 will affect 16.6 ha and 8.9 ha (53%) of open space. Access A4 will affect 15.8 ha, and the open space within that area is 6.4 ha (40%). In terms of absolute open space impact, alternative A1 has the lowest impact, followed by alternatives A4 and A2, while alternative A3 has the highest impact (see Figure 3).

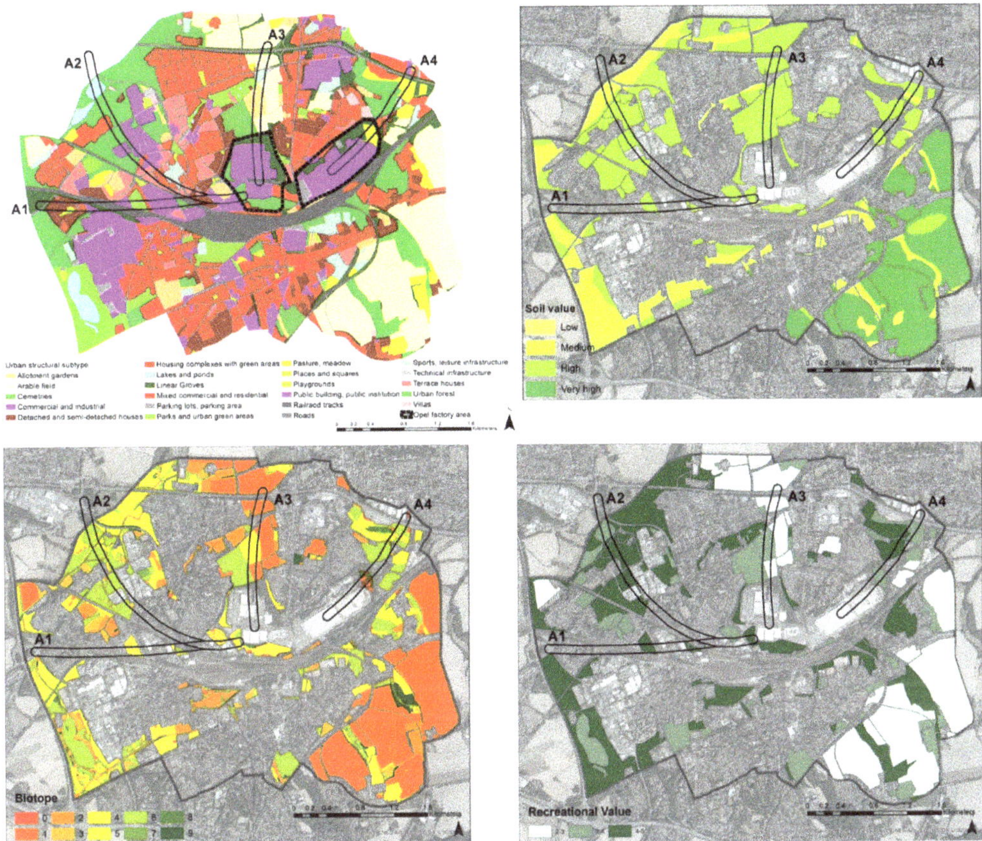

Figure 2. Urban structural subtypes in the study area (**upper left**). The areas of the European logistics centre for the distribution of spare car parts and the new industrial area are indicated with a black segmented line. The 50 m buffers of the four access roads are designated by the abbreviations A1, A2, A3 and A4. These buffers were used to calculate losses in the specific impacted areas. Soil values in open space (**upper right**). Biotopes values in open space (**down left**). Recreational values in open space (**down right**). Source: own elaboration and the assessment is based on information provided by the city of Bochum 2017 and Geological Survey NRW; satellite image as in Figure 1.

The USS most impacted by A1 was "commercial and industrial use", accounting for 5.6 ha and 22.4% of the total impacted area. In the case of A2, the USS most impacted was "urban forest", with 7.2 ha (28.3%) affected. A3 impacted 4.2 ha (25.5%) of "arable fields", while A4 also impacted "commercial and industrial uses" at a similar amount as that of A1 at 5.6 ha (33.6%). In terms of the impact on the USSs classified as open space, A2 had the highest impact due to the 7.2 ha of urban forest affected. The second-highest impact occurred with A3, due to the impact on "allotment gardens" and "arable fields". A4 impacted 3.6 ha of pastures and meadows, and, finally, A1 had less of an impact, with 1.7 ha of "urban forest" and 0.5 ha of "allotment gardens" affected.

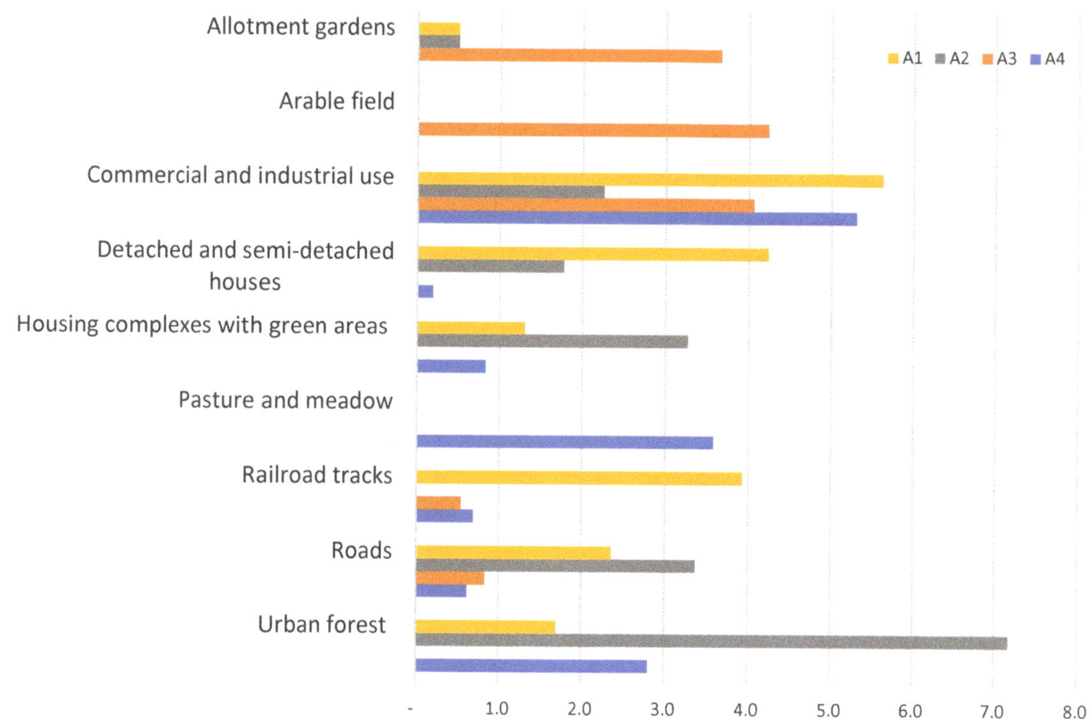

Figure 3. The USSs area lost from the development each of the access roads. Only the first three larger USSs in each alternative were plotted.

Impacts on Soil, Biotope and Recreation

When the assessment included the quality of the impacted areas in terms of soil, biotope and recreation, the situation was similar to that described above. Figure 4 shows the highest and the lowest impacts of the access roads, according to the respective losses in terms of hectares, of high-quality soil, biotope and recreation. The most severe losses of good quality soil, biotope, and recreation were connected to A2, which had substantial impacts. The lowest impact for the three analyzed variables was found again in A1. High-quality soil will be impacted most by A3.

3.2. Impacts on ES

In this section, we assess the impacts on ES, quantifying the impact of each of the access options in terms of P-ES, R-ES and C-ES (Figure 5). Similar to the previous analysis, A2 had the highest impact. However, the situation regarding the lowest impact changed, with A4 having the lowest impact. Considering the impacts in terms of ES bundles, A4 showed the lowest impact for P-ES and the second-lowest impact for R-ES and C-ES.

3.3. Analysis of Compensation Areas using ES

The spatially explicit assessment of ES using the hexagonal grid in the whole study area is presented in Figure 6, in addition to the analysis of hot and cold spots for provisioning (P-ES), regulating (R-ES), and cultural (C-ES) services. The results show a strong contrast between densely settled and industrial areas that produce cold spots, with the lowest ES values in the core, and the open space at the outskirts that concentrate the hot spots with the highest ES values.

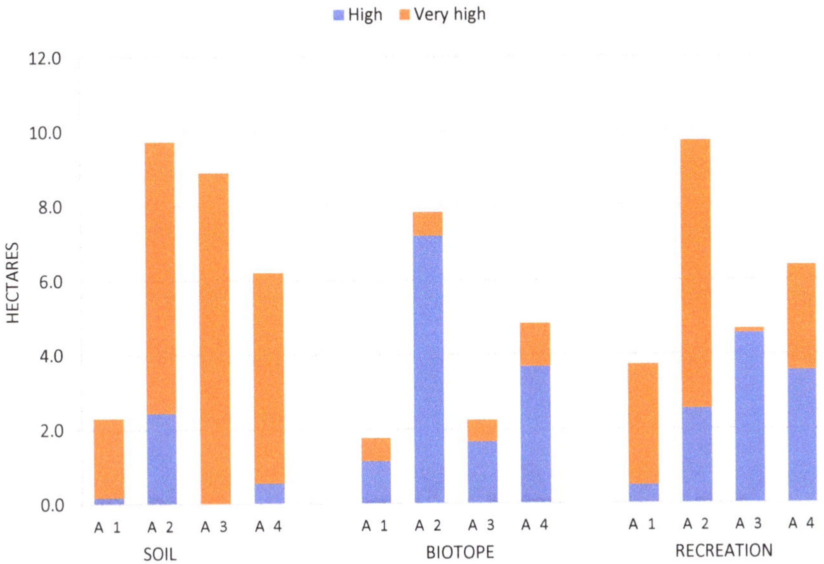

Figure 4. Area losses (hectares) of soil, biotope, and recreation for the four access road variants. The analysis considered only the areas providing high and very high values for each of the selected variables. In the case of soil, medium and high values were considered.

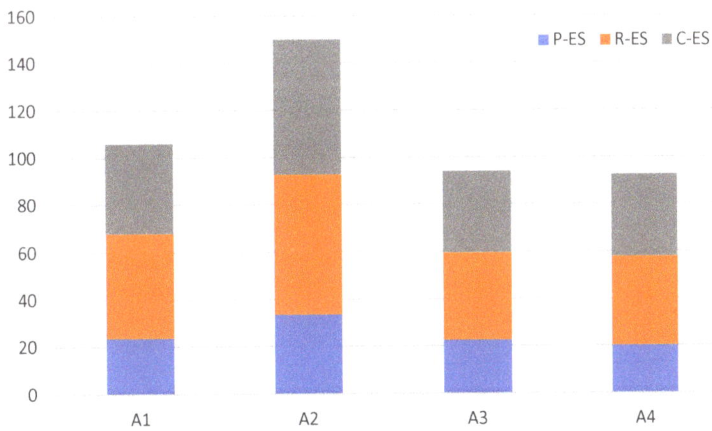

Figure 5. Impact of access roads in terms of ES. The impacts in terms of R-ES and C-ES for A4 are slightly higher than those for A3. Values for the P-ES, R-ES and C-ES were calculated using Equation (1) (E_k).

To illustrate our scheme of on-site compensation measures, we used A4, the alternative with less impact in terms of ES, according to our previous analysis. As a compensation area, we selected the hexagons present in a C-ES cold-spot cluster directly impacted by A4 (Figure 6). This cluster contained 63 hexagonal cells. We evenly distributed the amount of ES losses within these 63 cells. The amount to be compensated corresponded to the total sum product of the ES values per the USSs, P-ES, R-ES and C-ES, which are presented in the table in Figure 5. To fulfil the requirement of no net loss of important open space, we considered the replacement of the existing 3.6 ha of "pastures and meadows" and the 2.8 ha of "urban forest" that corresponded to the 6.4 ha of lost open space contained in the A4 buffer (Figures 3 and 5). Each hexagonal cell was 1 ha, which corresponded to four complete cells of a new urban forest, and three new cells of pastures and meadows. We discounted the supply of the new urban forest and pasture and meadow cells from the total ES amount to be compensated. The remainder was evenly distributed within the 56 target cells.

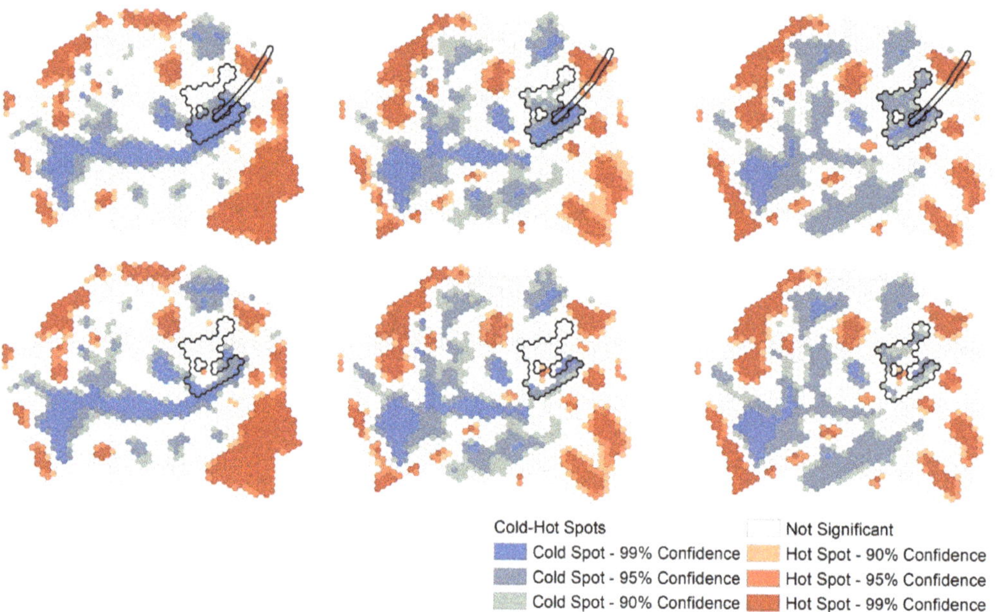

Figure 6. Compensation example for A4. Spatial distribution of ES. Upper row: total supply of provisioning ES (**left**), regulating ES (**center**), and cultural ES (**right**); center and lower rows: respective analyses of cold–hot spot for the current situation (**center**) and improvements after compensation (**lower**). The selected compensation area is indicated in black.

The central row in Figure 5 shows the analysis of cold–hot spot without compensation. The spatial structure of the ES describes a doughnut effect, with the outer areas forming a ring of hot spots and the inner areas containing cold spots. In the previous step, we selected alternative A4, which provided the amount of ES to be compensated. Once this compensation of ES was added to the selected cells, the existing cold spot was severely reduced. In the case of R-ES and C-ES, a new, small hotspot was generated in the area. This new small hot spot means that the improvement in the ES supply in the impacted area was, therefore, substantial.

4. Discussion

To be meaningful for society and decision-making, ES assessments should avoid the impulse to "distil the value of nature into a number (monetary or otherwise), and then communicate that number broadly" [40]. Alternative approaches to a single monetary value are better for considering what people care about in terms of specific decisions at stake, thus linking ES with social considerations. While ES science has evolved in meaningful ways to fulfil the promise of supporting decision-making, it is necessary to advance better characterizations of ES change, coupled with multimetric and qualitative context-specific valuations [40].

ES mapping has been increasingly used to assess and predict the expected impacts of urban development as a way to increase the quality of planning decisions [25,41]. The paradox of urban development is that attempts to increase the quality of urban environments can, at the same time, harm ES by sealing soil, fragmenting habitats, losing open space, and diminishing important vegetation structures, and, therefore, threaten human well-being [25,42].

Our empirical, methodological approach can counteract urban development shortcomings in a sound environmental compensation manner that accounts for the losses of ES. The proposed consideration of the spatial distribution of ES in a wider area surrounding a site that is affected by land-use change broadens the perspective. Our assessment gives an overall picture of an area's environmental situation, allowing multiple possible analyses of environmental performances, risks and strengths. We illustrated only three environmental aspects: soil, biotope and recreation. Similar studies have proven ES mapping a powerful input for scenario evaluation [3]. Here, we demonstrated that ES mapping using expert assessment can assist in evaluating possible impacts arising from urban development and in analyzing compensation. We presented two ways to analyze the impacts of urban development, with contrasting results. One way was based on the area-weighted loss of protected environmental goods, and the other way highlighted the impacts on ES. The merit of an ES assessment, as analyzed in the introduction (Section 1.2), is that it integrates into the decision-making process the hidden or underrepresented values of nature that could eventually be lost. Our analysis also included the full spectrum of USSs and the respective benefits that society receives from ecosystems in such areas. An analysis based on expert assessments of ES and the LULC matrix is highly correlated with biophysical measurements [16]. Therefore, our assessment is suitable for rapid preliminary evaluations and can be complemented or improved with biophysical measurements of key ES for a more complex analysis.

4.1. Limitations

There are two general limitations to consider when using our approach. The first limitation is that we used a general delineation of the planned roads. A more detailed assessment should use the exact delineation as detailed in engineering plans. The second limitation is that we did not assess in detail the area compensated in terms of USSs. In general, it is always possible to increase the supply of ES by including NbS in existing urban areas. However, a detailed estimation is necessary according to specific urban morphology.

4.2. The Role of Participatory Processes

Public participation can be greatly supported by using ES as a basis for guiding discussions and focusing on synergies and trade-offs between development options and stakeholder groups [1]. Our method allows the integration of civil society, a key aspect of participatory planning [1]. In the participatory process held in urban planning, argumentative lock-in-situations can block sustainable urban development. Environmentalists and environmental authorities remain in fierce opposition to municipal development agencies and developers. The former focuses on protection; the latter favours using open space for urban development while offering new economic possibilities and jobs. As a consequence, and due to powerful lobbies, city councils often overrule environmental concerns. Consequently, there is progress for one group of stakeholders only, at the expense of the other groups. We illustrated the procedure with an actual case study, with results that can feed a participatory process. Our method allows the use of participatory planning, where stakeholders and experts can identify development alternatives that could be judged by citizens assisting in a structured decision-making process that is value-focused, i.e., able to express what matters to people, and analytic, i.e., ascertains trade-offs between alternatives, as suggested by Chan [40]. Such an approach, where ES "values drive alternative scenarios and spatial analysis of benefits and costs" [40] holds the potential for reducing the social impacts brought on by urban development [1]. The theoretical basis of the ES framework is robust, but to put it into practice requires the involvement of key agents of development [43].

4.3. Environmental Encroachment Paradox

The assessment of environmental impacts induced by land-use change, such as the construction of road infrastructure, requires the inclusion of neighbouring areas that are spatially and functionally related to the directly affected site, and the taking into account of the urban context of urban transformations at larger scales. When ES are not assessed in areas where urban development takes place, the supply of ES might be reduced, affecting the well-being of residents. This local loss of ES can occur even if the overall ES supply remains equal or has been improved at larger scales.

The traditional approach of environmental impact assessments is to assess and judge possible encroachment on the environment expected at specific development sites. Mandatory compensation for the loss of nature (e.g., according to Directive 2011/92/EU) [44] is often restricted to replacing urban green spaces by improving rural or peri-urban green spaces, as in the case of German legislation [45]. According to the current legislation in Germany, compensation is restricted to the encroachment of legally protected goods (people and especially human health, animals, plants, biodiversity, soil, water, air, climate and landscape, cultural heritage and other properties, as well as interactions between protected goods) and must be carried out by balancing the loss of biotopes by increasing the quality of biotopes in other places. Typical compensation measures are planting trees, creating artificial wetlands, and transforming arable fields into species-rich meadows, or simply compensating by setting aside money for nature conservation purposes without any direct localised effect. These sites do not necessarily need to be close to the area of environmental encroachment. At the moment, even eco-accounting between distant locations and monetary compensation is legally possible. The highly valued vegetation will be placed far from the location of encroachment, making the compensation measures from such assessments lead to an upgrade in the quality of already existing open and green spaces elsewhere. Consequently, the environmental quality of the directly affected people, those living in the affected area, would not profit from such compensation. Thus, benefits may not be noticeable on or near the site impacted by the land-use change. Eventually, shifting compensation in distant areas means a deterioration of the supply of ES in the affected area.

Such legally binding, mandatory compensation for the loss of nature has several shortcomings: (i) compensation is carried out by replacing vegetation with vegetation; (ii) the loss of ES in areas to be developed is neglected; and (iii) the current environmental status is not improved. Vegetation and protection of species serve as the main indicators of the value of green elements. The degree of compensation is usually calculated by multiplying the biotope value with the area of the lost urban green space, neglecting other ES; and (iv) in practice, compensation is allowed to occur in distant areas, far from the affected site. In contrast, we suggest that compensation should include measures that maintain or eventually improve the environmental and living conditions in the affected area. Such measures could include countermeasures against urban heat islands, the realisation of nature-based solutions (NbS) against urban flash floods, reclamation of brownfields, river restoration, and other NbS. Using an ES assessment, it is possible to improve existing environmental quality in the affected areas. Our method goes beyond the calculation of impacts for each variant. We show how urban development can contribute to improving the overall environmental situation of areas subject to development, using the ES framework.

The consideration of the broad spectrum of NbS, along with ES supply, can open a vast range of possibilities for compensation. Admittedly, regulations that define compensation measures might have to be diversified, and practice in municipal administrations and consultancies needs to be adapted, for the ES concept to unfold its full advantage.

4.4. Estimating Urban Development Compensations with ES

To apply the ES framework in urban settings, it is necessary to include all types of urban areas belonging to the urban ecosystem, to consider the full spectrum of land-uses and not restrict the analysis only to the urban green space, as normally occurs [37]. Different degrees of urbanisation will deliver ES at different intensities, which can be higher for some bundles, such as cultural services, or lower for others, such as provisioning services [12,31,36]. Restricting the analyses only to green areas is a conceptual and methodological shortcoming which constrains the chances to ameliorate development alternatives in the places where it is needed. Furthermore, combining compensation measure determination with an ES assessment can improve preexisting situations, as we have indicated with our analysis.

Our approach ensures that the supply of ES will be maintained in the impacted areas, considering the broad spectrum of USSs. This means that the benefits of nature can be found everywhere. The relevance lies in the amount of ES—to maintain or increase when the amount is too low. Nature is not restricted to open space.

In our approach we considered no net loss of ES, i.e., maintaining the existing supply. It is also possible to apply this procedure to improve an existing situation. This could be realized by introducing a coefficient defining a target for ES improvement in the specific compensation area, or by targeting specific cold spots to transform them into hot spots.

4.5. ES and Urban Form: Spatial Distribution Matters

Enhancing the supply of ES in urban environments involves aspects of spatial distribution. In our approach, we maintained the same amount of preexisting ES that had been relocated to a nearby area previously identified as a cold spot for cultural ES. The analysis of cold and hot spots after the calculation of the compensation showed that the former cold spot was diminished in size, and partially transformed into a hot spot. This outcome indicated a substantial change in the spatial structure of the ES supply. It also showed that the supply of ES can be enhanced, even when no explicit quantitative improvements are considered, because a spatial structural assessment was involved in the supply of ES. Our results are in line with those in a similar study by Thomas [46], which showed that only by changing the spatial distribution of ES supply areas can the overall supply be altered, enhanced or diminished.

5. Conclusions

Our findings are relevant to research on the spatial structure of urban ES and the changes introduced by urban development. The peculiarities of the spatial structures found in urban areas, high heterogeneity, and complexity over relatively short distances, make the urban form a fundamental aspect to consider within ES assessments. Using USSs and a matrix-based approach for the assessment of ES allows the incorporation of the urban spatial structure that underpins the supply of ES. In our assessment, we illustrated how the selection of road alternatives can be improved by using an ES framework.

Our method illustrates how to improve planning procedures using an ES based approach. The suggested methodology is meant to counteract some of the shortcomings of traditional, legally binding regulations for environmental compensation due to urban development. Our method aims to strengthen urban resilience with a holistic understanding, using ES. The advantages of our method are the following: (i) not restricting compensation to merely accounting biotopes, but considering a wide range of ES; (ii) taking into account the overall spatial distribution of ES in the affected area in the search of legally mandatory compensation; (iii) highlighting synergies and trade-offs and allowing for NbS and strategic implementation of green infrastructure measures to leverage co-benefits [47]; and (iv) providing compensation near the affected locations.

Author Contributions: Conceptualization, H.Z. and L.I.; data curation, H.Z. and L.I.; formal analysis, L.I.; funding acquisition, H.Z. and L.I.; investigation, H.Z. and L.I.; methodology, H.Z. and L.I.; visualization, H.Z. and L.I.; writing—original draft, H.Z. and L.I.; writing—review and editing, H.Z. and L.I. All authors have read and agreed to the published version of the manuscript.

Funding: The finalization of this paper was supported by the research project, "Implementation of the Concept of Ecosystem Services in the Planning of Green Infrastructure to Strengthen the Resilience of the Metropolis Ruhr and Shanghai—research grant 01LE1805A", funded by the Bundesministerium für Bildung und Forschung (BMBF).

Data Availability Statement: Data available on request. The data presented in this study are available on request from the corresponding author.

Acknowledgments: Mapping was partly carried out in a transformation lab within the framework of the Double Degree Master Program, "Transformation of Urban Landscapes", jointly offered by Ruhr-Universität Bochum and Tongji-University, Shanghai.

Conflicts of Interest: The authors declare no conflict of interest.

References

1. Fürst, C.; Opdam, P.; Inostroza, L.; Luque, S. Evaluating the role of ecosystem services in participatory land use planning: Proposing a balanced score card. *Landsc. Ecol.* **2014**, *29*, 1435–1446. [CrossRef]
2. La Rosa, D.; Spyra, M.; Inostroza, L. Indicators of Cultural Ecosystem Services for urban planning: A review. *Ecol. Indic.* **2016**, *61*, 74–89. [CrossRef]
3. Mukul, S.A.; Sohel, M.S.I.; Herbohn, J.; Inostroza, L.; König, H. Integrating ecosystem services supply potential from future land-use scenarios in protected area management: A Bangladesh case study. *Ecosyst. Serv.* **2017**, *26*, 355–364. [CrossRef]
4. Gómez-Baggethun, E.; Barton, D.N. Classifying and valuing ecosystem services for urban planning. *Ecol. Econ.* **2013**, *86*, 235–245. [CrossRef]
5. Wolch, J.R.; Byrne, J.; Newell, J.P. Urban green space, public health, and environmental justice: The challenge of making cities 'just green enough'. *Landsc. Urban Plan.* **2014**, *125*, 234–244. [CrossRef]
6. Stępniewska, M.; Zwierzchowska, I.; Mizgajski, A. Capability of the Polish legal system to introduce the ecosystem services approach into environmental management. *Ecosyst. Serv.* **2018**, *29*, 271–281. [CrossRef]
7. European Union. *Green Infrastructure (GI)—Enhancing Europe's Natural Capital*; EU: Brussels, Belgium, 2013.
8. Costanza, R.; de Groot, R.; Braat, L.; Kubiszewski, I.; Fioramonti, L.; Sutton, P.; Farber, S.; Grasso, M. Twenty years of ecosystem services: How far have we come and how far do we still need to go? *Ecosyst. Serv.* **2017**, *28*, 1–16. [CrossRef]
9. Burkhard, B.; Maes, J. *Mapping Ecosystem Services*, 1st ed.; Burkhard, B., Maes, J., Eds.; PENSOFT Publishers: Sofia, Bulgaria, 2017. [CrossRef]
10. Haines-Young, R.; Potschin, M. Categorisation Systems: The Classification Challenge. 2016. Available online: https://www.nottingham.ac.uk/cem/pdf/CEM_Report15.pdf (accessed on 9 February 2021).
11. Haines-Young, R.; Potschin, M. *Common International Classification of Ecosystem Services (CICES) V5. 1. Guidance on the Application of the Revised Structure*; Fabis Consulting: Nottingham, UK, 2018.
12. Inostroza, L.; de la Barrera, F. Ecosystem Services and Urbanisation. A Spatially Explicit Assessment in Upper Silesia, Central Europe. *IOP Conf. Ser. Mater. Sci. Eng.* **2019**, *471*, 092028. [CrossRef]
13. Braat, L.C.; de Groot, R. The ecosystem services agenda: Bridging the worlds of natural science and economics, conservation and development, and public and private policy. *Ecosyst. Serv.* **2012**, *1*, 4–15. [CrossRef]
14. Montoya-Tangarife, C.; Barrera, F.D.e.; Salazar, A.; Inostroza, L.; de la Barrera, F.; Salazar, A.; Inostroza, L. Monitoring the Effects of Land Cover Change on the Supply of Ecosystem Services in an Urban Region: A Study of Santiago-Valparaı Chile. *PLoS ONE* **2017**, *12*, e0188117. [CrossRef]
15. Burkhard, B.; Kroll, F.; Nedkov, S.; Müller, F. Mapping ecosystem service supply, demand and budgets. *Ecol. Indic.* **2012**, *21*, 17–29. [CrossRef]
16. Roche, P.K.; Campagne, C.S. Are expert-based ecosystem services scores related to biophysical quantitative estimates? *Ecol. Indic.* **2019**, *106*, 105421. [CrossRef]
17. Hansen, R.; Frantzeskaki, N.; McPhearson, T.; Rall, E.L.; Kabisch, N.; Kaczorowska, A.; Kain, J.-H.; Artmann, M.; Pauleit, S. The uptake of the ecosystem services concept in planning discourses of European and American cities. *Ecosyst. Serv.* **2015**, *12*, 228–246. [CrossRef]
18. Cortinovis, C.; Geneletti, D. Ecosystem services in urban plans: What is there, and what is still needed for better decisions. *Land Use Policy* **2018**, *70*, 298–312. [CrossRef]
19. Geneletti, D.; Cortinovis, C.; Zardo, L.; Esmail, B.A. *Planning for Ecosystem Services in Cities*; Springer Nature: Cham, Switzerland, 2020. [CrossRef]
20. Baró, F.; Haase, D.; Gómez-Baggethun, E.; Frantzeskaki, N. Mismatches between ecosystem services supply and demand in urban areas: A quantitative assessment in five European cities. *Ecol. Indic.* **2015**, *55*, 146–158. [CrossRef]
21. Moore, S.A. Testing a Mature Hypothesis: Reflection on "Green Cities, Growing Cities, Just Cities: Urban Planning and the Contradiction of Sustainable Development". *J. Am. Plan. Assoc.* **2016**, *82*, 385–388. [CrossRef]
22. McKenzie, E.; Posner, S.; Tillmann, P.; Bernhardt, J.R.; Howard, K.; Rosenthal, A. Understanding the Use of Ecosystem Service Knowledge in Decision Making: Lessons from International Experiences of Spatial Planning. *Environ. Plan. C Gov. Policy* **2014**, *32*, 320–340. [CrossRef]

23. Wright, W.C.; Eppink, F.V.; Greenhalgh, S. Are ecosystem service studies presenting the right information for decision making? *Ecosyst. Serv.* **2017**, *25*, 128–139. [CrossRef]
24. Kaczorowska, A.; Kain, J.-H.; Kronenberg, J.; Haase, D. Ecosystem services in urban land use planning: Integration challenges in complex urban settings—Case of Stockholm. *Ecosyst. Serv.* **2016**, *22*, 204–212. [CrossRef]
25. Cortinovis, C.; Geneletti, D. A performance-based planning approach integrating supply and demand of urban ecosystem services. *Landsc. Urban Plan.* **2020**, *201*, 103842. [CrossRef]
26. European Commission. *EU Biodiversity Strategy for 2030 Bringing Nature Back into Our Lives*; European Commission: Brussels, Belgium, 2020.
27. European Union. *Territorial Agenda 2030. A Future for All Places*; Informal meeting of Ministers responsible for Spatial Planning and Territorial Development and/or Territorial Cohesion; EU: Brussels, Belgium, 2020.
28. European Commission. *EU Guidance on Integrating Ecosystems and Their Services into Decision—Part 1/3*; European Commission: Brussels, Belgium, 2019.
29. European Commission. *EU Guidance on Integrating Ecosystems and Their Services into Decision-Making—2/3*; European Commission: Brussels, Belgium, 2019.
30. Robertson, R. Glocalization. In *The International Encyclopedia of Anthropology*; John Wiley & Sons, Ltd.: Hoboken, NJ, USA, 2018. [CrossRef]
31. Zepp, H.; Andrzej, M.; Mess, C.; Zwierzchowska, I. A Preliminary Assessment of Urban Ecosystem Services in Central European Urban Areas. A Methodological Outline with Examples from Bochum (Germany) and Poznań (Poland). *Ber. Geogr. Landeskd.* **2016**, *90*, 67–84.
32. Bastian, O.L.A.F. Biotope Mapping and Evaluation as a Base of Nature Conservation and Landscape Planning. *Ekologia* **1996**, *15*, 5–17.
33. Schrey, H.P. *Bodenkarte von Nordrhein-Westfalen 1:50.000. Inhalt, Aufbau, Auswertung*; Geologischer Dienst Nordrhein-Westfalen: Krefeld, Germany, 2014.
34. Kopperoinen, L.; Itkonen, P.; Niemelä, J. Using expert knowledge in combining green infrastructure and ecosystem services in land use planning: An insight into a new place-based methodology. *Landsc. Ecol.* **2014**, *29*, 1361–1375. [CrossRef]
35. Rozas-Vásquez, D.; Fürst, C.; Geneletti, D.; Muñoz, F. Multi-actor involvement for integrating ecosystem services in strategic environmental assessment of spatial plans. *Environ. Impact Assess. Rev.* **2017**, *62*, 135–146. [CrossRef]
36. Inostroza, L. Clustering Spatially Explicit Bundles of Ecosystem Services in A Central European Region. *IOP Conf. Ser. Mater. Sci. Eng.* **2019**, *471*, 092027. [CrossRef]
37. Zepp, H.; Inostroza, L.; Sutcliffe, R.; Ahmed, S.; Moebus, S. Neighbourhood Environmental Contribution and Health. A novel indicator integrating urban form and urban green. *Chang. Adapt. Socio-Ecol. Syst.* **2018**, *4*, 46–51. [CrossRef]
38. Wentz, E.A.; York, A.M.; Alberti, M.; Conrow, L.; Fischer, H.; Inostroza, L.; Jantz, C.; Pickett, S.T.; Seto, K.C.; Taubenböck, H. Six fundamental aspects for conceptualizing multidimensional urban form: A spatial mapping perspective. *Landsc. Urban Plan.* **2018**, *179*, 55–62. [CrossRef]
39. Spyra, M.; Inostroza, L.; Hamerla, A.; Bondaruk, J. Ecosystem services deficits in cross-boundary landscapes: Spatial mismatches between green and grey systems. *Urban Ecosyst.* **2019**, *22*, 37–47. [CrossRef]
40. Chan, K.M.A.; Satterfield, T. The maturation of ecosystem services: Social and policy research expands, but whither biophysically informed valuation? *People Nat.* **2020**, *2*, 1021–1060. [CrossRef]
41. Grêt-Regamey, A.; Altwegg, J.; Sirén, E.A.; van Strien, M.J.; Weibel, B. Integrating ecosystem services into spatial planning—A spatial decision support tool. *Landsc. Urban Plan.* **2017**, *165*, 206–219. [CrossRef]
42. Alberti, M. The Effects of Urban Patterns on Ecosystem Function. *Int. Reg. Sci. Rev.* **2005**, *28*, 168–192. [CrossRef]
43. Inostroza, L.; König, H.J.; Pickard, B.; Zhen, L. Putting ecosystem services into practice: Trade-off assessment tools, indicators and decision support systems. *Ecosyst. Serv.* **2017**, *26*, 303–305. [CrossRef]
44. European Union. *Directive 2011/92/EU of the European Parliament and of the Council of 13 December 2011 on the Assessment of the Effects of Certain Public and Private Projects on the Environment*; European Union: Brussels, Belgium, 2011.
45. Federal Ministry of the Environment, Nature Conservation and Nuclear Safety. *Act on Nature Conservation and Landscape Management*; Federal Law Gazette [Bundesgesetzblatt] I p. 2542; Federal Ministry of the Environment, Nature Conservation and Nuclear Safety: Berlin, Germany, 2009; p. 74. [CrossRef]
46. Thomas, A.; Masante, D.; Jackson, B.; Cosby, B.; Emmett, B.; Jones, L. Fragmentation and Thresholds in Hydrological Flow-Based Ecosystem Services. *Ecol. Appl.* **2020**, *30*, 1–14. [CrossRef]
47. Meerow, S.; Newell, J.P. Spatial planning for multifunctional green infrastructure: Growing resilience in Detroit. *Landsc. Urban Plan.* **2017**, *159*, 62–75. [CrossRef]

Article

The Potential of Tram Networks in the Revitalization of the Warsaw Landscape

Jan Łukaszkiewicz [1], Beata Fortuna-Antoszkiewicz [1], Łukasz Oleszczuk [2] and Jitka Fialová [3,*]

1 Department of Landscape Architecture, Institute of Environmental Engineering, Warsaw University of Life Sciences-SGGW, UL. Nowoursynowska 159, 02-776 Warszaw, Poland; jan_lukaszkiewicz@sggw.edu.pl (J.Ł.); beata_fortuna_antoszkiewicz@sggw.edu.pl (B.F.-A.)
2 Legal and Analytical Services Department, 02-776 Warsaw, Poland; oleszczuk@zupa.org.pl
3 Department of Landscape Management, Faculty of Forestry and Wood Technology, Mendel University in Brno, 613 00 Brno, Czech Republic
* Correspondence: Jitka.fialova@mendelu.cz; Tel.: +420-545134096

Abstract: The current crisis of worldwide agglomeration and economic, spatial, and ownership factors, among others, mean that there is usually a shortage of new green areas, which are socially very beneficial. Therefore, various brownfields or degraded lands along public transport routes, e.g., tram lanes, are effectively transformed for this purpose. The significant potential of tram systems is that they can became a backbone of green corridors across the city. This case study of the Warsaw tram system (total length over 300 km of single tracks in service in 2019) enables us to simulate the potential growth of a biologically active area connected with an increasing share of greenery around tram lanes in Warsaw. Experience allows the authors to present the types of greenery systems based on existing and future tram corridors best suited for this city. The suggested usage of tram lanes as green corridors is in line with the generally-accepted concept of urban green infrastructure. Therefore, the aim of the authors is to present in a condensed fashion their views on a very important issue within the program of the revitalization of the Warsaw landscape by converting where possible the existing tram lines, as well as planning new ones according to the "green point of view".

Keywords: cityscape visual perception; green infrastructure; linear parks; sustainable landscape planning; tram lanes; Warsaw

check for updates

Citation: Łukaszkiewicz, J.; Fortuna-Antoszkiewicz, B.; Oleszczuk, Ł.; Fialová, J. The Potential of Tram Networks in the Revitalization of the Warsaw Landscape. *Land* **2021**, *10*, 375. https://doi.org/10.3390/land10040375

Academic Editors: Alessio Russo and Giuseppe T. Cirella

Received: 26 February 2021
Accepted: 1 April 2021
Published: 4 April 2021

Publisher's Note: MDPI stays neutral with regard to jurisdictional claims in published maps and institutional affiliations.

1. Introduction

Warsaw, Poland's capital, has a unique character as it was almost totally destroyed during the German occupation (the World War II period). Reconstruction of its infrastructure took decades and still is not complete. In addition, progression of the town's growth could not be based on the development of the existing structure, including the transportation system. The authors strongly believe that the growing demand for moving masses of people around the city is best fulfilled by the modernization and construction of the tram lines.

In the last few decades, there has been a fundamental change in the concept of designing tramway networks all over the world [1–5]. Concerning trams, passenger and environmentally friendly solutions are introduced by increasing accessibility, improving travel efficiency, and reducing energy consumption, thus reducing environmental costs [1–8]. The use of the multi-factor analysis method at each level of tram route planning and extensive promotional and information campaigns cause tramways to receive more and more social attention, as does the involvement of future passengers in the open debate at the design analysis stage (e.g., [5,7–11]). In this context, it is no exaggeration to say that the tramway as a form of public transport becomes a stimulus for developing and revitalizing cities [4,12]. Despite the multitude of solutions used in different countries, there are some similarities between them. One worth noting is the adaptation of degraded, fragmented, and marginalized space around buildings and elements of technical infrastructure within tram routes, and incorporating them again into the urban composition [1,2,4,5]. By introducing linear green systems accompanying tram lanes (insulating, ecological, and decorative functions) and new publicly accessible places for recreation and relaxation, a new quality in urban planning is achieved, which is also particularly important for the social quality during the COVID-19 pandemic [13–16].

The crisis of urban and industrial agglomerations observed worldwide at the turn of the 21st century manifested itself in several severe dysfunctions. Therefore, in the 21st century, there is a growing need to transform cities towards restoring the urban landscape's harmony and improving the inhabitants' living conditions. Among others, a skillful transformation with a humanistic approach to the entire complex cultural and natural structure of the urban fabric is needed (e.g., [17–22]). Phenomena such as the intensified global migration of people to cities (ca. 5.0 billion in 2030) [7], rising energy prices, and the degradation of the environment and landscapes—a light-hearted instrumental approach to natural resources [23]—overlap with the already existing various shortcomings of the existing functional and spatial solutions in cities, such as the density of buildings in city centers or the phenomenon of "urban sprawl" [13,14,20,22,24].

One of the ever-present problems is the inadequate development of urban public transport systems, which must continuously be adapted to the arising social needs in terms of quality and efficiency (e.g., [1,4,16,25–27]). It is already estimated that urban transport systems worldwide have such a significant impact on the environment that they are responsible for 20–25% of global energy consumption, CO_2 emissions to the atmosphere, gaseous pollutants (e.g., polycyclic aromatic hydrocarbons—PAHs), and dust (e.g., particulate matter—PM) [27–35]. For these reasons, urban transport based on private cars and diesel buses is gradually becoming a thing of the past [2,3,6,9]. The future of urban passenger transport involves four areas that are developing very rapidly: electrification, autonomy, connectivity, and sharing [8].

In line with the concept of "sustainable transport", which is increasingly used all over the world and is part of "smart cities" of the future, the aim is to create a public transport systems with a balanced impact on society, the environment, and climate [5,36]. There are attempts to combine specific engineering solutions, such as linear technical infrastructure with vegetation (e.g., green tram routes—Figure 1), aimed at crossing the border between the artificial and the natural in order to improve the quality of life on a social scale [20,21,37]. However, this is not a simple task, because it is known that urban mobility has two faces; on the one hand, it creates and stimulates economic growth, and on the other, it can also generate undesirable social, spatial, and environmental effects [3,4].

Figure 1. Tram lines as "green corridors" in the urban structure "smart city". Source: Smart City Blog, 29 June 2017 [38].

Simultaneously, urban planning formulates hypotheses (which have been confirmed many times in the past) that linear structures of public transport systems may support the revitalization of dysfunctional urban areas. Depending on the adopted priorities, strategies, and local spatial development policy, it is assumed that revitalizing activities will be concentrated along selected linear structures, necessary for the city and with a wide spatial range. It shall help connect the city's internal districts and bind the peripheries closer with the center [2,4,9,12,39]. For example, in the last several decades in urban planning in Europe and various regions of the world, a very positive, relatively new, and growing phenomenon has been observed, consisting of strengthening the integration of built-up areas thanks to the presence of tram systems [1,4,5,39].

Therefore, this publication makes a hypothesis that the tram—a necessary and environmentally friendly form of urban public transport—is a crucial urban tool enabling the simultaneous integration of dysfunctional and dispersed parts of the city and—no less important—an essential catalyst for a thorough restructuring of urban green systems and public urban spaces. This applies particularly to Warsaw, where very difficult and complex geological conditions make development of an underground transport system a very costly and time-consuming task.

This publication aims to present the importance and potential of a tram network for the reconstruction and development of the transport infrastructure in Warsaw. At the same time, based on the examples of other European cities that did not have as traumatic a past as Warsaw and that have developed in an evolutionary way, this paper tries to delineate how—drawing on the contemporary canons of sustainable development—the right proportions can be achieved in the design of the urban tram route surroundings, taking into account functional (optimum connection), environmental (resource enrichment), and landscape (the city's image) aspects.

2. Materials and Methods

The identification of the issues and the formulation of the main goals of the research allowed us to starting the first stage of work. An extensive literature search was conducted to compile examples of tram systems playing a key role on transformation of urban space. Combinations of keywords including "urban", "city", "tram lanes", "tramways", "revitalization", "green infrastructure", and "linear parks" were used in searching three online literature databases, including Scopus, ISI Web of Knowledge, and Google Scholar.

Issues identified during the literature review were divided synthetically in two major thematic groups:

- city policies (especially in Europe) concerning use of tram systems to stimulate urban development and revitalization of urban spaces;
- the importance of greenery used along tram routes for the urban environment and the quality of life.

The data collected at this stage of the research was from both the literature and the authors' own professional, scientific, and practical experience concerning the design and development of the tramway system in Warsaw. Such an example is the study and development concept of a tram route linking the Gocław and Saska Kępa districts in Warsaw, which consisted of elaboration of different initial variants of spatial solutions, and then open social consultations and elaborations of the final design.

The analysis of collected data allowed us to state that tram systems are an extremely vital tool for the revitalization of urbanized spaces, especially when they are combined with linear green systems (e.g., concerning remediation abilities of plants and the influence of green structures on air filtration in the city). This fact is of great importance not only for urban environments but, what is more crucial, for the urban society and the quality of life.

At the final stage of the work, the data collected from the literature and as part of the authors' own research formed the basis for the actual design work, enabling the development of a theoretical model for the transformation of the city tram system that would fit the specific conditions of Warsaw.

In general, it is a kind of forecast for the quantitative and qualitative development of a linear greenery layout associated with the city tram system. This form of presentation of the results of the authors' experiments was to show, in a model manner, selected aspects of designing the green forms in the vicinity of tram routes, based on the facts (collected and processed data). Case studies are an adequate and convenient scientific method used successfully in research in the fields of architecture, urban planning, and landscape architecture.

The obtained results in the form of estimated quantitative and qualitative indicators were compared synthetically with the literature in the discussion. On this basis, the final conclusions from the research were formulated.

3. Results—The Potential of Warsaw's Tram Network

Tram transport is very common in EU cities. Trams are a key part of EU public transportation, which is responsible for an annual rate of some 50 billion passengers (in 2018) [40]. All the major EU capitals have retained their original tram networks from the 19th century. Some of these networks have been upgraded to light rail standards, called Stadtbahn in Germany, premetros in Belgium, sneltram in the Netherlands, elétrico in Portugal, and fast trams in some other countries. Many city tramway networks extend over municipal boundaries. The city tram has always been efficient and one of the most environmentally-friendly forms of city mass transportation. It is competitive and resource-efficient, and it is typically characterized by low-carbon emissions. As an electric vehicle (EV), unlike other road transport means, it does not emit exhaust gases into the atmosphere, and its durability is many times greater than, for example, a city bus fleet [6]. In terms of the emissions of fine particulate matter (PM 2.5), which poses a great threat to human health and life, a tram is undoubtedly the least harmful compared to a bus (diesel engine) or even an underground [28,29,31,41]. Such transport, like low-emission trams, is one of elements of the traffic sector of the *Zero Pollution Action Plan* draft by the European Commission. The European Green Deal highlights the need for transport to become drastically less polluting in urban areas, emphasizing the importance of a combination of measures aimed at reducing emissions, mitigating urban congestion, and improving public transport options. The tramway as a form of public urban transport seems to be a perfect match for these expectations [16,25–27].

In Poland, tram networks are present in 14 cities. In 2018, the total length of tram lines in the country reached 2338.0 km (Table 1). Concerning individual Polish provinces (in 2016), the greatest length of tram lines was found in Silesia (405 km) and the second greatest in Masovia (363 km), of which Warsaw has the greatest share (Figure 2) [42].

Table 1. The total length of tram lines in Poland (km) changing in subsequent years (31 December 2019). Source of data: Statistical Yearbook of the Republic of Poland 2019 [43], compiled by J. Łukaszkiewicz, 2021.

Year	2010	2015	2017	2018
Km	2254.0	2425.0	2417.0	2338.0

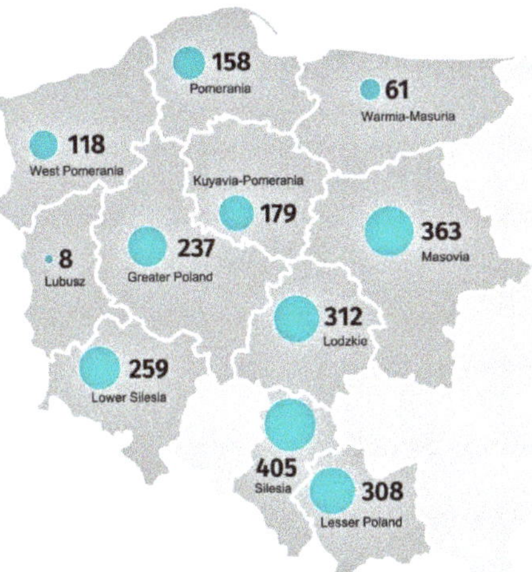

Figure 2. Trams in Poland in 2016—the total length (km) of tram lines in individual provinces (2408 km in total). Source of data: Statistics Poland [43,44], compiled by J. Łukaszkiewicz, 2021.

Concerning the introduction of green tram lanes, the Polish tradition in this matter dates back to the 1930s (Figure 3). Unfortunately, the catastrophic World War II and the following 45 years of communism caused this interesting idea to abandoned for many years, and the first solutions of this type were introduced again only at the end of the 1990s. Fifteen years later, in 2014, green lanes were present in 9 out of 14 Polish cities with tramway transportation systems [42,45]. The estimated average balance of the length of green lanes was only 3.6% then (most in Warsaw (4.0%) and Poznań (4.6%)), which accounted for approximately 66.5 km of the total length of the Polish tram network (ca. 1855.0 km) (Figure 4) [46,47].

Figure 3. Poland, Łódź, T. Kościuszko Avenue, 1930s/1940s (?)—a tram track in a spectacular floral setting, defining an already existing representative and recreational space (photo: author unknown, private collection).

Regarding Warsaw, introducing vegetation into streets and along tram tracks is not only a recent story. At the end of the 19th century, Warsaw used widespread greening and decorating of the streets with trees, lawns, and ornamental flower beds. Stefan Starzyński, the then distinguished president of Warsaw, transformed the capital into a modern European city under the slogan "Warsaw in flowers and greenery". The widespread introduction of vegetation served to raise the overall aesthetics, but it was also a way to improve the living conditions in this extensive and populous city. "...Greenery in the city is a matter of the health of the population. After all, health is the most precious human treasure..." [48] Thus, new parks and squares were established, and trees and vines were planted en masse along the streets together with flower beds and lawns. The lawns were also used for green tram tracks. "...In central districts, where compact buildings made it impossible to establish larger uniform units of greenery, efforts were made to exploit squares, roadways, and streets by establishing a significant number of new green areas and green belts, and by tree-lining several streets and introducing grass into tram tracks..." [49]. These were highly innovative activities, interrupted by the outbreak of World War II.

Figure 4. Poland, Katowice, W. Grassy tram lane along Kofranty Avenue. Photo: B. Fortuna-Antoszkiewicz, IV 2016.

In the second half of the 20th century, while rebuilding the city destroyed by war, efficient public transport was organized, including tram lines, which have been successfully used to this day. Main streets and important new arteries were given a carefully arranged floral settings. The accompanying linear spatial systems were arranged with rows of trees, hedges, and lawns [50,51]. Thanks to this, spatial order was introduced, and insulation and protection zones were consciously shaped, improving the city's climate and increasing street users' safety (Figure 5). Due to technological reasons, the greening of the tracks was abandoned in favor of lawns in parallel strips (Figure 6).

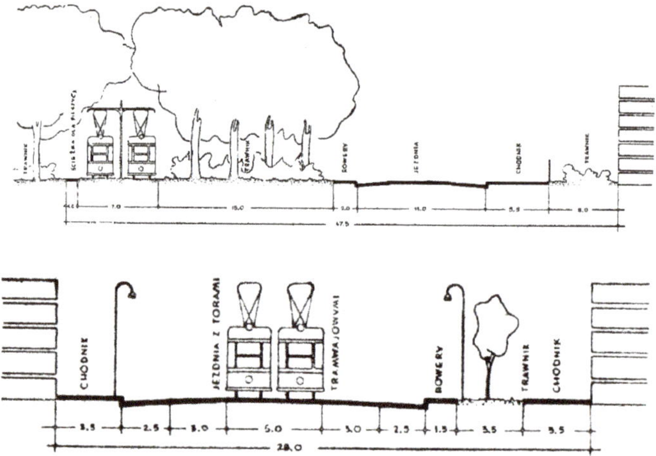

Figure 5. Cross-sections of streets with tram lanes surrounded by trees and greenery—standard from the 1950s [51].

Figure 6. Warsaw, 1950s. The newly arranged J. Marchlewskiego Avenue (currently Jana Pawła II Avenue)—a separate inner lane with a tram line surrounded by greenery (lawns with hedges) and regular rows of young trees on both sides of the road. Photo: Z. Siemaszko, collection: National Digital Archives [52].

Currently in Warsaw, around 70 years later, only fragmentary arrangements remain (dismal remnants !!!) of the old projects. Along with the development of car traffic, parking spaces appeared in place of gradually dying trees and degraded lawns in many cases. Such depletion of natural resources is particularly acute in the city center, where the process of building densification has been progressing over the last decades, often at the expense of small green enclaves (squares, undeveloped areas), and there the green track becomes a solution. Tramlines penetrate the highly urbanized center, making it possible to introduce a collision-free network of linear greenery [42,46,53,54].

At present (data from 2019) the length of green tram lanes in Warsaw has reached 25.0 km with a total length of approx. 433.0 km of tracks [46]. When renovating and constructing new tram routes, the Warsaw Trams have started to introduce green lanes as standard, implementing them both on concrete and ballast foundations [47] (Figures 7 and 8).

Figure 7. Reconstruction of a tram line, Rakowiecka Street, Warsaw. Photo: J. Łukaszkiewicz, III 2015.

The apparent advantage of green tram lanes today is the reduction of the noise level during the tram operation, the improvement of ecological aspects—increase in biologically effective urban areas—and the improved aesthetic experience of streets in cities [45]. In addition, skillfully-applied greenery allows the sound level from traffic to be reduced by 10.0 to 15.0 dB. This shows how much the sound level of the direct wave emitted by a tram decreases when the side of the tram lane is planted with large deciduous trees. The values of additional attenuation by the green belt range from 0.10 to 0.25 dB/m, depending on its type and configuration. A series of narrower belts produce more damping than one belt of the same width combined. The first lane, up to 50.0 m wide, is always the most critical. Tree belts with dense shrubs suppress noise by approx. 3.0 dB for every 30.0 m of width. Even a band of vegetation with a negligible acoustic attenuation changes the noise spectrum shape by dispersing and absorbing high-frequency components [55–61].

Figure 8. Reconstruction of a tram line—laying a lawn, Rakowiecka Street, Warsaw. Photo: J. Łukaszkiewicz, V 2015.

Additionally, vegetation reduces the speed of rising and falling of the sound level, which reduces the annoyance of noise [62]. According to various studies carried out in different conditions, the scope of noise reduction varies, but the share of the vegetation itself, especially high vegetation, remains unchallenged in this process. Therefore, instead of building acoustic screens in every situation, it is better to plant dense trees, which are incomparably more favorable for environmental and aesthetic reasons [63] (Figures 9 and 10).

Figure 9. Single trees along the tram tracks—the remains of protective plantings from the 1960s, Warsaw, Puławska Street. Photo: Warsaw Trams, Llc., 2020 [64].

In the years 2017–2020, the authors conducted research in Warsaw, aiming to identify the condition of city tram lines in terms of their quantity (length in km), their location (spatial context), and the form of the surrounding development (e.g., green or technical lanes). The obtained data show that as of 31 December 2019, the length of the single tracks reached 303.3 km (kmst—km of single tracks) (The measurement unit is 1.0 running meter of a single track (mst) or 1.0 running kilometer of a single track (kmst) [65]) including the following:

- utility tracks in depots—39.5 kmst;
- tracks used by passenger traffic—263.8 kmst.

Tracks applied for passenger traffic include the following:

- separated tracks—211.2 kmst; including green lanes with vegetation cover—approx. 25.5 kmst;
- not separated tracks available for cars, busses, and emergency vehicles—52.6 kmst [65].

Figure 10. Trees in a row at the tram-track, Warsaw, 2019. Photo: J. Bernacki [53].

Tramlines in Warsaw connect distant districts, mainly on the north-south and east-west axes (on both sides of the Vistula river), and they are concentrated within the city center with a high density of high-rise buildings (Figure 11).

Figure 11. Diagram of the tram system in Warsaw with the central zone of the city. Compiled by P. Wiśniewski, 2021.

In recent years, the Warsaw tram lines have been modernized, including the introduction of vegetation cover (turf or herbaceous plants). Currently, such green tracks account for approx. 8.0% of the total length of all tram lines in service, and approx. 12.1% in the category of separated tracks (as specified above) applied for passenger traffic. The green tracks are covered with turf or herbaceous vegetation from genera such as *Sempervivum* L. or *Sedum* L. (Figures 12–18).

Figure 12. Warsaw: Grochowska Street, Praga district. The vegetation cover of tracks, mainly of *Sempervivum* L. and *Sedum* L. species—a very ornamental green carpet with extensive maintenance. Photo shared with permission: Warsaw Trams Llc., 2020 [65].

Figure 13. Warsaw: Grochowska Street, Praga district. The vegetation cover of tracks, mainly of *Sempervivum* L. and *Sedum* L. species—the *Vicia cracca* visible in the middle as the ruderal species indicates the beginning of seminatural plant succession. Photo shared with permission: Warsaw Trams Llc., 2020 [65].

Figure 14. Warsaw: Puławska Street, Mokotów district. Smooth, grassy track cover gives a representative appearance to the main streets of the city. Photo shared with permission: Warsaw Trams Llc., 2020 [65].

Figure 15. Warsaw: Puławska Street, Mokotów district. Smooth, grassy track cover gives a representative appearance to the main streets of the city. Photo shared with permission: Tramwaje Warszawskie Llc., 2020 [65].

Figure 16. Warsaw: Rakowiecka Street, Mokotów district. The grassy track cover after mowing during summer. Photo: J. Łukaszkiewicz, VIII 2016 [65].

Figure 17. Warsaw: Zieleniecka Avenue, Praga district. Grassy tram lane nearby National Stadium is accomplished with a low hedge and a row of street trees on the left. Photo shared with permission: Warsaw Trams Llc., 2020 [65].

Figure 18. Warsaw: A. Mickiewicza Street, Żoliborz district. Grassy tram lane lined with rows of street trees on both sides. Photo shared with permission: Warsaw Trams Llc., 2020 [65].

The presented general data show that ca. 185.7 km of single tram tracks in Warsaw can be potentially transformed into a biologically active surface with turf or herbaceous vegetation cover. Some of these tracks run in separate corridors along the streets or independently across areas excluded for traffic, which offers the opportunity to introduce additional accompanying greenery (insulation and protective belts—one or double sided) of various widths.

Considering diverse spatial conditions and possible plant arrangements (different configurations of trees and shrubs connected with more or less extensive grassy areas), the authors introduced several model versions of natural protection zones (Figures 19–21). Linear vegetation belts make it possible to create insulation barriers of various densities of trees and shrubs, and in the broadest version—even to create structures like linear parks, available for direct use (added recreational function).

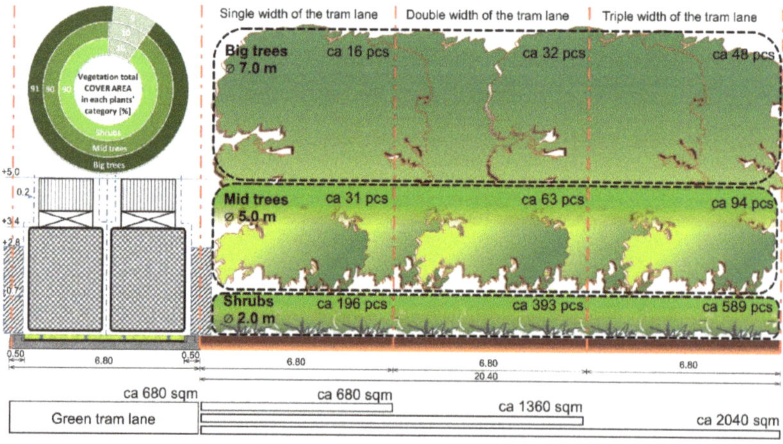

Figure 19. Model—Version 1: Land development along the tram route (one-sided intake)—a compact insulating green belt (regular triangular spacing, approx. 90% of the land area coverage). On the left: green tram route cross-section with dimensions—the pictogram after [66]. Compiled by J. Łukaszkiewicz.

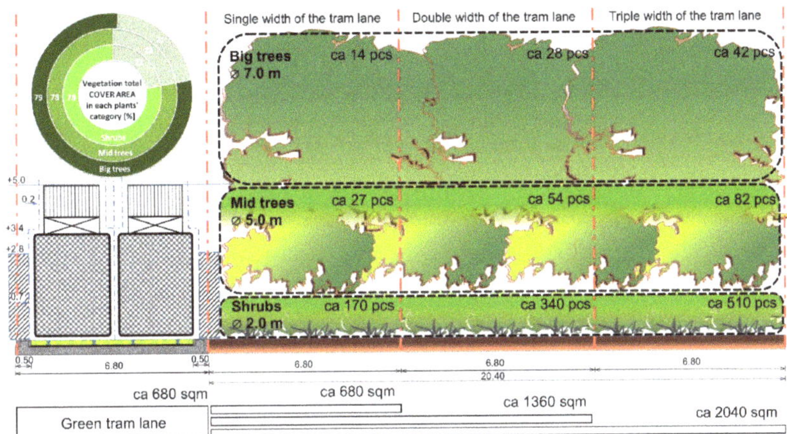

Figure 20. Model—Version 2: Land development along the tram route (one-sided intake)—a medium-dense insulating green belt (regular triangular spacing, approx. 80% of the land area coverage). On the left: green tram route cross-section with dimensions—the pictogram after [66]. Compiled by J. Łukaszkiewicz.

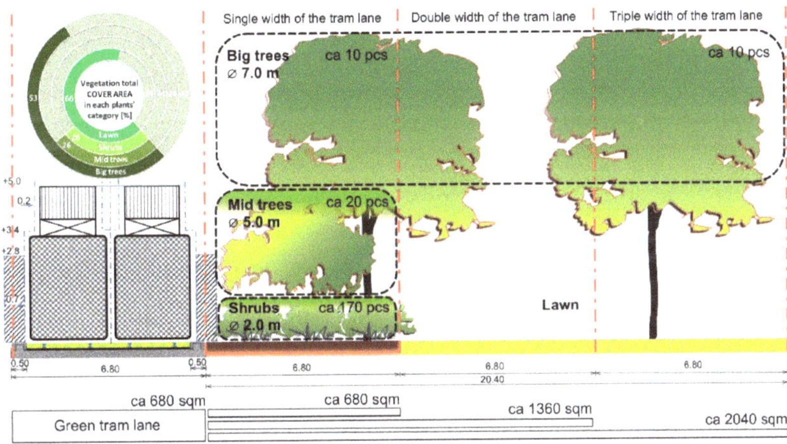

Figure 21. Model—Version 3: The linear park. Land development along the tram route with non-regular open forms of high and low greenery, providing for possible recreational use; the required width of the linear park must be at least 3x Tr. On the left: green tram route cross-section with dimensions—the pictogram after [66]. Compiled by J. Łukaszkiewicz.

3.1. The Model—Detailed Assumptions

1. The model assumes the hypothetical development of a track section, 100.0 m long, with a parallel linear strip of land of the same length; the purpose of developing the model is to visualize the possibilities of enriching the surroundings of tram routes with greenery as an additional natural resource for the urban environment.

2. The conventional unit of the width applied in the model (urban unit) for the measurement of the strip of land parallel to the trackway corresponds to the basic width of the double tram track in Warsaw—6.8 m (track without additional space for electric poles); the strip of land along the track may have a width of several times the width of the track; hence an additional strip of land with a width equal to a width of tracks lane (ca. 6.8 m) has a total area of ca. 680 m^2; a strip of land with a width equal a double width of tracks lane (ca. 13.6 m) = an area of ca. 1360 m^2; a strip of land with a

width equal to 3 times the width of the double tracks lane (ca. 20.4 m) = an area of ca. 2040.0 m^2

3. The model presents versions of vegetation cover for a strip of land with a total width equal to three times the track's width, i.e., ca. 20.0 m; research shows that the insulation green belts of this width are the most effective in stopping volatile and solid air pollutants, and at the same time, this width is sufficient to introduce local urban linear parks.

4. The model presents three versions of land development on one side of the tram double-track lane, assuming that the same development may occur on both sides or in a mosaic pattern.

5. It is assumed that the tram track itself has a green cover, i.e., grassy or herbaceous vegetation; the use of large trees (height > 10.0 m, Ø 7.0 m), smaller trees (height < 10.0 m, Ø 5.0 m), large shrubs (Ø 2.0 m), and lawns is assumed.

3.2. Model—Various Solutions

Version 1. Development of the area along the tram route in the form of a green barrier—a compact insulating green belt (regular triangular spacing, approx. 90% of the land area coverage); estimated numbers of plants in individual categories were calculated (large trees, h > 10.0 m/crown Ø approx. 7.0 m; smaller trees h < 10.0 m/crown Ø approx. 5.0 m; large shrubs Ø approx. 2.0 m) for a strip of land with a length of 100.0 m.

The width of the insulating green belt is a multiplication of the green double track width (Tr = 6.8 m); in the adopted version, the maximum width reaches the triple width of tracks lane exceeding 20.0 m and may be increased as far as possible and necessary, in accordance with the principle presented in the model (Figure 19).

Version 2. Development of the area along the tram route in the form of a green barrier—a medium-dense insulating green belt (regular square spacing, approx. 80% of the land area coverage); estimated numbers of plants in individual categories were calculated (large trees, h > 10.0 m/crown Ø approx. 7.0 m; smaller trees h < 10.0 m/crown Ø approx. 5.0 m; large shrubs Ø approx. 2.0 m) for a strip of land with a length of 100.0 m.

The width of the insulating green belt is a multiplication of the green double track width (Tr = 6.8 m); in the adopted version, the maximum width reaches the triple width of tracks lane exceeding 20.0 m and may be increased as far as possible and necessary, in accordance with the principle presented in the model (Figure 20).

Version 3. The linear park—land development along the tram route with non-regular open forms of high and low greenery, providing high luminosity beneath the canopy, with a significant share of open grassy areas providing for possible recreational use (an implementation of path layout, equipment, etc.); on the edge of the green tram lane, the greenery of the diverse height creates a compact, insulating physical and optical green barrier. Estimated numbers of plants in individual categories (large trees, h > 10.0 m/crown Ø approx. 7.0 m; smaller trees h < 10.0 m/crown Ø approx. 5.0 m; large shrubs Ø approx. 2.0 m) were determined for a strip of land with a length of 100.0 m. The width of the green belt arranged as a linear park is a multiple of the width of the green tram lane (Tr = 6.8 m); ... in the adopted version, the maximum width reaches the triple width of tracks lane exceeding 20.0 m (Figure 21).

Currently, green tram lanes in Warsaw provide approx. 8.9 ha of biologically active area (a total length of approx. 25.5 km of a single track with the mean track width of approx. 3.5 m), which is only approx. 1.78% of the total area of the ten largest parks in Warsaw (Table 2).

There is still approx. a total length of 185.7 kmst (single tram tracks) in Warsaw potentially suitable for transformation into green lanes. If at least some of these resources are available for direct use (after excluding "technical" sections, i.e., intersections, viaducts, stops, pedestrian crossings, etc.), the following can be assumed:

- the introduction of low vegetation cover (lawns or herbaceous plants) using only 50% of the resources of available single track length (approx. 92.85 km) with the mean-narrow version of track spacing (width of approx. 3.5 m) allows obtaining approx. 32.5 ha of biologically active area in total; connecting it with the area of already existing green tracks allows 41.4 ha of biologically active area to be obtained in the scale of the entire city's tramway system;

- the possibility of introducing additional protection green belts along the tram lanes using only 20% of available 185.7 kmst of single track length (due to restrictions resulting from spatial or technical conditions), gives approx. 37.14 km length of tracks in total; green belts of such a length and approx. 6.8 m average width would allow an additional biologically active area of approx. 25.26 ha or approx. 50.14 ha to be achieved—with green belts along tram lanes of approx. 13.5 m average width;
- the possibility of establishing a linear park in areas 20.0 m wide along the tram tracks would give an additional 2.0 ha of biologically active area for a 1.0 km long belt of greenery.

Table 2. The area of the 10 largest parks in Warsaw.

No.	Parks of Warsaw (10 Largest)	Area (ha)
1	Natolin	105.00
2	Pole Mokotowskie	100.00
3	Łazienki Królewskie	76.00
4	Skaryszewski	58.00
5	Marszałka E. Rydza-Śmigłego	53.00
6	Bródnowski	25.40
7	Wilanów	24.00
8	Dolinka Służewiecka	23.00
9	Moczydło	19.94
10	Morskie Oko—Promenada	17.90
	Total area of parks (ha)	502.24

In general, it can be assumed that the introduction of green tram lanes, including additional local green protection zones, would allow us to realistically obtain an additional biologically active area in the city, with a minimum size of approx. from 32.5 to 57.76 ha and even 82.64 ha—potentially enlarged by at least another 2.0 ha of "tram" linear parks. These numbers in total correspond to the average area of at least one large city park (from the group of largest in Warsaw) and an increase of ca. 6.5–16.5% in the combined area of all the largest parks in Warsaw. It is difficult to imagine introducing such a significant natural area into the dense and compact structure of the city in any other way.

Finally, the great importance of social participation in the design of green areas should be emphasized, especially linear parks in the vicinity of tram routes [10]. The authors' experiences in this area show that the recreational program of the planned park facility along the tram route should depend on public consultations. Residents of separate districts may have different preferences in terms of spending free time and using such places. The linear structure of the tram route crossing various parts of the city requires that the residents of these locations be able to comment on the recreational program. For example, this type of public consultation in Warsaw was carried out in the process of preparing an investment project—the construction of a new tram route: Saska Kępa-Gocław [10] (Figure 22).

Public consultations in Poland are anchored in the currently applicable legal provisions (the Environmental Protection Act), in relation to the Directive of the European Parliament and the Council of Europe (2011/92/EU). In the discussed case, the study group consisted of participants in consultation meetings—mainly the inhabitants of Gocław and Saska Kępa. In total, about 200 people participated in all four meetings (November–December 2017). It should be emphasized that over 1/3 of the participants (35.6%) called for the implementation of the widest developed strip of land, i.e., 60.0 m (Figure 23), because it allowed for the creation of new publicly accessible green areas. A deficit of places for sports such as running, rollerblading, and cycling was indicated. In general, the very assumption and approach to the problem of designing a tram route in a comprehensive manner, taking into account the natural, landscape, and spatial context, was accepted and positively assessed by the majority of the consultation participants. Ultimately, thanks to the social dialogue, the version of the linear park next to the tram route will provide a green isolation zone of the track up to a width of 28.0 m and balance the interests and needs of various social groups interested in the investment (see Figure 21; model—Version 3: the linear park).

Figure 22. Public consultations regarding the development of the surroundings of the new Saska Kępa-Gocław tram route in Warsaw, conducted in the form of open consultation points, where a wide team of specialists and moderators are at the disposal of interested people on selected days.

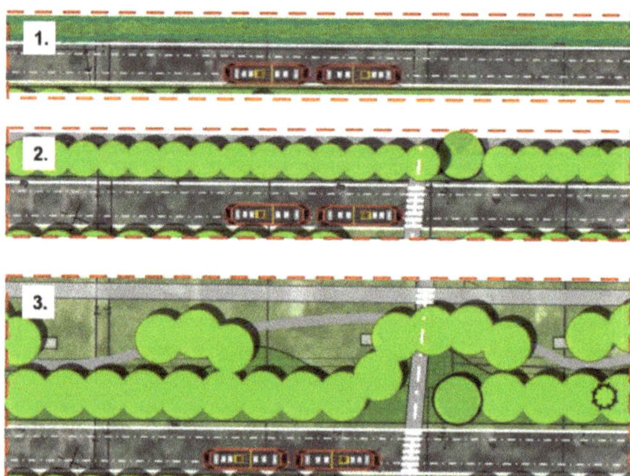

Figure 23. Insulating greenery along the planned Saska Kępa—Gocław tram route in Warsaw (three versions for public consultation). No 1. One-sided barrier of low and medium greenery (e.g., a strip of shrubs) along the tram route—with a width smaller than the width of the track itself—a frequent situation in densely built-up urban space. No 2. One-sided barrier of medium and high greenery (a belt of shrubs and trees), imitating the version of the 1st or 2nd model thanks to an additional green belt equal to one track width (the single width of tram lane). No 3. One-sided barrier of medium and high greenery—the green belt accompanying the tram reaches min. three times the width of the track (the triple width of tram lane), thanks to which it is possible to arrange greenery in a park style—linear layout—and to introduce a recreational program (roads, equipment, etc.). Graphics by J. Botwina. Compiled by J. Łukaszkiewicz [10].

4. Discussion

Warsaw, the capital of Poland, is a rapidly growing urban organism badly needing a general concept to organize this development to avoid city congestion and suffocation in the near future. One of the ideas seems to be the careful planning of the urban space growth along the tram lines as backbones of this development. The authors think that in light of the city's traumatic past and subsequent rapid development, President Starzyński's plan of building a radial structure has been lost, and now it is time to bring forward suitable plans before it is too late. Fortunately, there are examples to be followed, which are discussed here.

Sustainable development of the urban organisms involves the reduction of consumption of non-renewable spatial resources, assuming the reuse of wastelands or brownfields (e.g., [12,19–21]). Tram lines are often a perfect tool to integrate urban space, creating linear systems ("urban corridors"). In fact, the concept of development based on linear structures is not new in urban planning. The example is the idea of a linear city ("La Ciudad Lineal", ca. 1885) by the Spanish urbanist Arturo Soria y Mata (1844–1920) or the idea of "La Ville Radieuse" ("The Radiant City", 1920s) by Le Corbusier (1887–1965). Both theories later found further followers (e.g., [9,39]). For example, the concept of a "linear city" was applied in Curitiba (Curitiba), Brazil (approx. 3.0 million inhabitants) or in Australian cities (e.g., Melbourne, Coburg, and Sydney), where the tram is often an essential element of "urban corridors" [2,5].

Many examples from around the world indicate that a tramway can contribute to the improvement of the visual quality of urban public space and the promotion of the image of the city itself. For example, tram lines in France (France now has 240 city tram networks) are a symbol of modernity and an expression of French cities' pro-ecological aspirations: restoring the lost public spaces, rationalizing access to public transport, and limiting the traffic (e.g., Montpellier, Strasbourg or Bordeaux and many others) [1–5].

Similar to France, the promotion of many English cities is based on the local tram system, which is used to give the city center a new identity (e.g., Sheffield, Manchester, or Nottingham). In Germany, trams are also a tool for promoting a new public image of cities, accessibility, and good quality of city life (e.g., Heilbronn, Zwickau, Bad Wildbad, and others). In Oslo (Norway), the tram, which is part of the public transport, is promoted for its functionality—as a well-developed integrated system ensuring better city connections. In the Italian region of Calabria a modern tram line connecting the metropolis of Cosenza-Rende and the University of Calabria has been implemented since the 1980s [1,6,9].

The most interesting solutions involving trams in the urban spaces include the revitalization of the riverside area of Abandoibarra in Bilbao (Spain). The "Bilbao Effect" means a transformation of a mining and industrial port into a center of culture, art, and entertainment [67]. Similarly, a city-wide redevelopment of infrastructure was implemented in Petržalka district (Bratislava, Slovakia). Due to the introduction of a tram-train line running through the core of the district, the integration of the existing green spaces into a defined urban green system was achieved [68].

In the Czech Republic, the Prague municipality has approved an increase in the number of green tram lanes, applying mainly local, native species of drought-resistant plants, extensive drought-loving grasses, etc. [69–72]. The vegetation-modified tram lanes in Brno serve as additional urban greenery. Such green areas are often covered with herbaceous communities, transitioning between perennial beds and landscaped lawns [73].

It should be noted that in many cases, significant sections of tram routes run fully autonomously from the traffic zones (street, roads, etc.), combining fragmented parts of urban greenery, such as in Bordeaux or Strasbourg (France), Saarbrücken or Wolfartsweier Nord (Germany), Sheffield (England), Bratislava (Slovakia), and Bilbao (Spain). In practice, greenery appears in connection with tram tracks most often for two different reasons, which may partially complement each other. The first is the need to introduce insulating greenery to act as a barrier/cover instead of a classic fence, where the tracks are separated from other road users and pedestrians by linear shrub plantings, or where medium and high green plantings separate the entire length of the tracks from the traffic zone (e.g., Zwickau, Wolfartsweier, or Freiburg im Breisgau—Germany) (Figure 24). The second reason is the integration of the tram routes with the urban greenery system (existing and newly developed) thanks to coherent city-wide landscape strategies (e.g., Strasbourg, Bordeaux, and Montpellier—France, or Heilbronn—Germany) [1–3,12,20,21,67,68].

Figure 24. The tram train traveling in the "green corridor" consisting of grassy tracks, bordering trees, shrubs, and climbers attached to electric poles. Freiburg im Breisgau (Germany) has implemented numerous green tracks since 1980 [74].

Due to their often favorable location in the compact urban structure, tram lanes are predisposed to be transformed into green areas of phytoremediation, climatic, and biocenotic importance (e.g., [1–6,9,61,67–73,75]). For example, in Strasbourg, the skillfully shaped and maintained coherent continuity of green communication lanes outside the city center has resulted in biodiversity corridors—migration routes for small animals and plant seeds [1,3].

The vegetation along tram lanes is very welcome to mitigate the city's climate, e.g., the level of air pollution, the effects of climate change, etc. [23,28,61,75–79]. Naturally, the ability of plants to absorb pollutants and clean the air (phytoremediation) depends on their size and vitality—trees play the most significant role in this process. In general, research shows that the appropriate structure and the maintenance of tall greenery—both its arrangement (forms, spatial arrangements) and internal structure—are of greater importance for air quality improvement. The highest efficiency in the accumulation of dust particles is achieved by isolating tall greenery forms located 3.0–15.0 m from the source of transport emissions. It is relatively more than 2.5 times more than in the case of similar tree stands, but still growing further (e.g., [28–30,34,79,80]). Air purification of dust (PM) is optimal with a sufficiently loose tree crown structure (40% of density), reaching the optimal canopy density of approx. 35–70%. By increasing size or tree density of the smallest urban green spaces (e.g., around tram lanes), a real and most noticeable effect can be achieved in cleaning up the urban atmosphere (e.g., [28–30,33–35,80,81]).

In Warsaw, the prospective opportunity for the rapid development of the tram network system is essential and dictated by the economy rules as the environmental burden is smaller than in the case of the underground (no need to make deep excavations, deal with their negative effects on the hydro-groundwork conditions, use enormous amounts of building materials, transport masses of earth—an effect of earthmoving, disorganized public transport, closed streets, construction sites, boreholes, etc.).

Increasing the biologically active area in the center zone of Warsaw is now a necessary requirement. Although greenery is introduced in every possible way (green roofs, green walls, greenery on terraces, etc.), the main disadvantage of these forms of vegetation is their limited or complete lack of access to all users of urban public spaces. Such persons become only passive viewers (often only from a considerable distance) of such isolated arrangements. In terms of environmental advantages, forms such as green roofs, green walls, greenery on terraces, etc., undoubtedly play an important role—they serve to increase the biologically active surface and help to reduce the urban "heat island" phenomenon; however, in terms of social values—apart from a positive aesthetic effect—they do not have a more significant impact.

The greenery which is in Warsaw associated with tram lanes is also, to some extent, excluded from direct use (especially green tracks), but it is still visually closer as it is on the street level. Undoubtedly, this has a positive effect on the reception of a given space, which thus adopts a harmonious landscape appearance. In addition, the green tram lanes are real biocenotic corridors—they allow for unhindered existence and movement of small fauna (e.g., birds, insects, earthworms). Adding the technical considerations discussed earlier (protective function), it can be concluded that the vegetation accompanying tram routes is the simplest and most advantageous way of introducing greenery into the downtown area.

The model of the greenery version interrelated to tram tracks in Warsaw presented herein carries plenty of resemblance to solutions applied in other cities and countries. This is a concise conceptualization allowing for a choice of a specific greenery version in relation to the width of the lane of the accessible ground. In the widest version of the model (Version 3), the option to introduce linear parks along the tram tracks was also taken into account. In this case, the greenery is also a recreational space, besides having the isolating function. Such solutions are popular around the world [1–6,9,67–73], and this trend is also provided for in the design of the environment of tram route Saska Kępa—Gocław, in which the authors participated [10]. Furthermore, the conceptual research concerning the greenery structure accompanying the tram lines has shown great socioeconomic importance of the public consultations at the stage of the project preparation.

The presented case studies of cities show that a complex, coherent, and thoughtful approach to tramway lane design results in their successful integration with the urban surroundings [2,3,5,9,10,67]. The current world standard trams are environmentally friendly; they move almost noiselessly—often on surfaces covered with vegetation or in "green corridors"—without degrading visually valuable, often historic downtown areas. The tramways meet the expectations and carry out the communication tasks of modern cities, not only without generating adverse environmental effects but also positively influencing the process of shaping public spaces [1–4,39].

Moreover, one cannot forget that every—even the smallest—fragment of visible urban greenery is of great importance in relieving stress and mental disorders in city dwellers forced to stay in confinement during the global COVID-19 pandemic [13–15].

To sum up, those responsible for development of the future shape of a city and its public transport should take a leaf out of the book on other European cities, written by the decentralization and competition between cities across Europe and all over the world, showing that the evolution of urban theories makes it necessary to combine public transport with a high-standard multifunctional urban space. The tram system is a core of such projects, implementing not only communication functions without adverse effects on the environment (compared to other means of public transport), but also positively influencing the process of shaping urban public spaces by using "green thinking" as a key factor [1–3,5].

5. Conclusions

Trams are a key part of EU public transportation. The great importance of urban tram systems for the everyday transportation of masses of people is confirmed by the EU and Polish statistics data. The ecological advantages of trams are recognized in successive EU strategies and policies related to urban development and environmental protection. Globally, trams seem to be a vital tool for the development of urban space, which allow arranging and consolidating structures called "urban corridors". As additional green areas can be introduced into densely urbanized space through tramlines, the real social, environmental, and functional importance of this form of transport is finally tangible. The Warsaw tram system case study (total length over 300 km of single tracks in service in 2019) was implemented in order to simulate the potential growth of a biologically active area connected with an increasing share of greenery around tram lanes in Warsaw.

The suggested revitalization of the existing and designing the future tram lanes as green corridors is in line with the generally accepted concept of urban green infrastructure. Therefore, the authors aim to present their views on this significant issue in a condensed fashion, within the program of the revitalization of Warsaw landscape by converting the existing tram lines, where possible, and planning new ones according to the "green point of view".

The information used as a basis to analyze the relevant situation in Warsaw came from other studies, European as well as global, presenting the development of urban tram systems. The authors' experience in designing of the tram lane surroundings in Warsaw was of key importance for the research presented, especially in terms of greenery forms—their functionality, utility, and visual quality.

The synthesis of these issues is presented as a model with several versions, applied to Warsaw itself in the first place (as it is the case study of our research); however, the model could be applicable in other cities of the world. The development of such a model aims to indicate a potential solution to a specific problem, namely the shaping of tram route surroundings, especially with high greenery used. Our findings provide fundamental and valuable, yet overlooked, guidelines for urban tram system managers, urban policymakers, and local planners. The authors do not claim that the presented model is the only and best possible solution to the principles of designing and revitalization of greenery along tram routes. Nevertheless, it seems to be a viable overall proposition with which one can create detailed solutions adapted to the conditions of the specific site.

The paper indicates that in Warsaw—similar to other capital cities—some selected areas along the tram routes can be designed and arranged as linear parks with additional recreational functions, following the prevailing world trend. The linear park is, in a way, a contemporary approach to more traditional spatial linear forms such as a boulevard or a promenade. An excellent example of such a solution is the case of a planned tram line from Saska-Kępa to Gocław district in Warsaw. Moreover, also based on this example, this publication emphasizes the great importance of public consultations in planning new tram lines.

Taking into account all the results obtained so far, the main conclusions are as follows: (1) The presented case study (Warsaw) allows one to determine to what extent the introduction of green tracks (i.e., covered with herbaceous vegetation or additionally surrounded by tall vegetation) would realistically increase the biologically active area in the scale of the entire city; (2) With the adoption of minimum area parameters, the introduction of greenery associated with tram lines allows one to generate a biologically active area equal to a size of at least one large city park; in the case of Warsaw, it is an additional several dozen hectares of greenery in the densely built-up zone; (3) This is important especially in the central area—with dense, high-rise buildings and extensive technical infrastructure— where there is a shortage of space available for the introduction of new natural objects (parks, squares, promenades, etc.). Often, green tram tracks are the only rational way of introducing "ground-layer" vegetation into this zone; (4) The plans of city development should also consider the "green" tram lines leading to recreational areas on the outskirts, serving at the same time as "ventilation corridors"; (5) The case of Warsaw and the spatial simulations presented in this paper seem to confirm that this kind of approach, already successfully applied in many cities around the world, appears to be the best way of tackling some key environmental issues during the continuous development of Poland's capital.

Author Contributions: Conceptualization, methodology, and formal analysis: J.Ł. and B.F.-A.; investigation and resources: J.Ł., B.F.-A., and J.F.; data selection: J.Ł. and Ł.O.; writing—original draft preparation: J.Ł.; writing—review and editing: J.Ł., B.F.-A., and J.F.; visualization: J.Ł., B.F.-A., and Ł.O.; supervision: J.Ł., B.F.-A., and J.F. All authors have read and agreed to the published version of the manuscript.

Funding: This research received no external funding.

Institutional Review Board Statement: Not applicable.

Informed Consent Statement: Not applicable.

Acknowledgments: We would like to thank Věra Kolářová for her extremely competent help in language proofreading. We would like to thank our colleague Wiśniewski for his help in preparing graphic materials. The authors would like to thank the anonymous reviewers and editors for their constructive comments and suggestions.

Conflicts of Interest: The authors declare no conflict of interest.

References

1. Burns, M. HiTrans Best Practice Guide 3. Public Transport & Urban Design. 2005. Available online: https://www.crow.nl/downloads/documents/13360 (accessed on 15 January 2019).
2. Adams, R.A.M. Transforming Australian Cities. For a More Financially Viable and Sustainable Future. Transportation and Urban Design. City of Melbourne. 2009. Available online: http://www.transformingaustraliancities.com.au/wp-content/uploads/Transforming-Australian-Cities-Report.pdf (accessed on 21 September 2020).
3. Konopacki-Maciuk, Z. Trams as tools of urban transformation in French cities. *Tech. Trans. Archit.* **2014**, *10*, 61–79. [CrossRef]
4. Stipcic, M. Transit System as a Project of Urbanity. *Quad. Recer. Urban.* **2017**, *7*. Available online: https://www.raco.cat/index.php/QRU/article/view/321810 (accessed on 12 November 2020).
5. Lu, K.; Han, B.; Zhou, X. Smart urban transit systems: From integrated framework to interdisciplinary perspective. *Urban Rail Transit* **2018**, *4*, 49–67. [CrossRef]
6. Air Quality News Magazine—May 2020 by Spacehouse—Issuu. Available online: https://issuu.com/spacehouse/docs/final_aqn._mag._issue3.may2020 (accessed on 15 February 2021).
7. UNEP (United Nations Environmental Programme) Annual Report. 2007. Available online: http://wedocs.unep.org/bitstream/handle/20.500.11822/7647/-UNEP%202007%20Annual%20Report-2008806.pdf?sequence=5&isAllowed=y (accessed on 17 July 2018).
8. Bouton, S.; Hannon, E.; Haydamous, L.; Heid, B.; Knupfer, S.; Naucler, T.; Neuhaus, F.; Nijssen, J.T.; Ramanathan, S. An Integrated Perspective on the Future of Mobility. McKinsey Center for Business and Environment, September 2017. Available online: https://www.mckinsey.com/~{}/media/McKinsey/Business%20Functions/Sustainability/Our%20Insights/Urban%20commercial%20transport%20and%20the%20future%20of%20mobility/An-integrated-perspective-on-the-future-of-mobility.pdf (accessed on 20 January 2019).
9. Feudo, F.L.; Festa, C.D. A Tram-Train System to Connect the Urban Area of Cosenza to Its Province: A Simulation Model of Transport Demand Modal Split and a Territorial Analysis to Identify Adapted Transit Oriented Development Prospects. In Proceedings of the Building the Urban Future and Transit Oriented Development (BUFTOD), Champs-sur-Marne, France, 17 April 2012; HAL Id: Hal-00734634. Available online: https://hal.archives-ouvertes.fr/hal-00734634 (accessed on 4 November 2020).
10. Łukaszkiewicz, J.; Fortuna-Antoszkiewicz, B.; Botwina, J.; Oleszczuk, Ł.; Wiśniewski, P. Sustainable development of the city's transport infrastructure—A project of a new tram line with a linear park along the Exhibition Channel in Warsaw. *J. Environ. Sci. Eng. A* **2018**, *7*, 285–300. [CrossRef]
11. Jakimavičius, M.; Burinskienė, M. Multiple criteria assessment of a new tram line development scenario in Vilnius City public transport system. *Transport* **2013**, *28*, 431–437. [CrossRef]
12. Sas-Bojarska, A. Linear revitalization—Problems and challenges. Discursive article. *Urban Dev. Issues* **2017**, *53*, 5–19. [CrossRef]
13. Frumkin, H. Urban sprawl and public health. *Public Health Rep.* **2002**, *117*, 201–217. [CrossRef]
14. Ewing, R.; Meakins, G.; Hamidi, S.; Nelson, A.C. Relationship between urban sprawl and physical activity, obesity, and morbidity—Update and refinement. *Health Place* **2014**, *26*, 118–126. [CrossRef] [PubMed]
15. Soga, M.; Evans, M.J.; Tsuchiya, K.; Fukano, Y. A room with a green view: The importance of nearby nature for mental health during the COVID-19 pandemic. *Ecol. Appl.* **2020**, *31*, e02248. [CrossRef] [PubMed]
16. Althoff, J. The European Green Deal and the Future of Mobility. Available online: https://eu.boell.org/en/2020/07/20/european-green-deal-and-future-mobility (accessed on 20 March 2021).
17. Lynch, K. *The Image of the City*; MIT Press: Cambridge, MA, USA, 1960.
18. Mumford, L. *The City in History—Its Origins, Its Transformations, and Its Prospects*; Harcourt Brace & World: New York, NY, USA, 1961.
19. Gehl, J. *Cities for People*; Island Press: Washington, DC, USA, 2010.

20. Pluta, K. *Przestrzenie Publiczne Miast Europejskich Projektowanie Urbanistyczne [Public Spaces of European Urban Designs]*; Oficyna Wydawnicza Politechniki Warszawskiej: Warsaw, Poland, 2014.
21. Nyka, L. Architectural Research Addressing Societal Challenges. In *From Structures to Landscapes—Towards Re-Conceptualization of the Urban Condition*; Rodrigues, M.J., da Costa, C., Roseta, F., Lages, J.P., da Costa, S.C., Eds.; Taylor & Francis Group: London, UK, 2017; pp. 509–515.
22. Feliciantonio, D.; Salvati, L.; Sarantakou, E.; Rontos, K. Class diversification, economic growth and urban sprawl: Evidences from a pre-crisis European city. *Qual. Quant.* **2018**, *52*, 1501–1522. [CrossRef]
23. UNEP/GRID-Warsaw Centre. *Innovative Eco-City Conference 2018. COP24*; UNEP/GRID-Warsaw Centre: Katowice, Poland, 2018; pp. 1–16. Available online: https://www.gridw.pl/innowacyjneecomiasto18?lang=en (accessed on 14 December 2020).
24. Johnson, M.P. Environmental impacts of urban sprawl: A survey of the literature and proposed research agenda. *Environ. Plan. A Econ. Space* **2001**, *33*, 717–735. [CrossRef]
25. EUdebates Team. Which European Capital Has the Best Tram System? 2020. Available online: https://eudebates.tv/debates/eu-policies/transport-and-travel/which-european-capital-has-the-best-tram-system/ (accessed on 20 March 2021).
26. City Logistics. Air Quality in Europe Is Improving. 24 November 2020. Available online: http://www.citylogistics.info/research/air-quality-in-europe-is-improving/ (accessed on 20 March 2021).
27. Smith Kevin. UITP: EU's Green Deal Must Embrace Public Transport. 28 November 2019. Available online: https://www.railjournal.com/regions/europe/uitp-green-deal-embrace-public-transport/ (accessed on 20 March 2021).
28. Popek, R.; Gawrońska, H.; Gawroński, S.W. The level of particulate matter on foliage depends on the distance from the source of emission. *Int. J. Phytoremediat.* **2015**, *17*, 1262–1268. [CrossRef]
29. Gawroński, S.W. *Fitoremediacyjna Rola Roślin na Terenach Zurbanizowanych (Phytoremediation Role of Plants in Urbanized Areas)*; Nowak, G., Kubus, M., Sobisz, Z., Eds.; Drzewa i Krzewy w Rekultywacji (Trees and Shrubs in Environmental Reclamation), Mat. IX Zjazdu Polskiego Towarzystwa Dendrologicznego. Konferencja Naukowa, Wirty-Ustka, 19-22 września 2018 r. Wyd; Polskie Towarzystwo Dendrologiczne: Szczecin, Poland, 2018; pp. 19–27.
30. Sgrigna, G.; Sæbø, A.; Gawronski, S.; Popek, R.; Calfapietra, C. Particulate Matter deposition on Quercus ilex leaves in an industrial city of central Italy. *Environ. Pollut.* **2015**, *197*, 187–194. [CrossRef]
31. Moreno, T.; Reche, C.; Rivas, I.; Minguill'on, M.C.; Martins, V.; Vargas, C.; Buonanno, G.; Parga, J.; Pandolfi, M.; Brines, M.; et al. Urban air quality comparison for bus, tram, subway and pedestrian commutes in Barcelona. *Environ. Res.* **2015**, *142*, 495–510. [CrossRef]
32. Canales, D.; Bouton, S.; Trimble, E.; Thayne, J.; Da Silva, L.; Shastry, S.; Knupfer, S.; Powell, M. *Connected Urban Growth: Public-Private Collaborations for Transforming Urban Mobility*; Coalition for Urban Transitions: London, UK; Washington, DC, USA, 2017; Available online: http://newclimateeconomy.net/content/cities-working-papers (accessed on 15 October 2018).
33. Matos, P.; Vieira, J.; Rocha, B.; Branquinho, C.; Pinho, P. Modeling the provision of air-quality regulation ecosystem service provided by urban green spaces using lichens as ecological indicators. *Sci. Total Environ.* **2019**, *665*, 521–530. [CrossRef] [PubMed]
34. Jin, S.; Guo, J.; Wheeler, S.; Kan, L.; Che, S. Evaluation of impacts of trees on PM2.5 dispersion in urban streets. *Atmos. Environ.* **2014**, *99*, 277–287. [CrossRef]
35. Yang, J.; Chang, Y.; Yan, P. Ranking the suitability of common urban tree species for controlling PM2.5 pollution. *Atmos. Pollut. Res.* **2015**, *6*, 267–277. [CrossRef]
36. Reksins, M. (Ed.) Strategy for the sustainable development of the Warsaw transport system until 2015 and for the following years. In *Resolution No. LVIII/1749/2009 of the City Council of Warsaw of 9 July 2009*; City Council of Warsaw: Warsaw, Poland. (In Polish)
37. Nyka, L. *Przestrzeń Miejska Jako Krajobraz [Urban Space as Landscape]*; Czasopismo Techniczne 1-A; Architektura: Krakow, Poland, 2012; pp. 49–59.
38. Smart City Blog. Kolejne Środki Z Ue Na Transport Publiczny W Miastach [More Eu Funds for Public Transport in Cities]. Available online: https://smartcityblog.pl/transport-publiczny-w-miastach/ (accessed on 20 November 2020). (In Polish).
39. Furundzic, D.S.; Furundzic, B.S. Infrastructure Corridor as Linear City. In Proceedings of the 1st International Conference on Architecture & Urban Design, Epoka Unversity, Triana, Albania, 19–21 April 2012; pp. 721–728. Available online: https://core.ac.uk/download/pdf/152488845.pdf (accessed on 19 December 2020).
40. RailwayPro. Public Transport is a Must under the EU's Green Deal. 29 November 2019. Available online: https://www.railwaypro.com/wp/public-transport-is-a-must-under-the-eus-green-deal/ (accessed on 20 March 2021).
41. Quing, L.; Shinrin, Y. Pengiun Life. In *The Art and Science of Forest Bathing*; Insignis Media: Kraków, Poland, 2018.
42. Beister, M.; Górny, J.; Połom, M. Rozwój Infrastruktury Tramwajowej w Polsce w Okresie Członkostwa w Unii Europejskiej [The of the Publisherdevelopment of the Tramway Infrastructure in Poland during Accession to the European Union]. *Tech. Transp. Szyn.* **2015**, *22*, 20–36. Available online: http://www.tts.infotransport.pl/pl/ (accessed on 20 March 2021).
43. Statistics Poland. Statistical Products Department. *Statistical Yearbook of the Republic of Poland 2019*; Statistics Poland: Warsaw, Poland, 2019.
44. Statistics Poland. *Tramways in Poland*; Statistics Poland: Warsaw, Poland, 2018. Available online: https://stat.gov.pl/infografiki-widzety/infografiki/infografika-tramwaje,74,1.html (accessed on 20 March 2021).
45. Oleksiewicz, W.A. *Rozwój Zielonych Torowisk Tramwajowych w Polsce (Development of Green Tram Tracks in Poland)*; X Konferencja Naukowo-Techniczna "Miasto i Transport 2016"; Unpublished material-presentation.
46. Warsaw Trams, L. [Tramwaje Warszawskie sp. z o.o.]. Unpublished Report 2019. (In Polish).

47. Urbanowicz, W. Tramwaje Warszawskie Przetestują Zielone Torowisko na Tłuczniu. 2019. Available online: https://www.transport-publiczny.pl/mobile/tramwaje-warszawskie-przetestuja-zielone-torowisko-na-tluczniu-63017.html (accessed on 4 December 2020).
48. Starzyński, S. *Rozwój Stolicy (Odczyt Wygłoszony w Dniu 10 Czerwca 1938 r. na Zebraniu Urządzonym Przez Stołeczny Okręg Związku Rezerwistów). [Development of the Capital (Lecture Delivered on 10 June 1938 at a Meeting Organized by the Capital District of the Union of Reservists).]*; Wyd. Przez Stołeczny Okręg Związku Rezerwistów; [druk] Drukarnia Współczesna Sp. z o.o.: Warszawa, Poland, 1938. (In Polish)
49. Starzyński, S. *Sprawozdanie Prezydenta m. st. Warszawy Stefana Starzyńskiego za Okres od 3 III 1934 do 23 II 1939 Wygłoszone na 50 Posiedzeniu Tymczasowej Rady Miejskiej w Dniu 23 Lutego 1939 Roku. [The Report of the President of the Capital City of Warsaw, Stefan Starzyński, for the Period from 3 March 1934 to 23 February 1939, Delivered at the 50th Session of the Provisional City Council on 23 February 1939]*; Wyd. Zarząd Miejski w m. st. Warszawie; [druk] Drukarnia Miejska: Warszawa, Poland, 1939. (In Polish)
50. Fortuna-Antoszkiewicz, B. *Przemiany Formy Elementów I Układów Ogrodowych Wzdłuż Traktów Komunikacyjnych Na Przykładzie Traktu Królewskiego W Warszawie [Form's Transformations of Garden Elements and Arrangements Longwise Roads on the Example of Royal Road in Warsaw]*; SGGW: Warsaw, Poland, 2012.
51. Ptaszycka, A. *Przestrzenie Zielone W Miastach [Green Spaces in Cities]*; Ludowa Spółdzielnia Wydawnicza: Oddział w Poznaniu, Poznań, 1950.
52. Co Pamięta Warszawa. Available online: https://culture.pl/pl/artykul/co-pamieta-warszawa (accessed on 15 February 2021).
53. Klub Miłośników Komunikacji Miejskiej.Blog. 111 Lat Tramwaju Elektrycznego [111 Years of the Electric Tram] 10 April 2019. Available online: https://kmkm.waw.pl/111-lat-tramwaju-elektrycznego/ (accessed on 20 November 2020). (In Polish)
54. SISCOM. Available online: http://siskom.waw.pl/kp-tramwaje.htm (accessed on 20 March 2021).
55. Woźnicka, M.; Barszcz, P.; Janeczko, K.; Staniszewski, P.; Fortuna-Antoszkiewicz, B. Noise Level in Recreational Forest Management Areas on a Selected Example. In Proceedings of the Public Recreation and Landscape Protection—With nature hand in hand! Křtiny, Czech Republic, 2–4 May 2018; Fialová, J., Ed.; Wyd. Mendel University in Brno: Brno, Czechia, 2018; pp. 267–271.
56. Piechowicz, J.; Ozga, A.; Mleczko, D.; Kasprzak, C.; Stryczniewicz, L. *Ekologia Akustyczna na Obszarach Leśnych [Acoustic Ecology in Forest Areas]*; Wyd. Kat. Mechaniki i Wibroakustyki, Wydz. Inżynierii Mechanicznej i Robotyki, AGH im; Stanisława Staszica w Krakowie: Kraków, Poland, 2015. (In Polish)
57. Engel, Z. *Ochrona Środowiska Pracy Przed Drganiami i Hałasem [Environment Protection Against Vibration and Noise]*; Wydawnictwo Naukowe PWN: Warszawa, Poland, 1993. (In Polish)
58. Niemirski, W. *Kształtowanie Terenów Zieleni*; Arkady: Warszawa, Poland, 1973. (In Polish)
59. Sadowski, J.; Wodziński, L. *Akustyka Miasta [City Acoustics]*; z. 12.148: 1955; Architektura: Warsaw, Poland. (In Polish)
60. Sadowski, J. *Podstawy Akustyki Urbanistycznej [Basics of Urban Acoustics]*; Arkady: Warszawa, Poland, 1982. (In Polish)
61. Jakubcová, E.; Horváthová, E. Urban heat island mitigation—Microclimate regulation. *Sci. Agric. Bohem.* **2020**, *51*, 99–106.
62. Makarewicz, R. *Hałas w Środowisku [Noise in the Environment]*; Ośrodek Wydawnictw Naukowych: Poznań, Poland, 1996.
63. Fortuna-Antoszkiewicz, B. *Roślinność W Kompozycji Przestrzennej—Wartości i Zachowanie Dziedzictwa [Vegetation in Spatial Composition—Values and Preservation of Heritage]*; Wyd. SGGW: Warszawa, Poland, 2019. (In Polish)
64. *Single Trees by the Tram Tracks—The Remains of Protective Plantings from the 1960s*; Warsaw Trams, LLC.: Warsaw, Poland, 2020.
65. *Granted Permission for Sahing Photos, 2020–2021*; Warsaw Trams, LLC.: Warsaw, Poland, 2020.
66. Neufert, E. *Podręcznik Projektowania Architektoniczno—Budowlanego [Architectural and Construction Design Manual]*; Arkady: Warsaw, Poland, 1995.
67. Taplin, M. Bilbao: First Line is Just the Beginning. Available online: http://www.lrta.org/mag/articles/art0403.html (accessed on 25 September 2018).
68. Green Urban Axis—Petržalka Masterplan, Bratislava, Slovakia. Available online: http://markoandplacemakers.com/projects/green-urban-axis-petr-alka-masterplan-bratislava-slovakia (accessed on 15 February 2021).
69. Ozelěnění Tramvajových Pásů—TZB-Info. Available online: https://stavba.tzb-info.cz/terasy-a-zpevnene-plochy/21444-ozeleneni-tramvajovych-pasu (accessed on 15 February 2021).
70. Záchranářům Vadí Tráva na Kolejích. Available online: https://prazsky.denik.cz/zpravy_region/zachranarum-vadi-trava-na-kolejich-20160903.html (accessed on 15 February 2021).
71. Harciník, J. *Povrchy Tramvajových Tratí Hlavního Města Prahy*; Institut Plánování a Rozvoje hl. m: Prahy, Czech Republic, 2016; p. 44.
72. Zelené Tramvajové Pásy Oživí Prahu. Výzkumníci už Testují Nejvhodnější Rostliny—Pražský Deník. Available online: https://prazsky.denik.cz/zpravy_region/zeleny-tramvajovy-pas-doprava-klima-ekologie-rostlina-chytre-mesto.html (accessed on 15 February 2021).
73. Cejpková, K. *Principy Tvorby Veřejných Prostranství*; Kancelář architekta města Brna: Brno, Czech Republic, 2019; pp. 110–150.
74. Trams in Freiburg im Breisgau. Available online: https://en.wikipedia.org/wiki/Trams_in_Freiburg_im_Breisgau (accessed on 20 March 2021).
75. Šeptunová, Z.; Rieger, V.; Kundrata, M.; Hollan, J.; Gaillyová., Y.; Sedlák, R. *Adaptační Opatření na Zmírňování Vlivu Klimatickýxh Změn pro Město Brno*; Nadace Partnerství: Brno, Czech Republic, 2017; p. 8. Available online: https://www.lifetreecheck.eu/getattachment/c3688a57-a8eb-4925-ada7-47f1ef66c675/attachment%22 (accessed on 15 February 2021).
76. McPherson, G.E. Benefit-based tree valuation. *Arboric. Urban For.* **2007**, *33*, 1–11.

77. McPherson, G.E.; Simpson, J.R.; Scott, K.I. Actualizing microclimate and air quality benefits with parking lot tree shade ordinances. *Wetter Und Leben* **2001**, *50*, 353–369.
78. Chee Keng Lee, A.; Jordan, H.C.; Horsley, J. Value of urban green spaces in promoting healthy living and wellbeing: Prospects for planning. *Risk Manag. Healthc. Policy* **2015**, *8*, 131–137. [CrossRef]
79. Borowski, J.; Fortuna-Antoszkiewicz, B.; Łukaszkiewicz, J.; Rosłon-Szeryńska, E. Conditions for the effective development and protection of the resources of urban green infrastructure. *E3S Web Conf.* **2018**, *45*, 1–8. [CrossRef]
80. Sæbø, A.; Popek, R.; Nawrot, B.; Hanslin, H.M.; Gawrońska, H.; Gawroński, S.W. Plant species differences in particulate matter accumulation on leaf surfaces. *Sci. Total Environ.* **2012**, *427–428*, 347–354. [CrossRef]
81. Wang, X.; Teng, M.; Huang, C.; Zhou, Z.; Chen, X.; Xiang, Y. Canopy density effects on particulate matter attenuation coefficients in street canyons during summer in the Wuhan metropolitan area. *Atmos. Environ.* **2020**, *240*, 117739. [CrossRef]

 *land*

Article

Ecosystem Services Changes on Farmland in Response to Urbanization in the Guangdong–Hong Kong–Macao Greater Bay Area of China

Xuege Wang [1,2,3,4], Fengqin Yan [2,3,5], Yinwei Zeng [6], Ming Chen [2], Bin He [7], Lu Kang [2] and Fenzhen Su [1,2,3,7,*]

1   School of Geography and Ocean Sciences, Nanjing University, Nanjing 210023, China; wangxg@smail.nju.edu.cn
2   State Key Laboratory of Resources and Environmental Information System, Institute of Geographic Sciences and Natural Resources Research, Chinese Academy of Sciences, Beijing 100101, China; yanfq@lreis.ac.cn (F.Y.); chenming@lreis.ac.cn (M.C.); kangl@lreis.ac.cn (L.K.)
3   Collaborative Innovation Center for the South China Sea Studies, Nanjing University, Nanjing 210023, China
4   Department of Geography, Ghent University, 9000 Ghent, Belgium
5   Innovation Academy of South China Sea Ecology and Environmental Engineering, Chinese Academy of Sciences, Guangzhou 510301, China
6   Department of Plants and Crops, Ghent University, 9000 Ghent, Belgium; yinwei.zeng@ugent.be
7   Faculty of Geomatics, Lanzhou Jiaotong University, Lanzhou 730070, China; heb@lreis.ac.cn
*   Correspondence: sufz@lreis.ac.cn

Citation: Wang, X.; Yan, F.; Zeng, Y.; Chen, M.; He, B.; Kang, L.; Su, F. Ecosystem Services Changes on Farmland in Response to Urbanization in the Guangdong–Hong Kong–Macao Greater Bay Area of China. *Land* **2021**, *10*, 501. https://doi.org/10.3390/land10050501

Academic Editors: Alessio Russo and Giuseppe T. Cirella

Received: 22 March 2021
Accepted: 4 May 2021
Published: 8 May 2021

Publisher's Note: MDPI stays neutral with regard to jurisdictional claims in published maps and institutional affiliations.

Abstract: Extensive urbanization around the world has caused a great loss of farmland, which significantly impacts the ecosystem services provided by farmland. This study investigated the farmland loss due to urbanization in the Guangdong–Hong Kong–Macao Greater Bay Area (GBA) of China from 1980 to 2018 based on multiperiod datasets from the Land Use and Land Cover of China databases. Then, we calculated ecosystem service values (ESVs) of farmland using valuation methods to estimate the ecosystem service variations caused by urbanization in the study area. The results showed that 3711.3 km^2 of farmland disappeared because of urbanization, and paddy fields suffered much higher losses than dry farmland. Most of the farmland was converted to urban residential land from 1980 to 2018. In the past 38 years, the ESV of farmland decreased by 5036.7 million yuan due to urbanization, with the highest loss of 2177.5 million yuan from 2000–2010. The hydrological regulation, food production and gas regulation of farmland decreased the most due to urbanization. The top five cities that had the largest total ESV loss of farmland caused by urbanization were Guangzhou, Dongguan, Foshan, Shenzhen and Huizhou. This study revealed that urbanization has increasingly become the dominant reason for farmland loss in the GBA. Our study suggests that governments should increase the construction of ecological cities and attractive countryside to protect farmland and improve the regional ESV.

Keywords: ecosystem service value; farmland loss; construction land expansion; remote sensing

1. Introduction

Unprecedented urbanization has been occurring worldwide, which has caused the urban population to currently account for more than half of the world's population; this proportion is expected to increase to nearly three-quarters by 2050 [1]. Considerable urban expansion has mostly taken place in developed countries in the past, but it will mainly occur in developing countries in the coming decades. It is estimated that the urban area in developing countries will increase to 1.2 million km^2 by 2050, which is four times the urban area in 2000 [2]. As the largest developing country in the world, China has been experiencing unprecedented urban expansion after the implementation of the reform and opening up policy in 1978, with the increase of the urban population from 17.92% in 1978 to 58.5% in 2017, and a projected rise to 70% by 2030 [3,4]. Urban agglomerations in China are hotspots of urban expansion because of intensive anthropogenic activities and rapid economic growth, such as the Jing-Jin-Ji urban agglomerations, the Yangtze River Delta urban agglomerations, and the Pearl River Delta urban agglomerations [3,5,6].

Extensive urbanization has substantially encroached on farmland, wetlands and forestland, which has further led to decreases in the food supply, the loss of natural habitat, and the dramatic degradation of ecosystem services (ESs) [7–11]. Previous studies proved that urban expansion mainly originates from the conversion of farmland, and has also become the main source of farmland loss around the world [4,12–14]. In China, 74% of the lost farmland was converted into 85% of the newly increased urban land in the past decades [15]. Farmland preservation and urban expansion has become a hot topic. Urbanization has not only encroached substantially on farmland, but also impacted farmland quality and the intensity of agricultural activities [10,15,16]. Continued urban expansion will accelerate agricultural land expansion, putting high pressure on the natural ecosystems [10]. Moreover, urbanization has negative impacts on the ESs supported by farmland; e.g., food production ES, cultural heritage, wildlife habitat and open space [17–20]. Narduccia et al. showed that urban areas have more negative impacts on ESs than farmland [12]. The massive farmland loss due to urbanization is increasingly threatening farmland preservation and relative ESs [10,21–23]. Previous studies put more attention on the spatial and temporal relationship between urbanization and farmland, but they rarely explored the additional influences of urbanization on ESs and ecosystem functions supported by farmland. Therefore, it is necessary to explore the impacts of urbanization on farmland and associated ESs.

Estimate ESs are significant and alert decision makers and the public to the importance of these services. In 1997, Costanza et al. introduced the principle, methods and results of estimating the global ecosystem service value (ESV) by synthesizing previous studies based on a wide variety of methods, and they found that the global gross domestic product is much lower than the ESV [24]. In 2007, Xie et al. improved the valuation methods of Costanza et al. and developed a unit value adapted to China, which can be applied to assess ESs in specific land use areas [25]. In 2015, Xie et al. modified their original approach, developing a method for evaluating the value equivalent factor in the per unit area, and proposed an integrated method for the dynamic evaluation of the Chinese terrestrial ESV [26]. Xie et al.'s equivalent value per unit area method for the Chinese ESV has been widely adopted, but if it is directly applied to the regional scope, then the results will be biased. Xu et al. proposed a regional revision method, which entails revising the average ES equivalent values of the entire country to that of the study area based on the data of food products per unit area [27]. Xu et al.'s method has been widely implemented in different regions and has increased the accuracy of change detection [8,9,28].

The Guangdong–Hong Kong–Macao Greater Bay Area (GBA) is one of the most urbanized, industrialized and populous areas in China. However, the GBA used to be a predominantly agricultural region before 1980, was dominated by farms and rural villages and was highly suitable for agricultural development [29,30]. Extensive urbanization in the GBA has led to a large area of farmland loss, but few studies have examined the farmland loss in the GBA due to urbanization over recent decades. In previous studies, urbanization mostly refers to the expansion of urban land or urban areas, but it also includes the expansion of residential, road and industrial land [11,12,31,32]. To achieve a comprehensive understanding of urbanization, this study defines urbanization as construction land sprawl; namely, the expansion of residential areas, industrial areas, mining areas, transportation land and other construction land. Therefore, the changes in farmland and its ESV because of urbanization were examined in the GBA from 1980 to 2018 in this study. The objectives include the following: (1) to analyze the spatial and temporal changes in construction land in 1980, 1990, 2000, 2010 and 2018; (2) to examine the characteristics of farmland changes due to urbanization during the past 38 years; and (3) to evaluate the ESV of farmland loss under the effects of urbanization over the past four decades.

2. Materials and Methods

2.1. Study Area

The GBA is located in the south of China, covering a total area of approximately 55,000 km^2, with a population of approximately 45 million people in 2018 (Figure 1). The study area consists of 11 cities; specifically, Guangzhou, Shenzhen, Zhuhai, Foshan, Dongguan, Zhongshan, Jiangmen, Huizhou, Zhaoqing, Hong Kong and Macao. It has several typical geomorphic types, including mountains, hills, terraces and plains. The study area is located in a subtropical climate zone with an average annual temperature of 22.3 °C and an average annual rainfall of 1832 mm, and 83% of the rainfall occurs in the rainy season [33]. The study area has transformed from a predominantly agricultural region into one of the most developed urban agglomerations in the world since China's reform and opening-up policy [30,34]. As one of the regions in China with the highest degree of openness and a notable economic vitality, the GBA occupies an important strategic position in the overall national development situation.

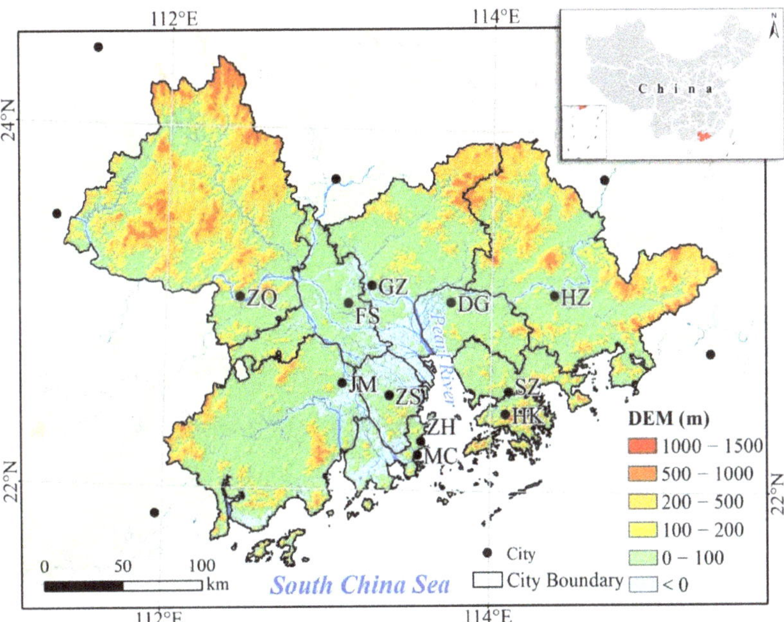

Figure 1. Location of the Greater Bay Area (GBA): from top to bottom, from left to right, cities are Hezhou (HZ), Wuzhou (WZ), Yangjiang (YJ), Zhaoqing (ZQ), Qingyuan (QY), Guangzhou (GZ), Foshan (FS), Jiangmen (JM), Zhongshan (ZS), Zhuhai (ZH), Macao (MC), Dongguan (DG), Shenzhen (SZ), Hong Kong (HK), Huizhou (HZ), Heyuan (HY), Shanwei (SW).

2.2. Data Source

The land use datasets of the farmland and construction land in this study related to 1980, 1990, 2000, 2010 and 2018 were extracted from the Land Use and Land Cover of China (CNLUCC) database [35]. The CNLUCC includes the land use and land cover (LULC) datasets for 1980, 1990, 1995, 2000, 2005, 2010, 2015 and 2018. These datasets were mainly produced by visual interpretation on the basis of a series of multitemporal Landsat images, including Thematic Mapper (TM), Enhanced Thematic Mapper (ETM) and Operational Land Imager (OLI) images. As one of the most accurate remote sensing monitoring data products in China, CNLUCC has played an important role in the investigation and research of national land resources, hydrology and ecology. In this study, the classification system included two primary classes and five subclasses, and the detailed information is listed in

Table 1. The two primary classes are farmland and construction land, and both were divided into several subclasses. Farmland is classified into two subclasses, namely paddy fields and dry farmland. Construction land includes three subclasses: urban residential land, rural residential land and other construction land. In addition, specific social statistical data, including the yield and area of the main grains in Guangdong Province and China, were obtained from the Statistical Yearbook of Guangdong Province and China of 1980, 1990, 2000, 2010 and 2018 and were downloaded from the China National Knowledge Infrastructure (CNKI) database (http://data.cnki.net/Yearbook/Navi?type=type&code=A (accessed on 20 May 2020)).

Table 1. Classification of farmland and construction land in this study.

Primary Classes	Subclasses	Implications
Farmland		This refers to land for planting crops, including cultivated land, newly opened wasteland, leisure land, rotation rest land, grassland rotation cropland, land for fruits such as mulberry, agriculture and forest areas mainly for planting crops, and beach and sea land areas cultivated for longer than three years.
	Paddy fields	This refers to cultivated land with guaranteed water sources and irrigation facilities, which are commonly irrigated in normal years and used to grow aquatic crops such as rice and lotus root, including cultivated land under rotation of rice and dry farmland crops.
	Dry farmland	This refers to cultivated land for dry crops without irrigation water sources and facilities, cultivated land for dry crops that contain water sources and irrigation facilities, which are commonly irrigated in normal years, the cultivated land used mainly for planting vegetables, and the leisure land and rotation land under normal rotation.
Construction land		This refers to urban and rural residential areas and industrial, mining, transportation and other land areas.
	Urban residential land	This refers to land for large, medium and small cities and built-up areas larger than counties and towns.
	Rural residential land	This refers to rural residential areas independent from construction land.
	Other construction land	This refers to land for factories, mines, large industrial areas, oil fields, salt fields, quarries, etc., and roads, airports and special land.

2.3. Methods

2.3.1. Analysis of the Changes in Land Use Types

To analyze the degree of changes in land use types, the annual change area (ACA, km^2/year) was calculated as follows:

$$ACA = \frac{A_2 - A_1}{t} \tag{1}$$

where A_1 and A_2 are the area of farmland or construction land at the start and end dates, respectively, and t is the time span of the study period.

2.3.2. Assessment of the Ecosystem Service Value of Farmland

The ESV was divided into four classes and eleven subclasses based on previous research [24,26,36]. The four classes were provision services, regulation services, support services and cultural services. Provision services include food production, primary production and water supply. Regulation services comprise gas regulation, climate regulation, environmental purification and hydrological regulation. Support services consist of soil conservation, nutrient cycling and biodiversity conservation. Cultural services include recreational and aesthetic values. Based on the research of Costanza et al. and

Xie et al. [24,26,37], the ESV of farmland in the GBA from 1980 to 2018 was quantitatively estimated with the ecosystem service value coefficient method as follows:

$$\mathrm{ESV} = \sum (A_k \times V_{ck}) \tag{2}$$

where ESV denotes the total annual value of the ESs, and A_k and V_{ck} are the area and value coefficient, respectively, for land use type k. In this study, the ESV of paddy fields and dry farmland were calculated first, then they were summed up to get the ESV of farmland.

To better reflect the ESV changes in the different study periods, the main grain yield in the GBA compared to that in all of China was used to revise the ecosystem services equivalent value [27]. The correction coefficient β_t was calculated as follows:

$$\beta_t = y_t / Y_t \tag{3}$$

where y_t and Y_t are the grain yields in Guangdong Province and China at time t, respectively. Because the data of grain yield in the GBA in the early study periods was not easy to acquire, the grain yield in the GBA was replaced by that of Guangdong Province in this study. Thus, β_{1980}, β_{1990}, β_{2000}, β_{2010} and β_{2018} were 1.34, 1.24, 1.38, 1.05 and 0.99, respectively.

The economic value of the ecosystem services equivalent value per unit area (Ea) was calculated as follows [8,28]:

$$E_a = 1/7 \times (P \times Y)/A \tag{4}$$

where Y and A are the total yield and area of the main grains in Guangdong Province in 2018, respectively; P is the average price of the main grains. In order to eliminate the impacts of changing values of the currency (yuan), P was given the value of 3.77 yuan/kg, which is the average price of the main grains in 2018, obtained from the Grain Net of South China (https://gdgrain.com/#!/list?params=%7B%22type%22:8%7D (accessed on 15 May 2020)).

3. Results

3.1. Urbanization in the GBA from 1980 to 2018

From 1980 to 2018, the area of construction land in the GBA showed a discernible growth trend, increasing from 2607.4 km^2 to 8243.5 km^2, a more than twofold increase (Figure 2a). A similar trend was observed in urban residential land and other construction land, which increased nearly sixfold (from 694.7 km^2 to 4778.5 km^2 and from 285.7 km^2 to 1991.2 km^2, respectively) (Figure 2b). The area of rural residential land peaked in 2000 with 1855.9 km^2 and then declined. By 2018, urban residential land occupied a dominant position among construction land (58.0%), followed by other construction land (24.2%) and rural residential land (17.9%). Regionally, construction land expanded rapidly in the 11 cities of the GBA in the past 40 years (Figure 2c). Among the 11 cities, only one had an area of construction land over 500 km^2 in 1980 (Guangzhou), but there were 8 cities by 2018. Among the four different periods, the highest ACA of construction land in the GBA was 279.7 km^2/year from 2000–2010, followed by 149.1 km^2/year from 1990–2000 and 108.7 km^2/year from 2010–2018 (Figure 2d). A similar trend also appeared in urban residential land and other construction land (Figure 2e). The ACA of rural residential land was negative in the final two study periods. The highest ACA of most cities was observed from 2000–2010 (Figure 2f). Lastly, although the urbanization and development degree of Hong Kong and Macao were much higher than that of the other cities in the GBA, their construction land expansion area and rate seemed much smaller than those of the other cities because their own administrative area was relatively small.

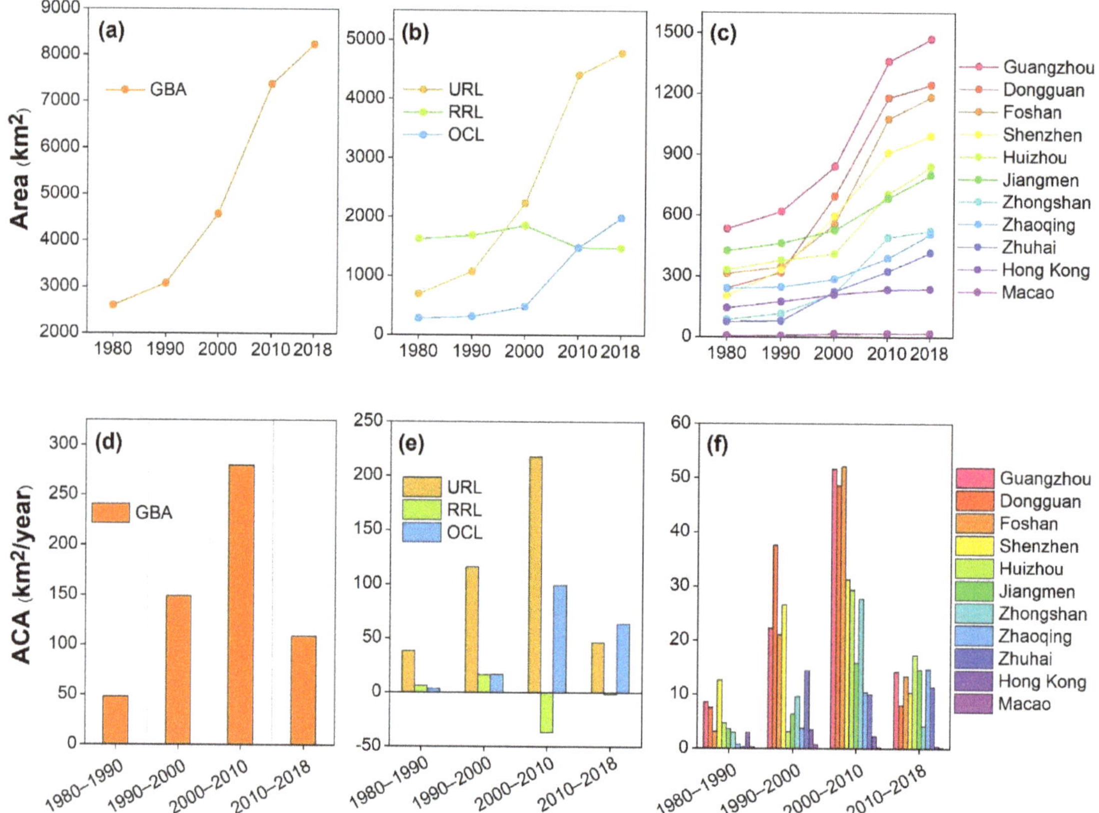

Figure 2. Construction land evolution of the GBA from 1980 to 2018: (**a–c**) are the growth of construction land for the total GBA, different types of construction land and 11 cities, respectively; (**d–f**) present the annual change area (ACA) of the construction land of the total GBA, different types and 11 cities, respectively. URL: urban residential land; RRL: rural residential land; OCL: other construction land.

3.2. Changes in Farmland in the GBA from 1980 to 2018

Over the past 38 years, the total area of the farmland in the GBA varied from 16,640.1 km^2 to 12,417.4 km^2, and more than a quarter of the total farmland area of 1980 had been lost (Figure 3a). An overall decreasing trend was also observed in the different categories of farmland and 11 cities (Figure 3b,c). The area of paddy fields was far larger than the area of dry farmland in the GBA; correspondingly, the paddy field loss (3086.3 km^2) was much greater than the dry farmland loss (1136.4 km^2) (Figure 3b). From 1980 to 2018, the top five farmland losses were in Guangzhou (−827.2 km^2), Dongguan (−744.7 km^2), Jiangmen (−452.9 km^2), Foshan (−428.2 km^2) and Shenzhen (−414.1 km^2) (Figure 3c). The highest ACA of farmland was −174.1 km^2/year from 2000–2010, followed by −158.0 km^2/year from 1990–2000, while the lowest ACA was −27.3 km^2/year from 2010–2018 (Figure 3d). The ACA of paddy fields, dry farmland and the farmland in 11 cites was also higher from 1990–2000 and 2000–2010 (Figure 3e,f).

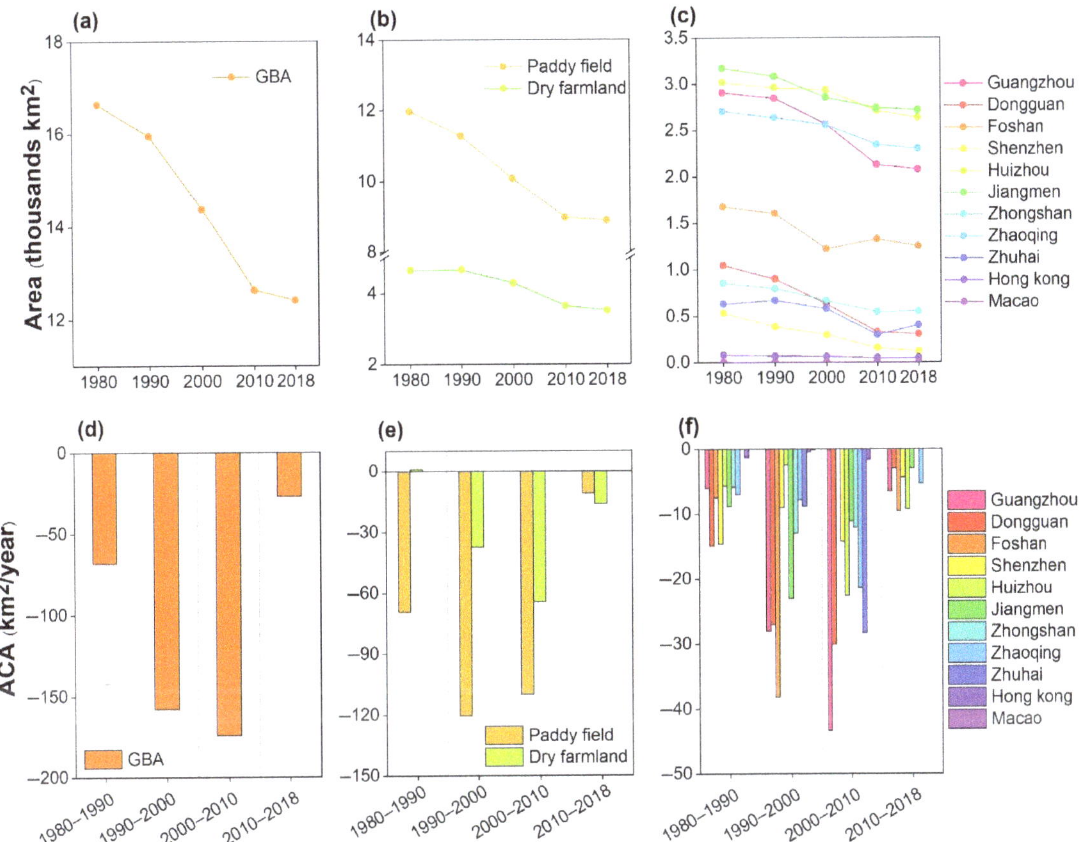

Figure 3. Farmland evolution of the GBA from 1980 to 2018: (**a–c**) are the change of farmland for the total GBA, different types and 11 cities, respectively; (**d–f**) present the ACA of farmland of the total GBA, different types and 11 cities, respectively.

From 1980 to 2018, the conversion from farmland to construction land was mainly distributed at the center of the GBA; e.g., Guangzhou, Dongguan, Shenzhen, Foshan and Zhongshan (Figure 4). Over the past 38 years, 3711.3 km^2 of farmland was converted into construction land. Among the four adjacent periods, the farmland loss due to urbanization increased drastically from the first study period (479.5 km^2) to the third study period (1772.4 km^2) but then dropped sharply in the last study period (439.2 km^2) (Figure 5a). However, the percentage of farmland loss due to urbanization in the total farmland loss exhibited a growth trend, increasing from 39.3% to 96.6% from the first study period to the last study period (Figure 5b), which demonstrates that urbanization increasingly became the dominant reason for farmland loss. From 1980 to 2018, the urbanization-triggered losses of paddy fields and dry farmland were 2599.4 km^2 and 1038.7 km^2, respectively (Figure 5c). The paddy field area lost due to urban sprawl was much more considerable than that of dry farmland over the past four decades. From 2000–2010, the losses of paddy fields and dry farmland due to construction land expansion were the highest, with areas of 1183.1 km^2 and 589.3 km^2, respectively. Over the past 38 years, 2086.7 km^2, 708.0 km^2 and 916.5 km^2 of farmland disappeared because of the expansion of urban residential land, rural residential land and other construction land, respectively (Figure 5d). The largest farmland loss due to urban residential land expansion was 1000.5 km^2 from 2000–2010, followed by 603.8 km^2 from 1990–2000 and 305.6 km^2 from 1980–1990. Farmland being converted to rural residential land mainly occurred from 1990–2000 with an area of 305.6 km², then from 2000–2010 (249.7 km²) and from 1980–1990 (132.6 km²). The largest farmland loss caused by other construction land was 522.2 km² from 2000–2010, followed by 242.4 km² from 2010–2018. In the past four decades, the top five cities that experienced farmland loss because of urbanization were Guangzhou (824.42 km²), Dongguan (653.34 km²), Foshan (591.03 km²), Shenzhen (371.35 km²) and Huizhou (362.22 km²) (Figure 5e). In most areas, this conversion was concentrated between 1990 and 2010.

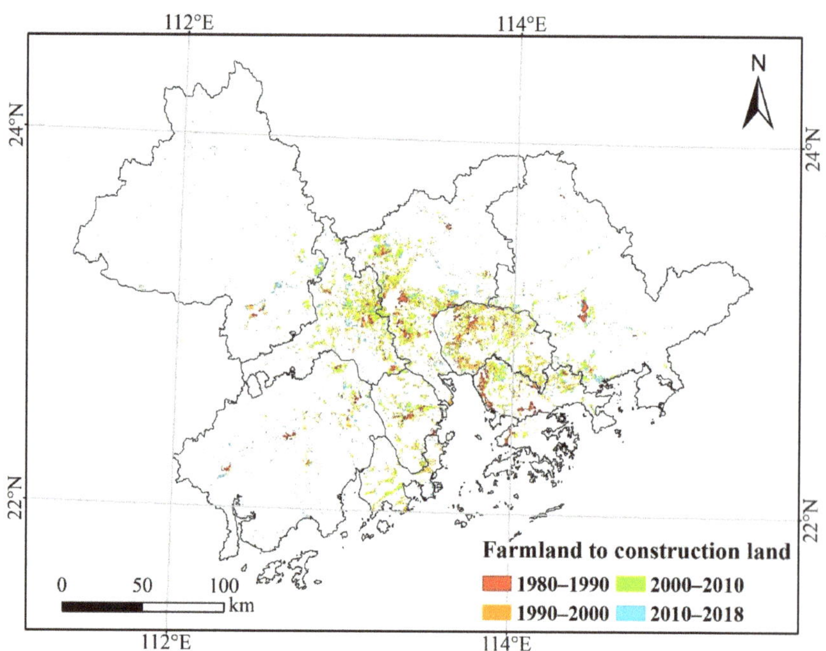

Figure 4. Spatial and temporal distribution of farmland loss due to urbanization in the GBA from 1980 to 2018.

3.3. The Impact of Urbanization on the Ecosystem Service Value of Farmland

An unprecedented expansion of construction land has led to a great loss of farmland, which has further impacted the ESs and functions of farmland in the GBA. In the past four decades, the ESV of farmland decreased by 5036.7 million yuan due to construction land encroachment, accounting for 43.1% of the total ESV loss of farmland (Table S1). The highest ESV loss of farmland because of urbanization was 2177.5 million yuan from 2000–2010, followed by 1655.1 million yuan from 1990–2000 (Figure 6a). For paddy fields and dry farmland, the ESV losses caused by urbanization were 3442.1 million yuan and 1594.7 million yuan, respectively (Figure 6b). The largest ESV losses of both paddy fields and dry farmland appeared from 2000–2010 (1438.8 million yuan and 738.7 million yuan, respectively), and the second-largest ESV losses were from 1990–2000 (1040.0 million yuan and 615.1 million yuan, respectively). Because of urbanization, the most affected ecosystem service function was hydrological regulation with a loss of 2514.2 million yuan, followed by food production and gas regulation with losses of 1541.4 million yuan and 1248.6 million yuan, respectively (Figure 6c). Specifically, the ESV of the water supply of farmland was positive under the effects of urbanization, because paddy fields need a large amount water for the growth of aquatic plants. To some degree, the loss of paddy fields is a benefit for the water supply. Regionally, the total ESV loss of farmland caused by urbanization in Guangzhou was the largest (1110.85 million yuan), followed by that in

Dongguan (921.8 million yuan), Foshan (791.6 million yuan), Shenzhen (518.1 million yuan) and Huizhou (460.4 million yuan). The largest ESV loss of most cites in most areas was also concentrated from 2000–2010 (Figure 6d). In addition, Figure 7 shows the temporal and spatial changes of ESV of farmland caused by urbanization in 11 cities. From 1980–1990 and 2010–2018, the ESV losses of farmland because of urbanization in 11 cities were mostly less than 100 million yuan (green and dark green). However, from 1990–2000 and 2000–2010, the ESV losses of farmland because of urbanization in Guangzhou, Foshan and Dongguan were larger, especially in Guanghzou from 2000–2010, showed with red color. Overall, the ESV losses of farmland due to urbanization in Zhaoqing, Zhuhai, Macao and Hong Kong were relatively less than in other areas.

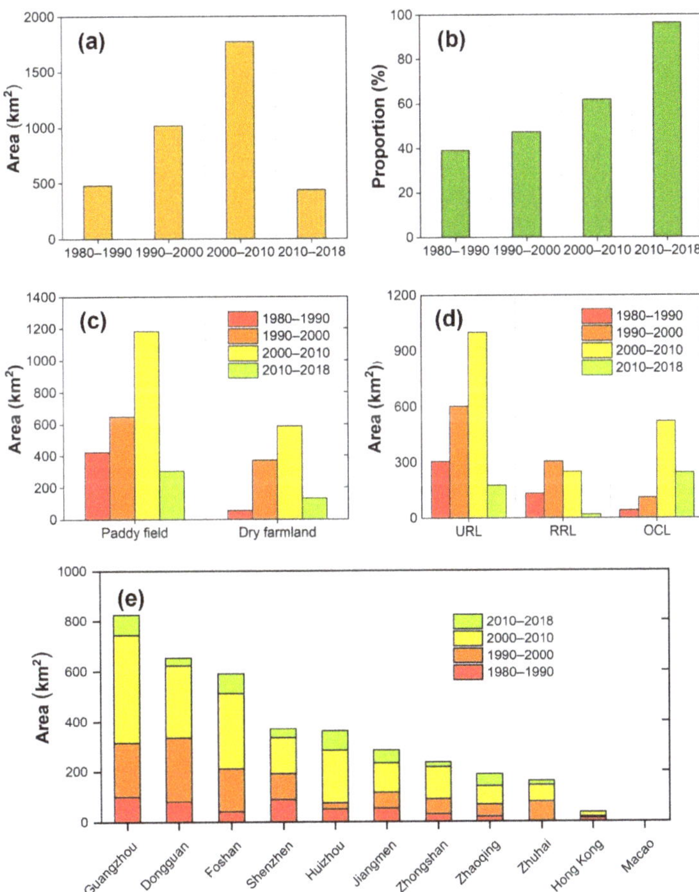

Figure 5. Farmland loss due to urbanization in the GBA during the four study periods: (**a**) is the area of farmland converted to construction land in the GBA during the four study periods; (**b**) presents the percentage of the area of farmland converted to construction land in the total area of farmland loss in the GBA during the four study periods; (**c**) shows the variation among the losses of different farmland categories due to urbanization in the GBA during the four study periods; (**d**) is the variations in the area of farmland converted into different categories of construction land in the GBA during the four study periods; and (**e**) shows the area of farmland converted to construction land in the 11 cities during the four study periods. URL: urban residential land; RRL: rural residential land; OCL: other construction land.

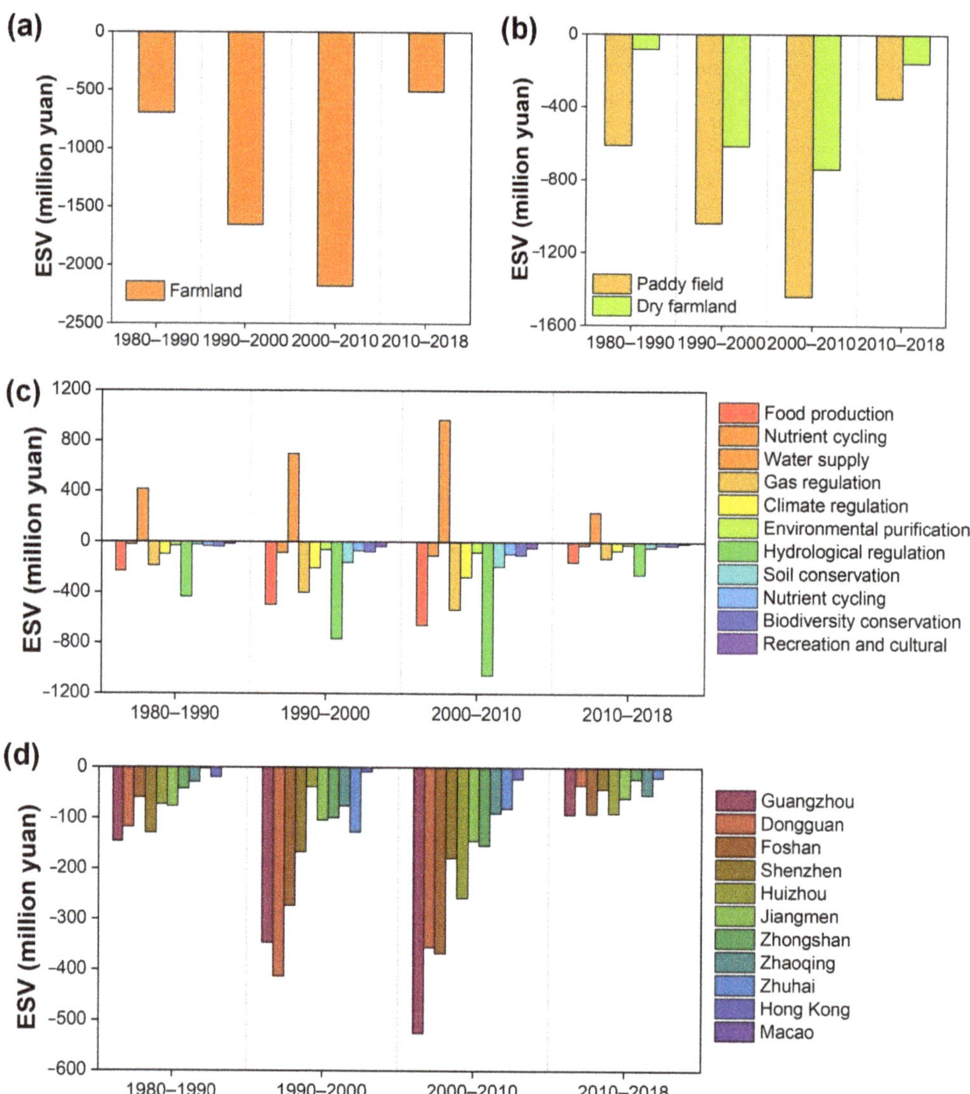

Figure 6. The ecosystem service value (ESV) loss of farmland encroached by urbanization in the GBA during the four study periods: (**a**) is the ESV loss of farmland converted to construction land in the GBA during the four study periods; (**b**) is the ESV loss of different farmland types converted to construction land in the GBA during the four study periods; (**c**) shows the variation of ecosystem service losses of farmland due to urbanization in the GBA during the four study periods; and (**d**) is the ESV loss of farmland converted to construction land in the 11 cities during the four study periods.

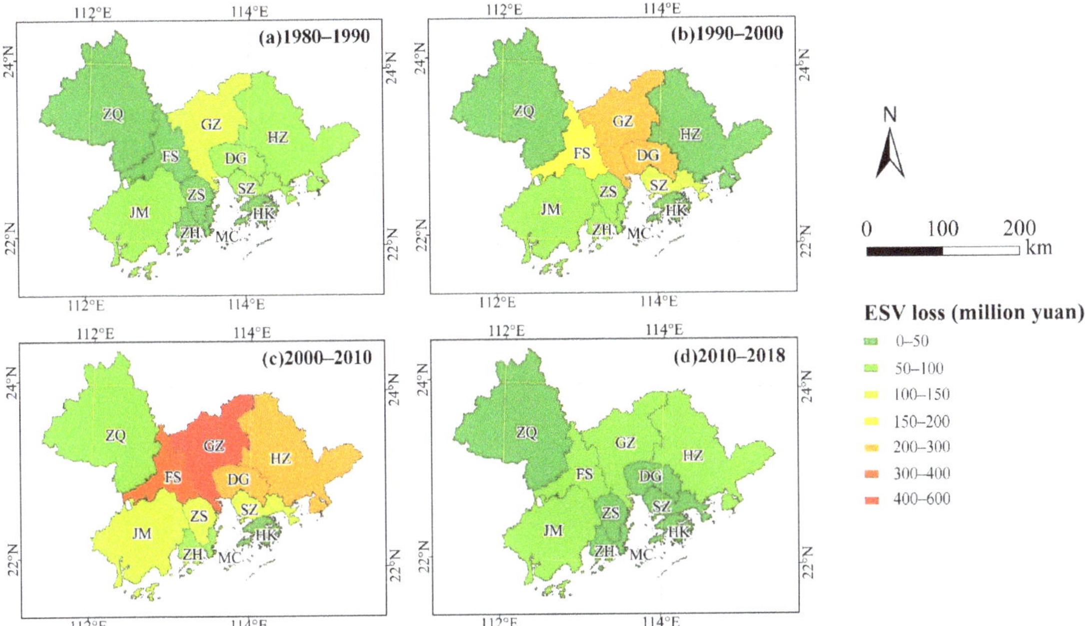

Figure 7. Spatial and temporal distribution of ESV loss of farmland because of urbanization in 11 cities in the four study periods.

4. Discussion

4.1. Farmland Changes Caused by Urbanization

The results of this study demonstrate that construction land in the GBA experienced dramatic changes in the overall extent, different types and regional extent, especially after 1990. Farmland is the land use type that contributed most to the expansion of construction land over the past four decades (Figure S1). Because of urbanization, 3711.3 km^2 of farmland disappeared in the GBA from 1980 to 2018. The highest farmland loss caused by urbanization was observed from 2000–2010 (1772.4 km^2), followed by that in the period from 1990–2000 (1020.1 km^2), but it was small from 1980–1990 (479.5 km^2) and from 2010–2018 (439.2 km^2). Our results are different from others showing the global conversion of farmland to other land use types; globally, farmland was mainly converted to grasslands and woodlands, which accounted for 57% and 36%, respectively [38]. These changes were strongly influenced by the social development background in China and local governments' behavior [39,40]. From 1980–1990, the first ten years after the implementation of the reform and opening-up policy in 1978, social and economic development was in the initial stage, production technologies lagged behind and infrastructure was relatively imperfect. All of these reasons contributed to a slower process of urbanization, which resulted in less farmland being converted to construction land. However, after ten years of development, the economy expanded to a certain extent, and the process of urbanization and industrialization also gradually accelerated, which led to farmland being intensively encroached on by construction land. Subsequently, China entered the World Trade Organization in 2001, which indicated that China's reform and opening up had entered a new stage in history. Rapid economic and scientific technology development impelled extensive urbanization, which caused the area of farmland loss to construction land to reach a peak from 2000–2010. As construction land expansion produced a great loss of farmland, a series of environmental problems appeared which threatened farmland preservation, food security, production capacity and social stability [41,42]. In addition, farmland conversion to construction land could also lead to intensification of farming, and abandonment and degradation of farmland [43,44]. Governments try their best to maintain a balance between urbanization development and farmland preservation by implementing much stricter policies for farmland conversion and ecological urbanization and construction [10,14]. As early as 2005, the central land policy for the preservation of 1.8 billion mu (1.2 million km^2) farmland was established, which is a key environmental policy to cope with China's land transformation crisis since the 1990s [45]. In addition, to improve the urbanization quality and to strengthen the protection of farmland and natural ecosystems, the National Plan on New Urbanization and the Opinions on Accelerating the Construction of Ecological Civilization were promulgated in 2014 and 2015, respectively [39]. Under these policies, farmland loss due to urbanization sharply declined after 2010. In addition, local governments' behavior has vital impacts on the conversion of farmland to construction land [14,46,47]. Driven by particular interests, local governments converted farmland to urban land at a low compensation and leased to developers at a much higher price [48], which significantly affected farmland change under rapid urbanization [14]. To preserve agricultural land and food security, a centralized fiscal reform in 1994 induced local governments' land financing behavior to significantly influence farmland conversion [14]. Although the central government limits the total construction land quotas that local governments can lease to developers, local governments still have sufficient autonomy to determine which parcel of land to lease out [14], and they resorted to land leasing to gain extra revenue to finance urban construction and balance fiscal expenditures [49]. This extra local revenue from land leasing was approximately 33.7% of the local revenues from 2007 to 2012, which mainly came from farmland conversion [50,51]. Lastly, the urbanization in Guangzhou, Dongguan, Foshan, Shenzhen and Huizhou was faster than in other cities over the past four decades (Figure 2); correspondingly, the area of farmland conversion to construction land was greater in these cities than in the other cities (Figure 5). The same is observed in other developing counties, like India, where farmland conversion to construction land predominantly occurred in the districts with high rates of economic growth and higher agricultural land suitability [22].

4.2. Ecosystem Service Value Changes due to Urbanization

Extensive urbanization has resulted in the degradation of farmland, which poses a great threat to food provision security, ecological environmental protection, ESs and regional sustainability [52–55]. In the past four decades, the ESV loss of farmland caused by urbanization was 5036.7 million yuan in the GBA, and the highest loss was 2177.5 million yuan from 2000–2010, followed by 1655.1 million yuan from 1990–2000. The change trend of the ESV loss of farmland is similar to the change trend of farmland lost because of urbanization, which is because the calculation of the ESV is based on area. In addition, our results indicate that ESs such as hydrological regulation, food production and gas regulation are particularly vulnerable to urbanization impacts (Figure 6c). As construction land expands, a large area of farmland is converted to impermeable surfaces, which causes rain and runoff to not penetrate the ground in time and participate in the natural water cycle, which further heavily affects hydrological regulation [56]. For food production, urbanization has multiple effects. Rapid urbanization in the GBA has led to an increase in the agricultural land use intensity, cropping frequency and chemical fertilizer use, a decline in the per capita availability of food grains, and soil and water pollution [10,22,57,58]. In this complex context, both food production and security are seriously influenced. The process of urbanization and industrialization inevitably causes environmental problems such as air pollution [59], which predominantly originates from industry, transport, power generation and construction [60]. Air pollution could lead to air quality deterioration and thus influence regional gas regulation [59,60].

4.3. Farmland Conservation and Ecosystem Services Protection

Governments have been striving to maintain a balance between urbanization development and farmland preservation [10,14], and the results of this study show that farmland loss due to urbanization sharply declined after 2010, but the results also indicate that construction land expansion has increasingly become the dominant reason for the loss of farmland in the GBA over the past 40 years. Therefore, it is still vital for governments to take more effective steps to regulate construction land expansion and to develop trade-offs and synergies among urban development, agricultural production and ecosystem preservation [14,39]. Based on the results of this study, we provide several recommendations for farmland preservation. First, to protect farmland and improve regional ESs, central and local governments should strictly control the area of new construction land and strengthen the construction of ecological cities and beautify countryside. In urban areas, the government could protect, plan and establish natural parks or green belts, including forests, grasses and wetlands. In rural areas, the government should promote reasonable planning for farmland use or exploitation. Governments can also 'green' abandoned industrial and mining land with forests, grasses or water bodies to improve ESs. Second, governments must strictly protect the red line of high-quality farmland and prohibit exploitation without rational reasons. Third, local governments should maintain the requisition–compensation balance of farmland, including the quantity, quality and ecological balance. For example, if the quality of compensatory farmland is lower than the quality of requisitioned farmland, then governments should invest more money to improve the quality of medium- and low-yield farmland or exploit new farmland rather than invest in urban construction [41]. Fourth, local governments should take measures to prevent environmental pollution, such as educating farmers to strictly control the use of agrochemicals and other soil additives to prevent soil pollution, and strictly prohibit the direct discharge of domestic sewage and industrial wastewater into rivers without treatment. Finally, the public should respond to government policies and protect the environment. For example, to protect the water and soil environment of farmland, enterprises and factories near farmland should strictly treat sewage according to regulations, and discharge the sewage only after it reaches the required standard. Farmers could reduce the use of pesticides and other chemicals to reduce the pollution of the farmland environment.

4.4. Limitations and Future Works

Estimates of ESV regionally and globally in monetary units play a critical role in heightening awareness and estimating the overall level of importance of ESs relative to and in combination with other contributors to sustainable human well-being [37]. Therefore, it is better for decision-makers and the public to consider the ESs as public goods or natural resources, and take these values into account when scenarios and policies are changed. However, this valuation method could not be used to examine the spatial changes in ESV. In addition, the ESV of construction land in this study was considered as zero, which is not appropriate in some degree and probably led to errors in the results. In our future work, we would like to combine the valuation methods and other evaluation methods to acquire more accurate results.

Farmland conservation and ecosystem services protection require cooperation in many aspects, including from governments, experts and the public. Therefore, it is necessary to build a system integrating geographic information system (GIS), spatial multicriteria evaluation (SME) and participatory GIS (PGIS) approaches, where decision-makers, experts and the public can participate and identify a range of ecosystem services [61–64]. Decision maker can easily collect and manage the results, which is useful for land use planning. That would be conducted in our future work.

5. Conclusions

This study explored the impacts of urbanization on farmland in the GBA from 1980 to 2018 based on multiple temporal land use datasets of the CNLUCC database. The ESV of farmland was also estimated by the valuation methods to study the ecosystem service changes caused by urbanization. Our results showed that the total area of farmland loss caused by urbanization was 3711.3 km^2 over the past 38 years, leading to a direct decline in total ESVs by 5036.7 million yuan. Paddy fields suffered much more losses than dry farmland because of urbanization. The expansion of construction land increasingly became the dominant reason for farmland loss in the GBA with the influence increasing from 39.3% to 96.6%. The value of hydrological regulation, food production and gas regulation showed much more decline. Guangzhou, Dongguan, Foshan, Shenzhen and Huizhou had the greatest total ESV loss of farmland caused by urbanization. The social development background in China and local governments' behavior played a vital role in the farmland conversion to construction land. To protect farmland and improve regional ESs, the central and local governments should strengthen the construction of ecological cities and beautify countryside by increasing capital investment, strengthening supervision, and raising public awareness of environmental protection, and the public should respond to land use policies and protect the environment.

Supplementary Materials: The following are available online at https://www.mdpi.com/article/10.3390/land10050501/s1, Figure S1: Conversion percentages of the different land use types into construction land in the GBA from 1980 to 2018, Table S1: ESV changes of the farmland in the GBA from 1980 to 2018 (million yuan).

Author Contributions: Conceptualization, X.W., F.Y. and F.S.; methodology, X.W. and F.Y.; formal analysis, X.W., M.C., B.H. and L.K.; writing—original draft preparation, X.W., Y.Z. and M.C.; writing—review and editing, X.W., F.Y., Y.Z. and F.S.; funding acquisition, X.W. and F.S. All authors have read and agreed to the published version of the manuscript.

Funding: This research was supported by the National Natural Science Foundation of China (41890854), the Program B for Outstanding Ph.D. Candidate of Nanjing University (No. 202002B087), and the China Scholarship Council (CSC) (No. 201906190120).

Institutional Review Board Statement: Not applicable.

Informed Consent Statement: Not applicable.

Data Availability Statement: Not applicable.

Acknowledgments: We appreciate critical and constructive comments and suggestions from the reviewers that helped improve the quality of this manuscript.

Conflicts of Interest: The authors declare no conflict of interest.

References

1. Desa, U.N. *World Urbanization Prospects: The 2014 Revision*; United Nations Department of Economics and Social Affairs, Population Division: New York, NY, USA, 2015.
2. Angel, S.; Parent, J.; Civco, D.L.; Blei, A.; Potere, D. The dimensions of global urban expansion: Estimates and projections for all countries, 2000–2050. *Prog. Plan.* **2011**, *75*, 53–107. [CrossRef]
3. Lin, M.; Lin, T.; Sun, C.; Jones, L.; Sui, J.; Zhao, Y.; Liu, J.; Xing, L.; Ye, H.; Zhang, G.; et al. Using the Eco-Erosion Index to assess regional ecological stress due to urbanization—A case study in the Yangtze River Delta urban agglomeration. *Ecol. Indic.* **2020**, *111*, 106028. [CrossRef]
4. Bai, X.; Shi, P.; Liu, Y. Society: Realizing China's urban dream. *Nature* **2014**, *509*, 158–160. [CrossRef] [PubMed]
5. Sun, Y.; Zhao, S. Spatiotemporal dynamics of urban expansion in 13 cities across the Jing-Jin-Ji Urban Agglomeration from 1978 to 2015. *Ecol. Indic.* **2018**, *87*, 302–313. [CrossRef]
6. Yang, C.; Li, Q.; Zhao, T.; Liu, H.; Gao, W.; Shi, T.; Guan, M.; Wu, G. Detecting Spatiotemporal Features and Rationalities of Urban Expansions within the Guangdong–Hong Kong–Macau Greater Bay Area of China from 1987 to 2017 Using Time-Series Landsat Images and Socioeconomic Data. *Remote Sens.* **2019**, *11*, 2215. [CrossRef]
7. Yang, X.J. China's Rapid Urbanization. *Science* **2013**, *342*, 310. [CrossRef] [PubMed]
8. Xiao, R.; Lin, M.; Fei, X.; Li, Y.; Zhang, Z.; Meng, Q. Exploring the interactive coercing relationship between urbanization and ecosystem service value in the Shanghai-Hangzhou Bay Metropolitan Region. *J. Clean. Prod.* **2020**, *253*, 119803. [CrossRef]
9. Wang, X.; Yan, F.; Su, F. Impacts of Urbanization on the Ecosystem Services in the Guangdong-Hong Kong-Macao Greater Bay Area, China. *Remote Sens.* **2020**, *12*, 3269. [CrossRef]

10. Jiang, L.; Deng, X.; Seto, K.C. The impact of urban expansion on agricultural land use intensity in China. *Land Use Policy* **2013**, *35*, 33–39. [CrossRef]

11. Mao, D.; Wang, Z.; Wu, J.; Wu, B.; Zeng, Y.; Song, K.; Yi, K.; Luo, L. China's wetlands loss to urban expansion. *Land Degrad. Dev.* **2018**, *29*, 2644–2657. [CrossRef]

12. Narducci, J.; Quintas-Soriano, C.; Castro, A.; Som-Castellano, R.; Brandt, J.S. Implications of urban growth and farmland loss for ecosystem services in the western United States. *Land Use Policy* **2019**, *86*, 1–11. [CrossRef]

13. Jones, N.; de Graaff, J.; Duarte, F.; Rodrigo, I.; Poortinga, A. Farming systems in two less favoured areas in Portugal: Their development from 1989 to 2009 and the implications for sustainable land management. *Land Degrad. Dev.* **2014**, *25*, 29–44. [CrossRef]

14. Huang, Z.; Du, X.; Castillo, C.S.Z. How does urbanization affect farmland protection? Evidence from China. *Resour. Conserv. Recycl.* **2019**, *145*, 139–147. [CrossRef]

15. Tu, Y.; Chen, B.; Yu, L.; Xin, Q.; Gong, P.; Xu, B. How does urban expansion interact with cropland loss? A comparison of 14 Chinese cities from 1980 to 2015. *Landsc. Ecol.* **2020**. [CrossRef]

16. Song, W.; Pijanowski, B.C.; Tayyebi, A. Urban expansion and its consumption of high-quality farmland in Beijing, China. *Ecol. Indic.* **2015**, *54*, 60–70. [CrossRef]

17. Chapin Iii, F.S.; Zavaleta, E.S.; Eviner, V.T.; Naylor, R.L.; Vitousek, P.M.; Reynolds, H.L.; Hooper, D.U.; Lavorel, S.; Sala, O.E.; Hobbie, S.E.; et al. Consequences of changing biodiversity. *Nature* **2000**, *405*, 234–242. [CrossRef]

18. Butler, S.J.; Vickery, J.A.; Norris, K. Farmland Biodiversity and the Footprint of Agriculture. *Science* **2007**, *315*, 381. [CrossRef]

19. Tilman, D.; Fargione, J.; Wolff, B.; Antonio, C.; Dobson, A.; Howarth, R.; Schindler, D.; Schlesinger, W.H.; Simberloff, D.; Swackhamer, D. Forecasting Agriculturally Driven Global Environmental Change. *Science* **2001**, *292*, 281. [CrossRef] [PubMed]

20. Gren, Å.; Andersson, E. Being efficient and green by rethinking the urban-rural divide—Combining urban expansion and food production by integrating an ecosystem service perspective into urban planning. *Sustain. Cities Soc.* **2018**, *40*, 75–82. [CrossRef]

21. Zhang, W.; Wang, W.; Li, X.; Ye, F. Economic development and farmland protection: An assessment of rewarded land conversion quotas trading in Zhejiang, China. *Land Use Policy* **2014**, *38*, 467–476. [CrossRef]

22. Pandey, B.; Seto, K.C. Urbanization and agricultural land loss in India: Comparing satellite estimates with census data. *J. Environ. Manag.* **2015**, *148*, 53–66. [CrossRef] [PubMed]

23. Sutton, M.A.; Oenema, O.; Erisman, J.W.; Leip, A.; van Grinsven, H.; Winiwarter, W. Too much of a good thing. *Nature* **2011**, *472*, 159–161. [CrossRef]

24. Costanza, R.; d'Arge, R.; de Groot, R.; Farber, S.; Grasso, M.; Hannon, B.; Limburg, K.; Naeem, S.; O'Neill, R.V.; Paruelo, J.; et al. The value of the world's ecosystem services and natural capital. *Nature* **1997**, *387*, 253–260. [CrossRef]

25. Xie, G.; Zhen, L.; Lu, C.; Xiao, Y.; Chen, C. Expert Knowledge Based Valuation Method of Ecosystem Services in China. *J. Nat. Resour.* **2008**, *23*, 911–919. [CrossRef]

26. Xie, G.; Zhang, C.; Zhang, L.; Chen, W.; Li, S. Improvement of the Evaluation Method for Ecosystem Service Value Based on Per Unit Area. *J. Nat. Resour.* **2015**, *30*, 1243–1254. [CrossRef]

27. Xu, L.; Xu, X.; Luo, T.; Zhu, G.; Ma, Z. Services based on land use: A case study of Bohai Rim. *Geogr. Res.* **2012**, *31*, 1775–1784.

28. Xiao, H.; Li, H.; Wang, L.; Chen, J.; Han, Y. Changes of Land Use and Ecosystem Service Value in the Guangdong-Hong Kong-Macao Greater Bay Area-A Case Study of Guangdong-Foshan-Zhaoqing. *Res. Soil. Water. Conserv.* **2020**, *27*, 290–297. [CrossRef]

29. Kang, L.; Ma, L.; Liu, Y. Evaluation of farmland losses from sea level rise and storm surges in the Pearl River Delta region under global climate change. *J. Geog. Sci.* **2016**, *26*, 439–456. [CrossRef]

30. Deuskar, C.; Baker, J.L.; Mason, D. *East Asia's Changing Urban Landscape: Measuring a Decade of Spatial Growth*; World Bank Publications: Washington, DC, USA, 2015.

31. Seto, K.C.; Güneralp, B.; Hutyra, L.R. Global forecasts of urban expansion to 2030 and direct impacts on biodiversity and carbon pools. *Proc. Natl. Acad. Sci. USA* **2012**, *109*, 16083. [CrossRef]

32. Wu, W.; Zhao, S.; Zhu, C.; Jiang, J. A comparative study of urban expansion in Beijing, Tianjin and Shijiazhuang over the past three decades. *Landsc. Urban Plan.* **2015**, *134*, 93–106. [CrossRef]

33. Zhang, J.; Li, H.; Zhou, Y.; Dou, L.; Cai, L.; Mo, L.; You, J. Bioavailability and soil-to-crop transfer of heavy metals in farmland soils: A case study in the Pearl River Delta, South China. *Environ. Pollut.* **2018**, *235*, 710–719. [CrossRef] [PubMed]

34. Yang, C.; Li, Q.; Hu, Z.; Chen, J.; Shi, T.; Ding, K.; Wu, G. Spatiotemporal evolution of urban agglomerations in four major bay areas of US, China and Japan from 1987 to 2017: Evidence from remote sensing images. *Sci. Total Environ.* **2019**, *671*, 232–247. [CrossRef] [PubMed]

35. Xu, X.; Liu, J.; Zhang, S.; Li, R.; Yan, C.; Wu, S. *China's Multi-Period Land Use Land Cover Remote Sensing Monitoring Data Set (CNLUCC)*; Resource and Environment Data Science and Data Center, Institute of Geographic Sciences and Natural Resources Research, Chinese Academy of Sciences: Beijing, China, 2018. [CrossRef]

36. Xie, G.; Li, W.; Xiao, Y.; Zhang, B.; Lu, C.; An, K.; Wang, J.; Xu, K.; Wang, J. Forest ecosystem services and their values in Beijing. *Chin. Geogr. Sci.* **2010**, *20*, 51–58. [CrossRef]

37. Costanza, R.; de Groot, R.; Sutton, P.; van der Ploeg, S.; Anderson, S.J.; Kubiszewski, I.; Farber, S.; Turner, R.K. Changes in the global value of ecosystem services. *Glob. Environ. Chang.* **2014**, *26*, 152–158. [CrossRef]

38. Yao, Z.; Zhang, L.; Tang, S.; Li, X.; Hao, T. The basic characteristics and spatial patterns of global cultivated land change since the 1980s. *J. Geog. Sci.* **2017**, *27*, 771–785. [CrossRef]
39. Wang, J.; Lin, Y.; Glendinning, A.; Xu, Y. Land-use changes and land policies evolution in China's urbanization processes. *Land Use Policy* **2018**, *75*, 375–387. [CrossRef]
40. Zhou, C.; Wang, Y.; Xu, Q.; Li, S. The new process of urbanization in the Pearl River Delta. *Geogr. Res.* **2019**, *38*, 45–63.
41. Liu, L.; Liu, Z.; Gong, J.; Wang, L.; Hu, Y. Quantifying the amount, heterogeneity, and pattern of farmland: Implications for China's requisition-compensation balance of farmland policy. *Land Use Policy* **2019**, *81*, 256–266. [CrossRef]
42. Cheng, L.; Jiang, P.; Chen, W.; Li, M.; Wang, L.; Gong, Y.; Pian, Y.; Xia, N.; Duan, Y.; Huang, Q. Farmland protection policies and rapid urbanization in China: A case study for Changzhou City. *Land Use Policy* **2015**, *48*, 552–566. [CrossRef]
43. Mishra, V. Population growth and intensification of land use in India. *Int. J. Popul. Geogr.* **2002**, *8*, 365–383. [CrossRef]
44. Reddy, V.R.; Reddy, B.S. Land alienation and local communities: Case studies in Hyderabad- Secunderabad. *Econ. Political Wkly.* **2007**, *42*, 3233–3240.
45. Chien, S. Local farmland loss and preservation in China—A perspective of quota territorialization. *Land Use Policy* **2015**, *49*, 65–74. [CrossRef]
46. Tian, L. Land use dynamics driven by rural industrialization and land finance in the peri-urban areas of China: "The examples of Jiangyin and Shunde". *Land Use Policy* **2015**, *45*, 117–127. [CrossRef]
47. Huang, Z.; Cao, J. Ergodicity and bifurcations for stochastic logistic equation with non-Gaussian Lévy noise. *Appl. Math. Comput.* **2018**, *330*, 1–10. [CrossRef]
48. Huang, Z.; Du, X. Government intervention and land misallocation: Evidence from China. *Cities* **2017**, *60*, 323–332. [CrossRef]
49. Cao, G.; Feng, C.; Tao, R. Local "Land Finance" in China's Urban Expansion: Challenges and Solutions. *China World Econ.* **2008**, *16*, 19–30. [CrossRef]
50. Cheng, X.; Zheng, H.; Duan, W.; Hao, X.; Duan, W. Fault restoration based on the path analysis for distribution grid. In Proceedings of the 2015 IEEE International Conference on Information and Automation, Beijing, China, 8–10 August 2015; pp. 552–556.
51. Tao, R.; Su, F.; Liu, M.; Cao, G. Land leasing and local public finance in China's regional development: Evidence from prefecture-level cities. *Urban Stud.* **2010**, *47*, 2217–2236. [CrossRef]
52. Jiang, L.; Deng, X.; Seto, K.C. Multi-level modeling of urban expansion and cultivated land conversion for urban hotspot counties in China. *Landsc. Urban Plan.* **2012**, *108*, 131–139. [CrossRef]
53. Musinguzi, P.; Ebanyat, P.; Tenywa, J.; Basamba, T.; Tenywa, M.; Mubiru, D. Precision of farmerbased fertility ratings and soil organic carbon for crop production on a Ferralsol. *Solid Earth* **2015**, *6*, 1063–1073. [CrossRef]
54. Kong, X. China must protect high-quality arable land. *Nature* **2014**, *506*, 7. [CrossRef]
55. Su, S.; Jiang, Z.; Zhang, Q.; Zhang, Y. Transformation of agricultural landscapes under rapid urbanization: A threat to sustainability in Hang-Jia-Hu region, China. *Appl. Geogr.* **2011**, *31*, 439–449. [CrossRef]
56. Bai, X.; McPhearson, T.; Cleugh, H.; Nagendra, H.; Tong, X.; Zhu, T.; Zhu, Y.-G. Linking Urbanization and the Environment: Conceptual and Empirical Advances. *Annu. Rev. Environ. Resour.* **2017**, *42*, 215–240. [CrossRef]
57. Boserup, E. *The Conditions of Agricultural Growth the Economics of Agrarian Change under Population Pressure*; Routledge: London, UK, 1965.
58. Hu, Y.; Liu, X.; Bai, J.; Shih, K.; Zeng, E.Y.; Cheng, H. Assessing heavy metal pollution in the surface soils of a region that had undergone three decades of intense industrialization and urbanization. *Environ. Sci. Pollut. Res.* **2013**, *20*, 6150–6159. [CrossRef] [PubMed]
59. Deshpande, A.; Mishra, P. Urbanization, air pollution and human health. *J. Environ. Res. Dev* **2007**, *1*, 13.
60. Xu, G.; Jiao, L.; Zhao, S.; Yuan, M.; Li, X.; Han, Y.; Zhang, B.; Dong, T. Examining the impacts of land use on air quality from a spatio-temporal perspective in Wuhan, China. *Atmosphere* **2016**, *7*, 62. [CrossRef]
61. Modica, G.; Zoccali, P.; Di Fazio, S. The e-Participation in Tranquillity Areas Identification as a Key Factor for Sustainable Landscape Planning. In *Computational Science and Its Applications—ICCSA 2013*; Springer: Berlin, Germany, 27 June 2013; pp. 550–565. [CrossRef]
62. Modica, G.; Pollino, M.; La Porta, L.; Di Fazio, S. Proposal of a Web-Based Multi-criteria Spatial Decision Support System (MC-SDSS) for Agriculture. In *Innovative Biosystems Engineering for Sustainable Agriculture, Forestry and Food Production*; Springer Nature: Cham, Switzerland, 2020; pp. 333–341. [CrossRef]
63. Jelokhani-Niaraki, M. Collaborative spatial multicriteria evaluation: A review and directions for future research. *Int. J. Geog. Inf. Sci.* **2020**, *35*, 9–42. [CrossRef]
64. Brown, G.; Fagerholm, N. Empirical PPGIS/PGIS mapping of ecosystem services: A review and evaluation. *Ecosyst. Serv.* **2015**, *13*, 119–133. [CrossRef]

land

Article

Applying the Evaluation of Cultural Ecosystem Services in Landscape Architecture Design: Challenges and Opportunities

Xin Cheng [1,*], Sylvie Van Damme [2] and Pieter Uyttenhove [3]

1   Department of Urban Planning and Landscape Architecture, Xihua University, Chengdu 610039, China
2   School of Arts, University College Ghent, 9000 Ghent, Belgium; sylvie.vandamme@hogent.be
3   Department of Architecture and Urban Planning, Ghent University, 9000 Ghent, Belgium; pieter.uyttenhove@ugent.be
*   Correspondence: xin.cheng@mail.xhu.edu.cn; Tel.: +86-159-0814-2467

Abstract: Landscape architects play a significant role in safeguarding urban landscapes and human well-being by means of design and they call for practical knowledge, skills, and methods to address increasing environmental pressure. Cultural ecosystem services (CES) are recognized as highly related to landscape architecture (LA) studies, and the outcomes of CES evaluations have the potential to support LA practice. However, few efforts have focused on systematically investigating CES in LA studies. Additionally, how CES evaluations are performed in LA studies is rarely researched. This study aims to identify the challenges and provide recommendations for applying CES evaluations to LA practice, focusing specifically on LA design. To conclude, three challenges are identified, namely a lack of consistent concepts (conceptual challenge); a lack of CES evaluation methods to inform designs (methodological challenge); and practical issues of transferring CES evaluations to LA design (practical challenge). Based on our findings, we highlight using CES as a common term to refer to socio-cultural values and encourage more CES evaluation methods to be developed and tested for LA design. In addition, we encourage more studies to explore the links of CES and landscape features and address other practical issues to better transfer CES evaluations onto LA designs.

Keywords: landscape architecture; cultural ecosystem services; design; evaluation

Citation: Cheng, X.; Van Damme, S.; Uyttenhove, P. Applying the Evaluation of Cultural Ecosystem Services in Landscape Architecture Design: Challenges and Opportunities. *Land* **2021**, *10*, 665. https://doi.org/10.3390/land10070665

Academic Editors: Alessio Russo and Giuseppe T. Cirella

Received: 18 May 2021
Accepted: 23 June 2021
Published: 24 June 2021

Publisher's Note: MDPI stays neutral with regard to jurisdictional claims in published maps and institutional affiliations.

1. Introduction

Cities experience increasing environmental stress given the sharp rise of city populations, and the consequent loss and degradation of urban green spaces. Landscape architects therefore play a significant role in maintaining the human environment and well-being by planning, design, and management of urban landscapes more than ever [1]. The Council of Europe [2] defines landscape as an area that is perceived by people, and the character is the result of the action and interaction of natural or human factors. Landscape architecture (LA), as a profession, provides site planning, design, and management advice to improve the character, quality, and experience of the landscape [3]. Design is the core activity of the LA practice, considering many factors, such as the landscape itself, the intentions of the clients, the interaction of the users and the design setting, the materials, and the processes of creative expression. To do this effectively, landscape architects apply solid design processes and implementation methods, elaborated on and explained in an increasing amount of studies and publications on those topics. For example, Langley et al. [4] and Yue and Shao [5] summarize the core knowledge domains of LA, including the evolution of analytical methods for LA planning and design. Milburn and Brown [6] and Lenzholzer et al. [7] focus on incorporating research into the LA design process. Some scholars focus on new tools, such as Li and Milburn [8] and Gu et al. [9], who work on geodesign as a design tool.

LA primarily aims to organize the complexity of the landscape into comprehensible, productive, and beautiful places to improve the function, health, and experience of life. However, landscape architects increasingly face new challenges within today's rapid urbanization, such as designing resilient landscapes for adjusting to a changing climate and controlling natural disasters, creating a sense of place, and managing the growth of informal settlement [10,11]. Meanwhile, clients and the general public are no longer only interested in the aesthetic value of landscape but also increasingly concerned about ecological functions and environment conservation [12]. To achieve these multiple, often-competing objectives, landscape architects need a clear understanding of the relationships between humans and the environment and the interaction processes on how humans shape the landscape they rely on and experience. Hence, there is an urgent call for practical knowledge, skills, and methods to address these challenges within the LA profession [12].

The concept of ecosystem services (ES) refers to the benefits that people obtain from ecosystems [13]. They highlight the contributions that environments and landscapes provide for society and the economy and, more generally, for human well-being. They are divided into provisioning services, regulating services, supporting services, and cultural services [13]. In the last decade, several researchers conducted studies on integrating ES evaluation into spatial planning policies such as land use/cover to foster sustainable development. More recently, the relationship between ES and urban landscapes has also been increasingly investigated. The outcomes of ES evaluations are recognized to have the potential to support practical application and decision-making [14–19]. Among ES, cultural ecosystem services (CES), namely the non-material benefits that people get from ecosystems, are regarded as directly influencing human-wellbeing and having the potential to motivate people's willingness to protect urban greens [14,20]. Whether or not people are familiar with the term, the concept resonates with nearly every human being, such as the experience of recreational activities and scenery appreciations. Although different definitions have been developed so far, CES are generally defined as the non-material benefits people obtain from ecosystems [13], which was adopted by this study. CES evaluation is one of the biggest challenges for further applying in practice. Evaluating what people obtain is a concise way of disseminating the importance of ecosystems or landscapes [14]. Various evaluation methods have been developed and they can generally be classified into monetary methods and non-monetary methods, depending on whether the evaluation outcomes are expressed by money or not [14]. The classification can be further divided into revealed preference and stated preference methods. The revealed preference method means observing the actual markets or human behaviors related to the CES to assess CES. The stated preference method means directly asking about one's values to assess CES [14]. CES evaluation studies and LA studies share many commonalities. First, they focus on the "human dimension" of landscapes [21]. For example, LA has long been concerned with questions of the human perception of and involvement with nature. Meanwhile, CES evaluations heavily depend on people's perceptions and preferences. Second, both have a broad knowledge about the assessment of aesthetic values, place identity, and cultural heritage on-site and in specific contexts [21]. Landscape architects may not be familiar with ES or CES; however, they are familiar with concepts such as aesthetics or heritage [22]. Third, both highlight the local scale of different landscape types, especially while studying urban green spaces like urban parks, gardens, and forests, and, fourth, both are concerned with the benefits provided by landscape features that can enhance the quality of the landscapes and the entire system [23]. For example, parks are recognized as providing a range of CES, and, in many areas, these relate to scientifically-assessed landscape features, such as the proportion of vegetation [24]. Similarly, De Valck et al. [25] test the different degrees of high or low vegetation that indicate outdoor recreation.

CES are highly related to LA studies, and the outcomes of CES evaluations have the potential to guide LA practices. Although they share many commonalities, few efforts have focused on systematically investigating CES in LA studies. Additionally, how CES evaluations are performed in LA studies is rarely investigated. This paper aims to identify the challenges and opportunities of integrating CES evaluations into LA designs. Specifically, it discusses the problems and challenges of existing studies, the examined CES categories, the implications and applications of evaluation methods, the links between CES and landscape features, and opportunities for future study.

With this in mind, we specifically ask which CES are studied, what evaluation methods are used, and how CES are connected to landscape features in LA studies. To achieve this, we focus on three areas:

1. First, we focus on the existing publications about CES within LA studies through bibliographic research. Looking at the existing publications from a scientific database helps with understanding the status quo of the CES in LA studies. This approach allows for investigating a topic comprehensively and covering a large timespan [26].

2. Second, we focus on how landscape architects express their ideas in design proposals related to CES through a systematic review of design proposals. Design proposals differ in scale and complexity. They involve various constituent elements for a variety of purposes and scales. They are often used to present design outcomes and communicate with clients. Focusing on the design proposals helps with understanding why CES are important, and how the abstract CES are expressed by the physical landscape features, according to designers.

3. Third, we focus on how landscape architects state their ideas and values related to CES through interviews. Not all of the information is shown in design proposals. The design process is complex, and the mechanisms behind the designs are unknown. Moreover, more practical issues can only be revealed by interviewing the designers as dealing with those issues is one of their daily tasks. Hence, we interviewed landscape architects directly to reveal more detailed information that might be missed, to reveal the real thoughts of designers and to get the primary data for further analysis.

Combining these three methods provided a comprehensive knowledge of CES evaluation in LA fields. In addition, this study focused on urban parks in China. As an important component of urban green spaces, urban parks are crucial for securing human well-being by providing diverse benefits, and they have been studied by ES and LA fields. Urban parks are important indicators used to estimate the quality of life in a city. Moreover, studying urban parks is vitally important for their design, planning, and management, especially in the case of China, which is experiencing increasing social and environmental pressure caused by the rapid population growth in cities. Hence, it is an urgent call for valuable tools to aid urban park design.

2. Methods

We combined three methods to investigate the challenges and opportunities of applying CES evaluations in LA studies. Table 1 shows the study process. Specifically, we first reviewed existing scientific articles to have a general overview of the research on urban parks and identify items related to CES and the links between these terms. Then, we conducted a systematic review of selected park proposals based on a guideline (Table A1 in Appendix A). Subsequently, we interviewed landscape architects directly to know their thoughts about CES evaluations and other practical issues (Table A2).

Table 1. Overview of the study process.

Study Process	Data Collection	Data Analysis
(1) Review journal articles, dissertations, and conference papers from CNKI database	5273 publications	Quantitative
(2) Systematic review of park proposals	83 proposals	Quantitative
(3) Interviews of landscape architects	12 interviewees	Qualitative

CNKI is a key national information and knowledge database that includes journals, doctoral dissertations, master's theses, proceedings, newspapers, yearbooks, statistical yearbooks, patents, standards, etc. in China.

2.1. Review of Keywords and Abstract of Scientific Publications

To review the existing publications about CES within LA studies, we used key terms and phrases based on reviewing the abstracts and keywords of scientific publications. This technique can be used to quantify a large sample [26,27]. We first searched "landscape architecture design*" and "urban park*" in the China National Knowledge Infrastructure (CNKI) database, including journals, dissertations, and conference papers. We set the timespan up to November 2019, which resulted in 5273 items. Then, we screened the key terms using CiteSpace.5.5.R2 (parameter settings: Years Per Slice: 2; Node Types: Keyword). The frequency of the keywords could reflect hot topics and core content over a set period [28]. Subsequently, we counted the occurrence frequency of key terms in relation to CES. There are different definitions and classifications of CES, and different authors use different terms based on their study focus. Here, we recorded the terms that the authors reported. Finally, other terms were also presented and counted if they were linked to CES terms.

2.2. Review of Design Proposals

A systematic review was conducted to review the design proposals of urban parks in China. More specifically, we first collected park proposals from LA companies, government agents, and internet searches. A balance of park types and scales were also noticed, and, finally obtained, 83 items were obtained. Second, we asked a set of questions when we reviewed each park proposal and recorded the answers (Table A1), including the time, park scales, park types, CES types, evaluation methods, and the relationships with landscape features. More specifically, we set the classification of the answers of identified proposals as shown in Table 2. There are several different classification mechanisms of urban parks, and it is beyond the scope of this study to give a comprehensive account of the various classifications. In this study, we based on the Standard for Classification of Urban Green Space of China.

Table 2. Review of design proposals.

Question Categories	Answer Descriptions
(1) Time	The year the design proposal came out.
(2) Scale	The size of the park.
(3) Type	The urban park was classified into four major categories: comprehensive park, community park, topic park (children's park, the cultural relics park, the commemorative park, zoological garden, sports park, etc.), and belt-shaped park. Among them, comprehensive parks include a large area of green land and numerous public service facilities, which are the main locations for residents to recreate, and a significant public open space [29].
(4) CES types	We identified CES types by focusing on the non-material characteristics, meaning there was no pre-set classification before reviewing each proposal. We recorded the original terms that authors reported in proposals and similar terms were grouped.
(5) Evaluation of CES methods	We recorded if the designers introduced CES evaluation methods in their design proposals.
(6) The relationships with other landscape elements/features (vegetation, benches, etc.)	The classifications of characteristics were selected from the literature that focused on the park characteristics, including Bertram and Rehdanz [30], Campbell et al. [31], Chiesura [32], and Hegetschweiler et al. [33]. Based on this classification, we recorded and grouped the terms that the authors reported in their designs. For the detailed types, see Table A1.

2.3. Interviews of Landscape Architects

To reveal more detailed information from designers, 12 semi-structured phone interviews were conducted between November, 2019 and January, 2020. The interviewees were landscape architects with rich experience in park design in China and consisted of three groups. The first group consisted of designers in the government's LA design institute. The second group was made up of professors working at the university who were also doing LA projects, and the third group contained landscape architects from LA design companies. The interview started with a brief description of this study. Subsequently, a set of questions was asked based on guidelines that directed towards our research questions (Table A2). The interviewees were prompted with a talk-generating question, and the structure of the interview was adjusted to their statements. Each interview was conducted over 30 to 90 min, depending on the interviewee's interests and schedule. In addition, the results were qualitatively analyzed according to the research goals. For the presentation of our results, quotations were translated from Chinese to English. The interview texts were summarized when interviewees shared the same ideas. The individual ideas were recorded independently.

3. Results

3.1. CES in Scientific Publications

This section presents the high-frequency keywords assigned by the CNKI database. A minimum of two occurrences of each keyword was included in the network, resulting in 208 keywords and 1266 links. Notably, we merged similar terms, because the original language was Chinese, and some terms share the same meaning, and the final results contained 30 keywords (Table 3). Specifically, the results revealed that regional/local culture (frequency = 212), humanization/user-friendly (85), and cultural characteristic (70) were mentioned far more than other keywords. Other high-frequency keywords included historical culture (16), urban culture (9), theme culture (8), leisure and recreation (8), and place spirit (5). All of these keywords served as a reference point for finding and understanding the possibilities of applying CES in the landscape design of urban parks.

Table 3. List of keywords in relation to CES.

Keyword	Frequency	Keyword	Frequency
Regional/local culture	212	Expression	4
Humanization-/user-friendly	85	Experience	4
Culture characteristic	70	Art	4
Historical culture	16	Creative	3
Urban culture	9	Nature education	3
Theme culture	8	Urban characteristic	2
Leisure and recreation	8	Culture heritage	2
Place spirit	5	Social life	2
Ecological aesthetic	4	Tradition	2

In addition, the intensity of the link between two keywords was computed based on the number of times they were mentioned together. Sixty-three keywords were found linked to the CES keywords (Table A3). The main connecting keywords were landscape architecture design (23), landscape architecture (23), urban park (18), theme park (11), and ecology (8). The first three keywords were expected as they were our search terms. Other highly linked keywords included design (7), plant design (7), local culture (7), urban culture (6), historical culture (6), waterfront landscape (6), landscape renovation (5), humanization/user-friendly (5), urban wetland park (5), and comprehensive park (5). Other items, such as sculpture, environmental infrastructure, and plant community, were related to landscape features. There were surprisingly no terms directly about evaluation or evaluation methods.

3.2. CES in Park Proposal

The sizes of the 83 reviewed parks ranged from 2.5 ha to 1500 ha between 2003 and 2019. The parks included 31 comprehensive parks, 37 topic parks, 7 community parks, and 8 belt-shaped parks (Figure 1). Topic parks included the wetland park (6), the sports park (6), the expo park (5), the children's park (3), the water park (3), the amusement park (2), the botanical garden (2), the shopping park (2), the ecological park (2), the cultural park(2), the lake park (1), the cultural relics park (1), the commemorative park (1), and the zoological garden (1).

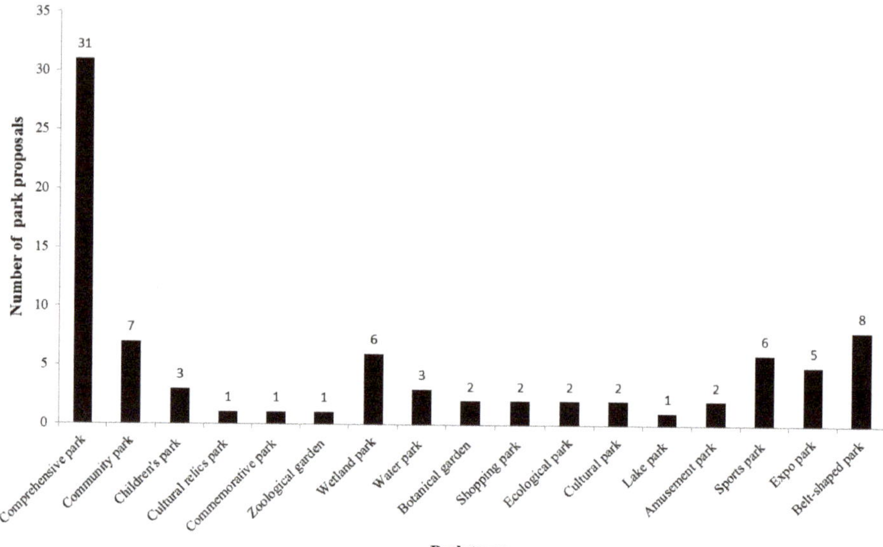

Figure 1. The types of the park.

The concept of CES was not found in park proposals, but 50 terms were identified in relation to CES (Table 4). The most mentioned types were recreation (73), education (26), and aesthetic (13), followed by traditional culture (12), sense of place (11), social interaction (10), history (8), Shan-Shui culture (8), sense of nature (7), art (7), tourism (6), and experience (6). Recreation included walking, running, boating, fishing, biking, bird watching, tea dinking, flower watching, farming, bird watching, fishing, skating, etc.

The majority of proposals (74) mentioned that they used a document research approach through analyzing materials such as landform, existing information and data, and master plan. Among them, 56 proposals analyzed successful case studies in China and other countries. Twelve proposals stated that they conducted field surveys, and four used questionnaires to ask people how they perceived the existing parks and what they liked.

Landscape features were wildly used to link CES to designs, and 26 features were found in proposals (Table A4). Plants (41) were the most mentioned feature in park proposals, followed by architecture (28), playground (26), sculpture (21), and sports-/fitness facilities (20). Other features, such as signs (19), furniture (12), recreation facilities (10), path (9), water (8), and landform (7), were also mentioned often.

3.3. CES in the View of Landscape Architects

Three interviewees who were also involved in landscape education in a university stated that they knew the concept of CES, but they had not used it in their designs. The other interviewees stated this was the first time to hear about this concept. However, the landscape architects were familiar with related terms, such as socio-cultural value. They stated that it is essential that parks are designed for recreation, aesthetics, and education purposes. They agreed on the importance of socio-cultural values and non-material benefits in LA designs. However, they also stated that it is difficult to be aware of them and capture them because of their abstract and intangible characteristics, which often depend on designer's experience, values, and preferences.

Table 4. CES categories in park proposals.

CES Category	Number	CES Category	Number
Ethnic culture	1	Spiritual	4
Education	26	knowledge	1
Recreation	73	Tourism	6
Spiritual and religious value	3	Shopping experience	1
Cultural diversity	3	Identity	3
Traditional culture	12	Respect of nature	1
History	8	Love	2
Sense of nature	7	Traditional philosophy	1
Life style	1	Peace	1
Local culture	3	Agricultural culture	3
Experience	6	Cultural interaction	1
Sense of place	11	Fairy tale	1
Social interaction	10	Tranquil	1
Social equity	1	Human friendly	1
Folk culture	5	Historical celebrities	1
Memory	1	Local identity	1
Aesthetic	13	Commerce culture	1
Art	7	Local connection	1
Patriotism	1	Industry culture	1
Cultural heritage	2	Surname culture	1
Celebrity culture	1	Minority culture	1
Food culture	3	Ancient drama culture	1
Social relations	5	Ancient architecture culture	1
Shan-Shui culture	8	Adventure	1
Health	2	Technological culture	1

On the one hand, landscape architects recognized the huge potential of CES evaluations to support their practices. They claimed that it may help them capture cultural benefits and better understand human and environmental processes and further integrate them into the ES framework to achieve multifunctional landscapes, not single-function landscapes. Designers stated that it is difficult to find an overarching value that would allow aesthetic, social, economic, and environmental values to be measured against one another [34]. Moreover, it should be a useful tool to communicate with clients and stakeholders by showing them potential benefits. One designer stated, "It's very important for communicating with clients who make the final choice. Therefore, you're actually selling your ideas to them and somehow persuade them to trust your designs. So more supporting proofs are needed to strengthen your design plans."

However, designers also stated that there are still many issues to be addressed. The biggest challenges are the methodological and practical issues. For example, the interviewees asked series of questions: "when do I need to evaluate CES and in which stages (before design, during design, or after design)?" "What methods do I need, and how?" "What kind of expertise is needed for this?" "Do I need CES experts to aid me?" "Who is involved in this evaluation, and how can it be facilitated?" "What is the difference when dealing with different parks of different scales? For example, the design of children parks is totally different from the zoological garden." "Does it influence the designer's creative expression or artistry?"

When asked how they transfer the CES to their designs, designers stated that recreation, aesthetics, or educational values are easier to express. For example, recreational values are often expressed by creating spaces for recreation and equipping them with recreational facilities. As for the other more abstract CES, designers stated that they are often inspired by history, a story, a culture, or a celebrity, and they often use physical landscape features to create the cultural atmosphere. For example, Shan-Shui culture was highlighted by interviewees; one stated, "Shan-Shui refers to mountains and rivers and broadly refers to nature in general, which is an important source of inspiration for creating places." Another example is the use of plants: "Bamboo, for example, is often used in designs to emphasize the elegance of the environment because, in Chinese culture, it is a symbol representing the character of moral integrity, modesty, and loyalty." Sculptures are also widely used to commemorate a celebrity or historical event or person. Designers also stated that the process is complex and depends on sites and specific contexts.

4. Discussion: Challenges and Opportunities

This study focuses on which CES are studied, what evaluation methods are used, and how the concepts connected to physical landscape featured in LA studies. In this section, we highlight three key challenges based on our findings and further discuss opportunities for landscape architects to contribute to apply CES evaluations into LA studies in the future.

4.1. Conceptual Challenges: A Lack of Consistent Concepts

Although the term CES was not been used, various socio-cultural values were found in the LA publications and design proposals. Local culture and historical culture were widely highlighted, benefiting from the fruitful history and culture of China. Landscape architects did not use the term CES because the concept of CES is relatively new, proposed in recent decades. Chinese scholars seemed to be unaware of this progress until recent years, and the number of studies focusing particularly on CES is still very limited [35]. In addition, some interviewees stated that they are unaware of the latest research trends or methods because they do not read international research publications as their English is poor. Another landscape architect stated that both CES and other terms he often uses such as historical culture or Shan-Shui culture all refer to the social-cultural values, so it does not matter whether he uses the term CES or not. The landscape architects also highlighted that the key is how to evaluate them and use them to guide the LA design practice.

Although CES have a range of definitions based on diverging views, leading to alternative classification schemes, they are generally described as socio-cultural values or non-material benefits. There are several mainstream classifications of CES in ES studies, such as the Millennium Ecosystem Assessment classification, Economics of Ecosystems and Biodiversity, Common International Classification of Ecosystem Services, Final Ecosystem Goods and Services Classification System, and, more recently, the Intergovernmental Platform on Biodiversity and Ecosystem Services. Diverse classifications were developed to clarify CES, and they are still in progress [36,37]. However, a consistent concept is necessary as this gives an opportunity to organize a dialogue and cooperation between LA and CES studies [21]. CES have the potential to be a common term referring to socio-cultural benefits. The results show that many CES are referenced and studied in LA research and design proposals but not by explicitly using the term CES. Hence, an opportunity is missed by not highlighting the link between the environment and humans that implicitly exists within the research and proposals. Given that this anthropocentric rationale can be effective at generating support for environmental conservation [38], the explicit inclusion of the CES concept may also lead to better ecosystem protection in general [39]. Moreover, practitioners who lack basic knowledge about CES concepts may be less aware of what could benefit from their designs. Hence, the inclusion of a common term, concept, or definition for CES in the LA studies and design proposals and other similar documents is needed to promote LA practice. Furthermore, the existing CES evaluation studies by many other countries and their various methods and indicators can inspire LA studies in China. However, a concern about the use of CES evaluations by designers is that the application of CES as a scientific concept might cause a loss of creativity and artistic thinking. It is a challenge for designers to balance the evaluation and creativity, which requires designers to have a full understanding of the knowledge of CES and LA design.

4.2. Methodological Challenges: Lack of Evaluation Methods to Inform Designs

Most design proposals use document research, followed by questionnaires which are based on people's preferences and perceptions. In reality, expert-based methods are the most used, even when this is not mentioned in the proposals. Indeed, the designs highly depend on the experiences, skills, and values of the designers as experts, and evidence-based design is widely used for LA designs in China. An evidence-based design highlights the process of design by critically and appropriately integrating various aspects, such as credible data, practitioner design expertise, client needs, and resources, in order to achieve project objectives [40]. Although evidence-based design has been widely used, it has been simultaneously critiqued for rigidity and misapplication in China. Hence, landscape architects gravitate to the support of research by integrating "research-informed design" as a broader term with concepts, fields, tools, or methods to support their practice. In this study, CES is regarded as a research tool that has the potential to aid LA practices. It is clear that, although CES are difficult to evaluate, a systematic evaluation of CES could guide the designers to capture and maximize them.

The diversity of CES study methods provides a rich inventory for scholars to assess CES. Instruments for assessing CES, including quantification, valuing, mapping, and modeling, are increasingly studied in CES and ES research. Many evaluation methods provide an opportunity to evaluate CES in LA studies. The evaluation methods are generally divided into monetary and non-monetary methods or revealed preference and stated preference methods as introduced in the first section. Cheng et al. [14]; Hirons et al. [41]; and Spangenberg and Settele [42] summarize the evaluation methods. In urban green space studies, non-monetary methods were used more, especially questionnaires and interviews, which emphasize the preferences and perceptions of people. It highlights the importance of the role that humans play in interaction with landscapes. Participatory mapping can be used to evaluate the crucial locations and settings for an urban park. Such assessments can provide valuable information for designers to increase CES provisions in the urban landscape. A social media-based method was used to reveal people's preference of CES based on the social media data from various resources such as Flickr or Instagram. Monetary methods, such as willingness to pay (WTP), can be used to know people's preferences for specific park settings. Both monetary and non-monetary methods and their combinations are encouraged to be used in LA studies. To achieve this, combing economics with other disciplines such as social or behavioral sciences is significant to know how to shift human aspects to environment setting, which is encouraged in future studies.

4.3. Practical Challenges: Practical Issues of Transferring CES to LA Design

In addition to the conceptual and methodological challenges, there are also more practical issues. They are: the relationships between CES and landscape features, the evaluation scopes (i.e., park types and size), data collection, and operation.

4.3.1. CES and Landscape Features

CES are highly related to landscape features. Sculpture, environmental infrastructure, and plant community were identified in the publications review. In addition, plants, architecture, playground, sculpture, and sports-/fitness facilities were the features used most to indicate CES in the reviewed design proposals. Landscape architects also highlighted the importance of the landscape features to express their ideas about CES. Linking the CES to physical landscape features is regarded as an efficient way to ground the abstract concept into the designs. The link between CES and landscape features provided an opportunity to help designers easily grasp CES and create essential links between ecosystem processes and management actions [43]. For example, some studies have tried to link CES to a location by means of participatory mapping, which allows people to mark the sites of CES (see Brown and Hausner [44]; Bryan et al. [45]; and Plieninger et al. [46]). Such studies provide designers with insight into the location of CES supplies and their correlations with specific features [30]. De Valck et al. [25] test the influence of different degrees of high or low vegetation on recreation and how such specific connections have the potential to support the vegetation design and landscape management. Benches are also regarded as an indicator of aesthetic values that can guide LA designs [47]. However, the number of studies linking CES and landscape features are still limited, and future studies should focus more on the links between CES and landscape features to guide LA designs.

4.3.2. Scopes, Data Collection, and Implementation

Many other factors influence the transfer of the concept of CES into LA design. The design process is complicated, and landscape architects need to coordinate a series of tasks to be performed by a number of people over a set period. They need to consider factors such as scale, types, layout, and features and organize the landscape to create a functional, comprehensible, and beautiful place. In addition, landscape architects need to consider the data collection and operation process when integrating CES evaluation into this process.

Specifically, the first issue is the evaluation scope including park scales and types. Parks with different types and scales have different expressions. For instance, sports parks emphasize recreation more than commemorative parks, which highlight spiritual values as stated by designers, and, in this case, landscape architects often have to trade off different CES. Meanwhile, different parks with different sizes add complexity. Hence, defining types and scales is crucial at the beginning to clarify the study scope and further select evaluation methods.

The second issue is how to collect the data. Data collection is significant to support practices, and depends on what methods are selected. The data can be divided into primary data or secondary data and quantitative data or qualitative data. For example, LA designs often depend on the secondary data derived from the document research, and primary data derived from the interviews. The primary data often engage stakeholders because CES evaluation is recognized to rely heavily on people's preferences and perceptions. LA is a user-inspired and user-useful discipline that requires designers to achieve the requirements of the clients and collaborate with them, making it more complex. CES evaluation could contribute to the increased congruence between different stakeholders. It is a useful tool for communicating with clients by showing what benefits they can have based on design proposals.

The third issue is the operation of the CES evaluation, such as who is involved in the evaluation and how it can be implemented. There is often a lack of local research capacity to undertake valuation research [48], which requires designers to have the capacity to act as enumerators or facilitators. As most designers do not know about CES, training on how to evaluate CES is essential for designers. Additionally, the development of evaluation methods, toolboxes, and practical guidelines is important to aid designers. Meanwhile, cooperation with ecologists or other experts is also highly encouraged at the beginning.

We take account of the complexity of practical issues and emphasize that there is not one simple and straightforward way to address practical issues. This study presented assumptions and discussions for tackling the complexity of involving CES evaluation in landscape design.

5. Conclusions

Landscape architects are facing increasing pressure due to rapid urbanization and call for practical knowledge, skills, and methods, etc. to support their designs. This study proposes that the concept of CES could have the potential to address their pressures by integrating CES evaluation in LA designs. This study identified three challenges, including consistent concepts, methods for evaluating CES, and practical issues of transferring CES to LA design. We further provided recommendations about how to deal with these challenges by highlighting opportunities for designers to contribute to LA and CES research in the future. (1) We highlighted developing consistent concepts and highlighted using CES as a common term to refer to socio-cultural values. (2) We encouraged using more evaluation methods to assess CES in LA studies, including monetary and non-monetary methods, such as WTP, participatory mapping, and the social media-based method. In addition, (3) we encouraged more studies addressing various practical issues to better guide LA designs. The first issue is to define park types and scales, which is crucial at the beginning to clarify the study scope and further select evaluation methods. The second issue is to collect data, especially the primary data that are often ignored in LA studies. The third issue is to develop evaluation methods, toolboxes, and practical guidelines to aid designers. In addition, training on how to evaluate CES is also essential for designers.

Author Contributions: X.C.: Conceptualization, Methodology, Investigation, Formal analysis, Writing—Original draft preparation. S.V.D.: Methodology, Writing—Reviewing and Editing. P.U.: Writing—Reviewing and Editing. All authors have read and agreed to the published version of the manuscript.

Funding: This research was funded by The Talent Introduction Program of Xihua University.

Institutional Review Board Statement: Not applicable.

Informed Consent Statement: Not applicable.

Data Availability Statement: The data presented in this study are contained within this article.

Conflicts of Interest: We declare that we have no conflicts of interest to this work. We have no financial and personal relationships with other people or organizations that can inappropriately influence our work.

Appendix A. Supplementary Data

This appendix contains four tables.

Table A1. Set of questions asked for every design proposal reviewed.

Question	Response Categories
(1) Year	
(2) Park types	Comprehensiveness park, community park, topic park (children's park, the cultural relics park, the commemorative park, the zoological garden, and the botanical garden, etc.), and belt-shaped park.
(3) Park scale	Size
(4) In relation to CES categories	Cultural diversity, spiritual and religious values, knowledge systems, educational values, inspiration, aesthetic values, social relations, recreation and ecotourism, cultural heritage values, and sense of place, etc.
(5) Evaluation methods If yes, what?	Yes/No Document research, questionnaires, and interviews, etc.
(6) In relation to landscape features If yes, what?	Yes/No Landform, plant, recreation facilities (camping, picnic, BBQ, Bird/fish feeder, birdbath, bird box, etc.); water, architecture, benches, sculptures, etc.

Table A2. Example of interview guidelines.

Interviewer: Interviewee:	Date (MM/DD/YYYY):	Start Time:	Stop Time:
Park once involved:			
Q1. Have you ever heard about the concept of CES, ES, or services, etc.? If yes, how do you think about this concept? If no, the interviewers will explain the concept to interviewees.			
Q2. Do you mention CES/ social cultural values (aesthetic values, recreation, etc.) in your park designs? How? Could you give me some examples? If no, why?			
Q3. What challenges can you imagine for transferring CES into your design?			
Q4. What the opportunities can you think of for applying this concept in your daily design?			

The interview begins with an informed consent about the recording and an explanation about the confidentiality of the interview. Following is a rough and easy to understand description of the aim of our study.

Table A3. Keywords link to CES.

Keyword	Frequency	Keyword	Frequency	Keyword	Frequency
landscape architecture design	23	forest park	2	experience	1
landscape architecture	23	urban	2	protection	1
urban park	18	environmental infrastructure	2	renew	1
theme park	11	function	2	sustainable development	1
ecology	8	geology park	2	community park	1
plant design	7	old people	2	mountain park	1
design	7	ecological design	2	rural landscape	1
local culture	7	development	2	country park	1
historical culture	6	place/cite/space	2	design point	1
waterfront Landscape	6	landscape development	2	expression	1
humanization/user-friendly	5	ecological landscape	2	ecological restoration	1
urban wetland park	5	leisure and recreation	2	creative	1
landscape renovation	5	integrate	2	exemplary	1
comprehensive park	5	recreational space	2	environment	1
nature	4	urban culture	1	design principle	1
urban green	4	urbanization	1	plant community	1
open space	4	park design	1	urban design	1
culture	4	point, line, and surface	1	recreational behavior	1
design strategy	3	characteristic	1	influence	1
wetland	3	application	1	modern	1
architecture design	3	cultural heritage	1	sculpture	1

Table A4. Landscape feature categories in park proposals.

Code	Category	Number	Code	Category	Number
1	Plants	41	14	Bridge	4
2	Architecture	28	15	Bench	5
3	Playground	26	16	Market square	2
4	Sculpture	21	17	Children playground	2
5	Sports-/fitness facilities	20	18	Lawn	2
6	Signs	19	19	Local stone	2
7	Furniture	12	20	Fountain	2
8	Recreation facilities	10	21	Camping facilities	1
9	Path	9	22	Shopping street	1
10	Water	8	23	Stores	1
11	Landform	7	24	Viewing platform	1
12	Lighting	5	25	Lighting tower	1
13	Tower	5	26	Watch tower	1

References

1. Chen, X.; Wu, J. Sustainable landscape architecture: Implications of the Chinese philosophy of "unity of man with nature" and beyond. *Landsc. Ecol.* **2009**, *24*, 1015–1026. [CrossRef]
2. *European Landscape Convention*; Report and Convention; Council of Europe: Strasbourg, France, 2000.
3. Deveikienė, V. Methodological guidelines for optimizing the interaction between landscape architecture and urban planning. *Sci. J. Latv. Univ. Life Sci. Technol. Landsc. Archit. Art* **2018**, *12*, 7–21. [CrossRef]
4. Langley, W.N.; Corry, R.C.; Brown, R.D. Core Knowledge Domains of Landscape Architecture. *Landsc. J.* **2018**, *37*, 9–21. [CrossRef]
5. Yue, B.-R.; Shao, S.-L. The evolution and trends of analytical methods for landscape planning and design. *J. Xi'an Univ. Archit. Technol. Nat. Sci. Ed.* **2010**, *42*, 690–695.

6. Milburn, L.-A.S.; Brown, R.D. The relationship between research and design in landscape architecture. *Landsc. Urban Plan.* **2003**, *64*, 47–66. [CrossRef]

7. Lenzholzer, S.; Duchhart, I.; Koh, J. 'Research through designing'in landscape architecture. *Landsc. Urban Plan.* **2013**, *113*, 120–127. [CrossRef]

8. Li, W.; Milburn, L.-A. The evolution of geodesign as a design and planning tool. *Landsc. Urban Plan.* **2016**, *156*, 5–8. [CrossRef]

9. Gu, Y.; Deal, B.; Larsen, L. Geodesign processes and ecological systems thinking in a coupled human-environment context: An integrated framework for landscape architecture. *Sustainability* **2018**, *10*, 3306. [CrossRef]

10. Alizadeh, B.; Hitchmough, J. A review of urban landscape adaptation to the challenge of climate change. *Int. J. Clim. Chang. Strateg. Manag.* **2019**, *11*, 178–194. [CrossRef]

11. Steiner, F. Frontiers in urban ecological design and planning research. *Landsc. Urban Plan.* **2014**, *125*, 304–311. [CrossRef]

12. Grose, M.; Frisby, M. Mixing ecological science into landscape architecture. *Front. Ecol. Environ.* **2019**, *17*, 296–297. [CrossRef]

13. MEA. *Millennium Ecosystem Assessment: Ecosystems and Human Well-being: Synthesis*; Island Press: Washington, DC, USA, 2005.

14. Cheng, X.; Van Damme, S.; Li, L.; Uyttenhove, P. Evaluation of cultural ecosystem services: A review of methods. *Ecosyst. Serv.* **2019**, *37*, 100925. [CrossRef]

15. De Groot, R. Function-analysis and valuation as a tool to assess land use conflicts in planning for sustainable, multi-functional landscapes. *Landsc. Urban Plan.* **2006**, *75*, 175–186. [CrossRef]

16. De Groot, R.; Hein, L. *Concept and Valuation of Landscape Functions at Different Scales, in Multifunctional Land Use*; Springer: Berlin/Heidelberg, Germany, 2007; pp. 15–36.

17. Lautenbach, S.; Kugel, C.; Lausch, A.; Seppelt, R. Analysis of historic changes in regional ecosystem service provisioning using land use data. *Ecol. Indic.* **2011**, *11*, 676–687. [CrossRef]

18. Maes, J.; Egoh, B.; Willemen, L.; Liquete, C.; Vihervaara, P.; Schägner, J.P.; Grizzetti, B.; Drakou, E.; La Notte, A.; Zulian, G.; et al. Mapping ecosystem services for policy support and decision making in the European Union. *Ecosyst. Serv.* **2012**, *1*, 31–39. [CrossRef]

19. Mooney, P. A systematic approach to incorporating multiple ecosystem services in landscape planning and design. *Landsc. J.* **2014**, *33*, 141–171. [CrossRef]

20. Dickinson, D.C.; Hobbs, R.J. Cultural ecosystem services: Characteristics, challenges and lessons for urban green space research. *Ecosyst. Serv.* **2017**, *25*, 179–194. [CrossRef]

21. Schaich, H.; Bieling, C.; Plieninger, T. Linking Ecosystem Services with Cultural Landscape Research. *GAIA Ecol. Perspect. Sci. Soc.* **2010**, *19*, 269–277. [CrossRef]

22. Riechers, M.; Barkmann, J.; Tscharntke, T. Perceptions of cultural ecosystem services from urban green. *Ecosyst. Serv.* **2016**, *17*, 33–39. [CrossRef]

23. Lovell, S.; Johnston, D.M. Designing landscapes for performance based on emerging principles in landscape ecology. *Ecol. Soc.* **2009**, *14*, 44. [CrossRef]

24. Plieninger, T.; Bieling, C.; Fagerholm, N.; Byg, A.; Hartel, T.; Hurley, P.; López-Santiago, C.A.; Nagabhatla, N.; Oteros-Rozas, E.; Raymond, C.M.; et al. The role of cultural ecosystem services in landscape management and planning. *Curr. Opin. Environ. Sustain.* **2015**, *14*, 28–33. [CrossRef]

25. De Valck, J.; Landuyt, D.; Broekx, S.; Liekens, I.; De Nocker, L.; Vranken, L. Outdoor recreation in various landscapes: Which site characteristics really matter? *Land Use Policy* **2017**, *65*, 186–197. [CrossRef]

26. Scholte, S.S.; van Teeffelen, A.J.; Verburg, P.H. Integrating socio-cultural perspectives into ecosystem service valuation: A review of concepts and methods. *Ecol. Econ.* **2015**, *114* (Suppl. C), 67–78. [CrossRef]

27. Beck, A.C.; Campbell, D.; Shrives, P.J. Content analysis in environmental reporting research: Enrichment and rehearsal of the method in a British–German context. *Br. Account. Rev.* **2010**, *42*, 207–222. [CrossRef]

28. Shrivastava, R.; Mahajan, P. Artificial intelligence research in India: A scientometric analysis. *Sci. Technol. Libr.* **2016**, *35*, 136–151. [CrossRef]

29. Li, H.M.; Jin, Y.M. A Construction of Landscape Evaluation Indicator System for the Urban Comprehensive Park Based on the Theory about Three Elements. *Appl. Mech. Mater.* **2011**, *99–100*, 496–500. [CrossRef]

30. Bertram, C.; Rehdanz, K. Preferences for cultural urban ecosystem services: Comparing attitudes, perception, and use. *Ecosyst. Serv.* **2015**, *12*, 187–199. [CrossRef]

31. Campbell, L.K.; Svendsen, E.S.; Sonti, N.F.; Johnson, M.L. A social assessment of urban parkland: Analyzing park use and meaning to inform management and resilience planning. *Environ. Sci. Policy* **2016**, *62*, 34–44. [CrossRef]

32. Chiesura, A. The role of urban parks for the sustainable city. *Landsc. Urban Plan.* **2004**, *68*, 129–138. [CrossRef]

33. Hegetschweiler, K.T.; de Vries, S.; Arnberger, A.; Bell, S.; Brennan, M.; Siter, N.; Olafsson, A.S.; Voigt, A.; Hunziker, M. Linking demand and supply factors in identifying cultural ecosystem services of urban green infrastructures: A review of European studies. *Urban For. Urban Green.* **2017**, *21*, 48–59. [CrossRef]

34. Thompson, I. Aesthetic, social and ecological values in landscape architecture: A discourse analysis. *Ethics Place Environ.* **2000**, *3*, 269–287. [CrossRef]

35. Jiang, W. Ecosystem services research in China: A critical review. *Ecosyst. Serv.* **2017**, *26*, 10–16. [CrossRef]

36. Costanza, R.; De Groot, R.; Braat, L.; Kubiszewski, I.; Fioramonti, L.; Sutton, P.; Farber, S.; Grasso, M. Twenty years of ecosystem services: How far have we come and how far do we still need to go? *Ecosyst. Serv.* **2017**, *28*, 1–16. [CrossRef]

37. Czúcz, B.; Arany, I.; Potschin-Young, M.; Bereczki, K.; Kertész, M.; Kiss, M.; Aszalós, R.; Haines-Young, R. Where concepts meet the real world: A systematic review of ecosystem service indicators and their classification using CICES. *Ecosyst. Serv.* **2018**, *29*, 145–157. [CrossRef]

38. Schröter, M.; Van Der Zanden, E.H.; van Oudenhoven, A.; Remme, R.; Serna-Chavez, H.M.; De Groot, R.S.; Opdam, P. Ecosystem services as a contested concept: A synthesis of critique and counter-arguments. *Conserv. Lett.* **2014**, *7*, 514–523. [CrossRef]

39. Lam, S.T.; Conway, T.M. Ecosystem services in urban land use planning policies: A case study of Ontario municipalities. *Land Use Policy* **2018**, *77*, 641–651. [CrossRef]

40. Peavey, E.; Vander Wyst, K.B. Evidence-based design and research-informed design: What's the difference? Conceptual definitions and comparative analysis. *HERD Health Environ. Res. Des. J.* **2017**, *10*, 143–156. [CrossRef]

41. Hirons, M.; Comberti, C.; Dunford, R. Valuing Cultural Ecosystem Services. *Annu. Rev. Environ. Resour.* **2016**, *41*, 545–574. [CrossRef]

42. Spangenberg, J.H.; Settele, J. Precisely incorrect? Monetising the value of ecosystem services. *Ecol. Complex.* **2010**, *7*, 327–337. [CrossRef]

43. Kupschus, S.; Schratzberger, M.; Righton, D. Practical implementation of ecosystem monitoring for the ecosystem approach to management. *J. Appl. Ecol.* **2016**, *53*, 1236–1247. [CrossRef]

44. Brown, G.; Hausner, V.H. An empirical analysis of cultural ecosystem values in coastal landscapes. *Ocean Coast. Manag.* **2017**, *142*, 49–60. [CrossRef]

45. Bryan, B.A.; Raymond, C.M.; Crossman, N.; MacDonald, D.H. Targeting the management of ecosystem services based on social values: Where, what, and how? *Landsc. Urban Plan.* **2010**, *97*, 111–122. [CrossRef]

46. Plieninger, T.; Dijks, S.; Oteros-Rozas, E.; Bieling, C. Assessing, mapping, and quantifying cultural ecosystem services at community level. *Land Use Policy* **2013**, *33*, 118–129. [CrossRef]

47. Bieling, C.; Plieninger, T. Recording Manifestations of Cultural Ecosystem Services in the Landscape. *Landsc. Res.* **2013**, *38*, 649–667. [CrossRef]

48. Christie, M.; Fazey, I.; Cooper, R.; Hyde, T.; Kenter, J.O. An evaluation of monetary and non-monetary techniques for assessing the importance of biodiversity and ecosystem services to people in countries with developing economies. *Ecol. Econ.* **2012**, *83*, 67–78. [CrossRef]

 *land*

Communication

Estimating Air Pollution Removal and Monetary Value for Urban Green Infrastructure Strategies Using Web-Based Applications

Alessio Russo [1,*], Wing Tung Chan [1] and Giuseppe T. Cirella [2]

1   School of Arts, Francis Close Hall Campus, University of Gloucestershire, Swindon Road, Cheltenham GL50 4AZ, UK; wtung919@gmail.com
2   Faculty of Economics, University of Gdansk, 81-824 Sopot, Poland; gt.cirella@ug.edu.pl
*   Correspondence: arusso@glos.ac.uk; Tel.: +44-(0)1242-714557

Abstract: More communities around the world are recognizing the benefits of green infrastructure (GI) and are planting millions of trees to improve air quality and overall well-being in cities. However, there is a need for accurate tools that can measure and value these benefits whilst also informing the community and city managers. In recent years, several online tools have been developed to assess ecosystem services. However, the reliability of such tools depends on the incorporation of local or regional data and site-specific inputs. In this communication, we have reviewed two of the freely available tools (i.e., i-Tree Canopy and the United Kingdom Office for National Statistics) using Bristol City Centre as an example. We have also discussed strengths and weaknesses for their use and, as tree planting strategy tools, explored further developments of such tools in a European context. Results show that both tools can easily calculate ecosystem services such as air pollutant removal and monetary values and at the same time be used to support GI strategies in compact cities. These tools, however, can only be partially utilized for tree planting design as they do not consider soil and root space, nor do they include drawing and painting futures. Our evaluation also highlights major gaps in the current tools, suggesting areas where more research is needed.

Keywords: urban ecosystem services; urban tree planting; i-Tree Canopy; Office for National Statistics; health damage costs; United Kingdom

Citation: Russo, A.; Chan, W.T.; Cirella, G.T. Estimating Air Pollution Removal and Monetary Value for Urban Green Infrastructure Strategies Using Web-Based Applications. *Land* **2021**, *10*, 788. https://doi.org/10.3390/land10080788

Academic Editor: Thomas Panagopoulos

Received: 22 June 2021
Accepted: 25 July 2021
Published: 27 July 2021

1. Introduction

Air pollution caused by the growth of urbanization and industrialization continues to plague societies in the twenty-first century [1]. Urbanization plays a major role in worsening ventilation conditions and increases the emissions of pollutants [2]. The transformation of land use, caused by urbanization, reduces ventilation quality via building morphology and, indirectly, the urban wind velocity [3]. Air pollution derived from human activities comes from both indoor and outdoor environments [4]. It causes harm to health, decreases economic growth, and augments social problems (i.e., by way of knock-on societal effects) [5]. In 2015, the World Health Organization [6] estimated 4.3 million deaths occurred due to indoor air pollution and 3.7 million due to outdoor air pollution (i.e., 8 million for the year). Data published by the United Kingdom (UK) Royal College of Physicians demonstrates that there are around 40,000 fatalities each year due to air pollution [7].

Several studies show that green infrastructure (GI) can improve air quality in cities [8–11]. In particular, urban vegetation provides several ecosystem services and plays a vital role in air pollutant removal, heavy metal removal, rainwater interception, and microclimatic improvements [12–14].

Planting millions, billions, or even trillions of trees as a simple solution to air pollution and other major environmental problems is being proposed by an increasing number of global, regional, and national projects [15]. Large tree planting can also improve life satisfaction [16,17]. According to Jones' [17] research, life satisfaction among NYC residents improved by 0.018 points on a 4-point scale during the first three years of the Million Trees NYC initiative, when over 400,000 trees were planted. This is a USD 505 increase in per capita monthly family income, or a 6.5% gain. According to the existing literature on the benefits of urban trees, the observed increases in life satisfaction following Million Trees NYC could be due to improved air quality, lower ambient temperatures during the spring and summer, lower crime rates, improved recreation and exercise opportunities, or greater social and community cohesion [17]. Tree planting, however, is a lot more difficult than it appears [15]. It takes between 15 and 40 years for a tree to grow a sufficiently large canopy to offer several ecosystem services (e.g., aesthetics, reducing air pollution, controlling rainwater, and carbon storage) [18]. However, street tree growth is influenced by critical landscape design issues that affect access of the tree roots to water, air, and nutrients [19]. Landscape architects and urban foresters should consider the concept of "optimal planting," which includes several factors such as the extent of rooting space and the quality of urban soils for supporting trees [19,20]. Therefore, there is a need for tools that can aid in the process of tree planting as well as the implementation of landscape design in order to guarantee healthy trees can provide sufficient ecosystem services in the built environment [21].

In addition to tree planting campaigns, several nations and towns throughout the world have made deliberate pledges to provide high-quality GI [22,23]. In particular, GI strategy (which outlines which GI and ecosystem service assets already exist and how they can be improved [24]) serves as the foundation for policies and decisions on development proposals in cities to avoid loss or harm before considering mitigation or compensatory measures [25]. However, an issue raised in the scientific literature and by stakeholders is a lack of reliable friendly user models with local data for assessing ecosystem services that support GI strategies [26], as well as strong evidence on the most cost-effective and sustainable models and procedures for long-term management and maintenance of high-quality GI [22]. Therefore, rapid ecosystem services evaluation tools and models have sparked widespread interest across all sectors; nonetheless, it is widely acknowledged that systematic use of ecosystem services in decision- and policy-making necessitates a level of accuracy that is seldom achieved in practice [27,28]. Experts from the disciplines of forestry, agriculture, urban planning, and environmental engineering must collaborate to develop accurate tools that can simulate plant-built environment interactions [29]. Fortunately, numerous models that simulate and quantify energy, water flows, and ecosystem services in various ecosystems already exist [29]. For example, the last few years have seen an increase in the use of the United States Department of Agriculture, Forest Service i-Tree tools in the American and international market (e.g., Australia, Canada, Mexico, South Korea, much of the European Union, and the UK). Even though the science and development of the i-Tree tools date back to the mid-1990s, the software suite was released as a framework for science delivery in 2006 [30]. Today, i-Tree tools include several desktop applications (e.g., i-Tree Eco and i-Tree Hydro) and web-based applications (e.g., i-Tree Canopy, i-Tree Country, i-Tree Design, and i-Tree Species) that provide baseline data for tree benefits and planning over time [26,31].

In the UK, for instance, several ecosystem services provided by GI that specifically target air pollutant removal have been calculated using i-Tree tools. For example, dating back to 2011, the i-Tree Eco project started in Torbay, England that has now been introduced in more than 20 cities and towns [32,33]. In 2013, Natural England [34] evaluated three of the tools (i.e., i-Tree Design, i-Tree Eco, and i-Tree Streets) for applications nationwide. i-Tree Eco uses data collected using standardized time-consuming field methods that require professional foresters and arboriculturists [26,33] in which data on the number and health of trees assess their quantity and monetary value (i.e., in terms of air purification, carbon storage, and carbon sequestration [32]). Similarly, ready-made GI valuation tools available online can be used by those with little to no ecological background or training, offering low-budget alternatives for applications and assessments [34]. In particular, i-Tree

Canopy (i.e., estimating tree canopy levels using aerial photography) as well as the UK's Office for National Statistics tool (i.e., framed by using postcodes)—both based on different spatial parameters and methods—have been mostly used to assess GI benefits and the UK natural account. Specifically, the Office for National Statistics tool was created in response to the government's commitment to incorporate natural capital accounting in the UK Environmental Accounts by 2020 [35]. In addition, the UK Government has pledged to boost tree planting rates across the country to 30,000 ha per year. Between 2020 and 2025, they have allocated over GBP 500 million of the GBP 640 million Nature for Climate Fund on trees and forests in England to assist this goal [36]. To meet these ambitious goals in the coming years, evaluation tools and new guidance through the National Model Design Code on how trees can be included in the built environment (including design parameters for street tree placement) are required [36].

Moreover, such publicly-funded planting efforts rarely receive formal or even informal benefit-cost analyses, implying that large amounts of resources (i.e., financial, labor, etc.) are being deployed without a clear understanding of their returns, preventing comparisons of the net benefits per penny spent on afforestation to other potential urban improvement projects, such as early childhood education [17]. Thus, understanding the net benefits of urban trees is essential for justifying public-funding planting efforts or just allocating money to maintain existing urban trees on public land [37]. In this effort, this communication examines tree cover and relating ecosystem service utility using the Bristol City Center as an example by: (1) illustrating the main features of free user-friendly web applications (i.e., i-Tree Canopy and the Office for National Statistics tool), and (2) comparing i-Tree Canopy Version 6.1 (i.e., using American quantified datasets), Version 7.1 (i.e., local UK quantified datasets), and the Office for National Statistics website in the context of their use and as tree planting strategy tools in Europe. The tools are centered on aiding policymakers to best understand the benefits of maintaining trees and GI in terms of a balanced urban ecosystem services output.

2. Materials and Methods

2.1. Study Area

Bristol is the largest city in South West England with an estimated population of 463,400 people [38]. In 2019, Bristol's Council Cabinet approved a GBP 4 million, five-year management contract for preserving the city's trees, with the goal of doubling the city's tree canopy [39]. A commissioned report from the City Council showed that around 300 deaths each year (i.e., 8.5% of total deaths) in the City of Bristol had been attributed to air pollution [40], making it crucial to control and reduce air pollution in certain areas. The study area is comprised of six areas in Bristol City Centre according to the postcode BS1 (i.e., Bristol Central, see https://www.streetlist.co.uk/bs/bs1, accessed on 1 June 2021) with a population of 11,991 inhabitants living between Broadmead and Wapping Wharf (Figure 1). The choice was supported by: (1) preliminary desk research using ArcGIS Version 10.5.1 and the EDINA Digimap web-based mapping service that evaluated the physical BS1 zones, which took into account the location of air pollutant monitors, population density, and NO_2 concentrations and found that NO_2 is above the UK legal limits within postcodes BS1-2 and BS1-3 [41], and (2) according to council figures, it has planted approximately 6000 trees in each of the last four years. However, far too many of these are younger, smaller trees that are not in the city center, where they are most needed, and will take decades to reach maturity [42]. The six areas (i.e., letters A–F) represent the postcode sectors within the BS1 district—each with an area of 1 km^2.

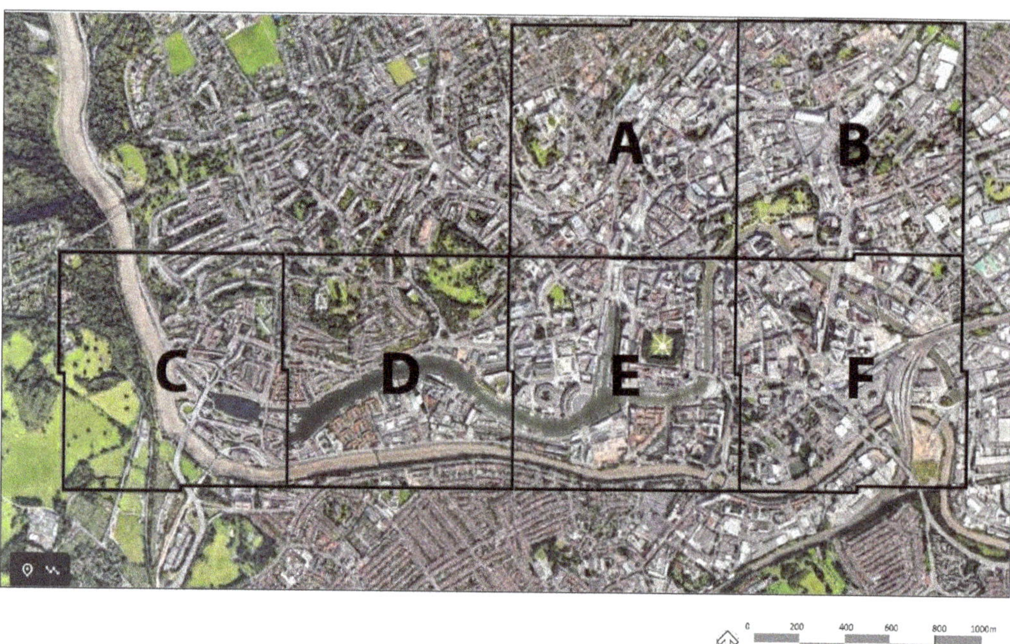

Figure 1. The study area made up of the six quadrants, i.e., **A–F**, included in the postcode BS1. Source: Google Earth.

2.2. Office for National Statistics Web-based Application

Quantified pollution removed by vegetation (i.e., per kg) and avoided health damage costs (i.e., GBP per person) in each area is calculated using the Office for National Statistics website [43]—an online tool developed by the Centre for Ecology and Hydrology. The tool is available online at: https://www.ons.gov.uk/economy/environmentalaccounts/articles/ukairpollutionremovalhowmuchpollutiondoesvegetationremoveinyourarea/2018-07-30, accessed on 1 June 2021. This calculator also provides the avoidable health damage costs per person within postcodes and compares it to the UK average. In 2020, to calculate pollution removed by vegetation as well as the avoided health damage costs for the selected areas, we have entered postcodes within the BS1 district into the Office for National Statistics tool (i.e., area A = BS1-1 and BS1-2, area B = BS1-2 and BS1-6, area C = BS1-6, area D = BS1-6, area E = BS1-4 and BS1-6, and area F = 1-6). The Office for National Statistics website's methods are being developed to incorporate the values within the UK's natural capital accounts [44]. Air pollution removal by urban green and blue infrastructure is calculated using the EMEP4UK model which is a dynamic atmospheric chemistry transport model [44,45]. Table 1 illustrates pollutants removed by urban green and blue infrastructure (i.e., urban trees and woodland, urban grassland, and urban water) as dry pollutant deposition (i.e., in terms of kilo tonnes per year) throughout the UK in 2015. The negative values for several pollutants removed by urban water, which are legitimate outputs of the scenario comparison, imply that dry deposition to water would be higher if there were no woodland or grassland [44].

The monetary account's economic and health calculations are based on damage cost per unit of exposure, with the economic benefit calculated directly from mortality and morbidity statistics for each local authority in the UK, as well as the receiving population's change in pollutant exposure [44]. Detailed methods are given in Jones et al. [46].

Table 1. UK pollutants removed by GI as dry pollutant deposition (kilo tonnes per year), 2015 [46].

Pollutant	UGBI [†]	Year 2015
PM_{10}	Urban trees and woodland	1.23
	Urban grassland	1.45
	Urban water	−0.004
$PM_{2.5}$	Urban trees and woodland	0.70
	Urban grassland	0.31
	Urban water	−0.003
SO_2	Urban trees and woodland	0.59
	Urban grassland	1.00
	Urban water	−0.049
NH_3	Urban trees and woodland	0.44
	Urban grassland	0.95
	Urban water	−0.045
NO_2	Urban trees and woodland	0.41
	Urban grassland	1.61
	Urban water	0.000
O_3	Urban trees and woodland	4.97
	Urban grassland	16.94
	Urban water	−0.003

[†] urban green and blue infrastructure.

2.3. i-Tree Canopy

To calculate the tree cover and the monetary value for the selected areas in Bristol, the free online tool i-Tree Canopy Version 6.1 in 2020 and Version 7.1 in 2021 was used. To compare the data from the Office for National Statistics, the Pollution Removal GeoPackage (i.e., found at: http://geoportal.statistics.gov.uk/search?q=Geopackage, accessed on 1 June 2021) was used to create six ESRI shapefiles in ArcGIS Version 10.5.1. The boundary of each postcode area and the ESRI shapefiles were imported into the i-Tree Canopy tool. Each boundary was 1 km^2. A total of 500 random points (i.e., with a standard error (SE) < 3%) were photo interpreted for each area for a total of 3000 points. Within each area, the percentage of each cover class (i.e., 'p') was calculated as the number of sample points (i.e., 'x') hitting the cover attribute divided by the total number of interpretable sample points (i.e., 'n') within the area of analysis (i.e., $p = x/n$). The SE of the estimate is calculated using Equation (1) [47,48].

$$SE = \frac{\sqrt{p\,(1-p)}}{n} \tag{1}$$

where p = percentage of each cover class, n = total number of interpretable sample points.

For the photo interpretation, two photo interpreters with a background in landscape architecture and urban forestry classified each point using three cover classes: two default classes (i.e., tree and non-tree) and grass. i-Tree Canopy Version 6.1 calculates air pollutant removal and monetary values using the default values (i.e., the multipliers) of air pollutant removal rates (i.e., g/m^2/year) and monetary values (i.e., USD m^{-2} year^{-1}) for a unit tree cover derived from i-Tree Eco projects across the United States [49]. In this version for international projects (i.e., outside the United States), the default values are derived from the United States' total removal amount and monetary values used from American urban areas [49]. The monetary values are in USD and the tool calculates currency values from the online currency exchange tool at: https://www.openexchangerates.org, accessed on 2 June 2021. On the other hand, i-Tree Canopy Version 7.1 estimates the ecosystem service rates using i-Tree Eco batch runs as well as using local pollution and weather data [50]. A description of the metadata used in the model is available in the i-Tree Canopy metadata and data sources [50]. Furthermore, the monetary values for ecosystem services in the UK are provided by Treeconomics, which are available online at: https://www.itreetools.org/documents/734/UK_Benefit_Prices_from_Danielle_Hill_Treeconomics_-_Benefits_Prices_by_County_Final_1.xlsx, accessed on 10 June 2021. In this study, both versions were run. Specific to Version 7.1, the tool used the UK's average data as well local data in urban areas, i.e., the South West data (Table 2).

Table 2. Multipliers derived from i-Tree Eco projects in the UK and using South West data [50].

Pollutant	Removal Rate (g/m^2/Year) *	Monetary Value (GBP/t/Year) *	Removal Rate (g/m^2/Year) **	Monetary Value (GBP/t/Year) **
CO	0.148	956.63	0.072	956.63
NO$_2$	3.065	187.91	2.037	114.41
O$_3$	10.304	928.15	9.06	770.40
PM$_{10}$	2.08	33,713.00	2.033	33,713.00
PM$_{2.5}$	0.521	30,654.87	0.567	26,838.42
SO$_2$	0.405	64.93	0.251	41.56

* UK average data; ** South West data; Metric units: g = grams, m = meters, t = metric tons.

2.4. Comparison of the Office for National Statistics and i-Tree Canopy Tools

In terms of indicators, the i-Tree Canopy toolset contains six common air pollutants (i.e., carbon monoxide (CO), nitrogen dioxide (NO$_2$), ozone (O$_3$), particulate matter less than 2.5 microns (PM$_{2.5}$), particulate matter greater than 2.5 microns and less than 10 microns (PM$_{10}$), and sulfur dioxide (SO$_2$) [49]) while the Office for National Statistics also contains six common air pollutants (i.e., ammonia (NH$_3$), NO$_2$, O$_3$, PM$_{2.5}$, PM$_{10}$, and SO$_2$). In order to compare the two tools, we have not included CO from i-Tree and NH$_3$ from the Office for National Statistics. Furthermore, to compare the total pollutant removal amounts between i-Tree Canopy Version 7.1 with local data and the Office for National Statistics' findings, the Office for National Statistics values were converted from kg/km^2 to kg/ha.

3. Results and Discussion

3.1. Tree Cover Results Using i-Tree Canopy

The percentage of tree cover in the six areas are given in Figure 2. Area E and F have the lowest coverage of 12.2 ± 1.46%, while the range of tree cover in the six study areas is between 12.2 ± 1.46% and 35.6 ± 2.14%. The lowest value is higher than in a previous study using i-Tree Canopy in Bristol, which found a value of 10% tree cover in the city center in 2018 [51]. This difference is due to the fact that the Bristol Tree Canopy Cover Survey in 2018 [51] assessed each area according to city council wards, e.g., Bristol Central was calculated using the total area of 223.14 ha with a tree canopy of 10.0 ± 1.62% using i-Tree Canopy and a tree canopy of 6.5% using i-Tree Eco. Furthermore, the UK tree cover, in general, was found to be lower than in other European and American cities [52]. Doick et al. [53] suggest that UK towns and cities strive to achieve a 20% tree canopy cover as a minimum standard while towns and cities with at least 20% cover should increase their tree cover by at least 5% within the next 10 to 20 years [53]. Unless supplemented by more comprehensive criteria, the canopy cover targets cannot give a true representation of the structure, health, and function of GI [54].

Studies have suggested that increasing tree canopy may provide more support for mental health [55]. Recently, Marselle et al. [56] found that people with a poor socioeconomic level who lived in an area with a high density of street trees within 100 m of their home had a lower chance of being given antidepressants. In a study conducted by Kondo et al. [57], it was found that a five-percentage-point increase in tree canopy might result in a 302-death decrease every year, worth USD 29 billion in Philadelphia. Moreover, a 10% increase in canopy over the city was linked to a USD 36 billion reduction in mortality. If Philadelphia achieves its objective of raising tree canopy cover to 30% by 2025, 403 premature adult deaths (i.e., 3% of total mortality) might be avoided annually [57]. On the contrary, UK city councils have raised concerns about the possible impact of increased tree cover in urban parks on crime and personal safety, as well as the fact that leaf fall from deciduous trees can obstruct urban run-off drains [58]. Furthermore, while increasing tree cover is associated with better pollution reduction, local-scale trees and forest design, it can also influence local-scale pollution concentrations [59]. In Baltimore, Troy et al. [60] found that a 10% increase in tree canopy was linked to a 12% reduction in crime. These conflicting findings between the American and British indicate a clear policy difference at the local level with different scientific viewpoints used to support their case.

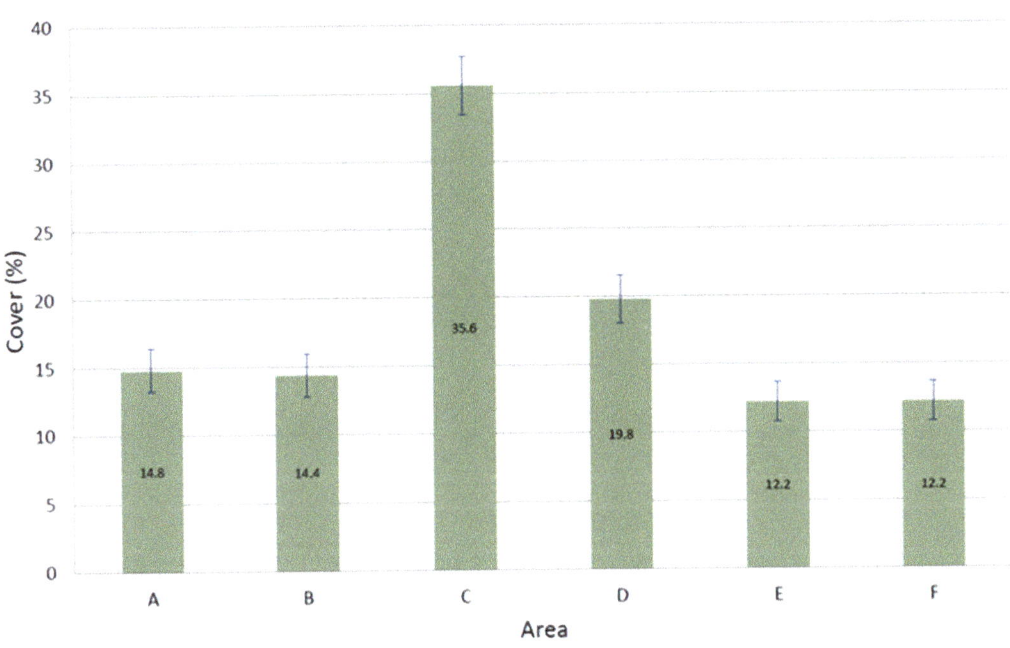

Figure 2. Tree cover in the six study areas shown in percentage (error bars represent SE).

3.2. Air Pollution Removal, Monetary Value, and Tool Evaluation

The range of pollutants removed is between 1006.8 ± 120.79 to 2935.18 ± 129.58 kg using i-Tree Canopy Version 6.1 (i.e., calculated using the US average), while i-Tree Canopy Version 7.1 (i.e., using the UK average data) calculated a reduction between 2231.72 ± 249.68 and 5347.27 ± 341.57 kg. More specifically, at the local level, the results yielded a range between 2063.22 ± 221.29 and 4944.86 ± 298.57 kg (Figure 3). Area C recorded the highest removal of pollutants, which is nearly threefold compared to the lowest in area E and F. The difference between the US and the UK average data is more than double.

Using i-Tree Canopy Version 7.1 with the UK data, the data overestimated the reduction of pollutants in area A and underestimated it in area E by a total of 103 kg when compared to the local South West data. Figure 4 shows a comparison of the total pollutant removal amount between i-Tree Canopy Version 7.1 with the local data and the Office for National Statistics' findings.

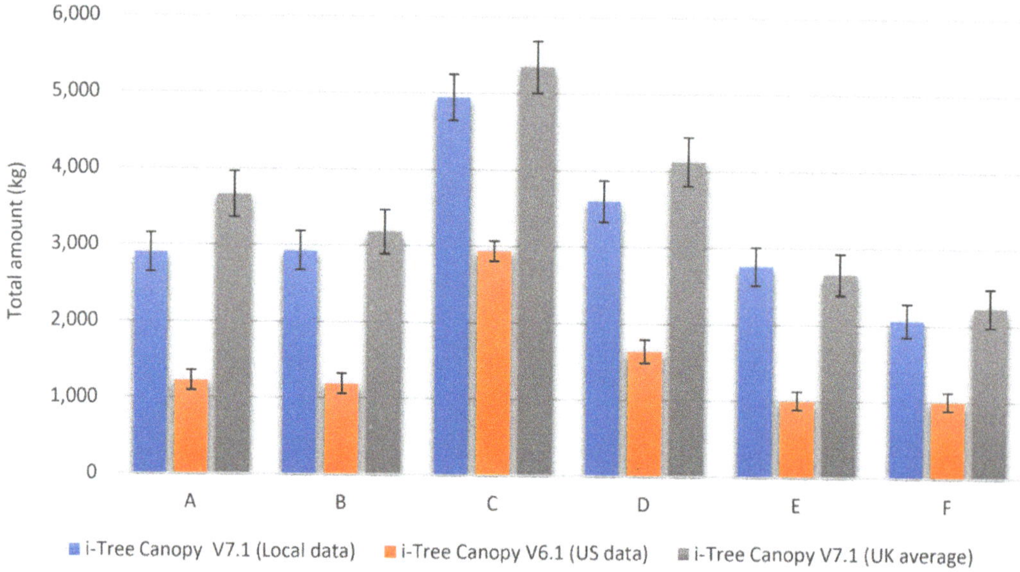

Figure 3. Total amount of air pollutants (i.e., NO_2, O_3, $PM_{2.5}$, PM_{10}, and SO_2) removed by trees in a year estimated using i-Tree Canopy Version 6.1 and i-Tree Canopy Version 7.1 with local and UK average data (error bars represent SE, calculated from SEs of sampled and classified points).

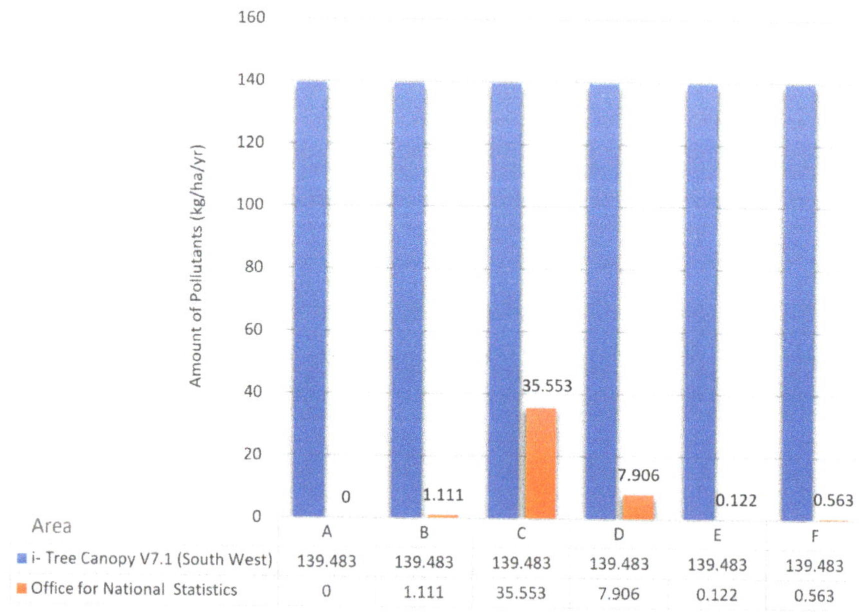

Area	A	B	C	D	E	F
i- Tree Canopy V7.1 (South West)	139.483	139.483	139.483	139.483	139.483	139.483
Office for National Statistics	0	1.111	35.553	7.906	0.122	0.563

Figure 4. Comparison of the estimates of air pollutants (i.e., NO_2, O_3, $PM_{2.5}$, PM_{10}, and SO_2) removed per ha by trees using i-Tree Canopy Version 7.1 (local Data) and the urban green and blue infrastructure using the Office for National Statistics. Metric units: kg = kilograms, ha = hectares; Office for National Statistics values were converted from kg/km^2 to kg/ha; i-Tree Canopy Version 7.1 values are based on the removal multipliers (i.e., NO_2 = 20.375, O_3 = 90.596, $PM_{2.5}$ = 5.670, PM_{10} = 20.329, and SO_2 = 2.513) in kg/ha/year from South West data.

There is a very strong contrast between the results using either tool. Throughout the entire study area, the i-Tree Canopy results recorded a constant value of 139.483 kg/ha/year, while the Office for National Statistics did not record any pollutant removal for area A and near zero for areas B, E, and F. In area C, however, it recorded its highest pollutant removal amount at 35.553 kg/ha/year. Figure 5 shows the amount of each pollutant removed annually using i-Tree Canopy Version 7.1 with local data. Among the five pollutants, O_3 had the highest removal amount and NO_2 ranged between 301.38 ± 32.34 and 722.32 ± 43.61 kg. In contrast, the Office for National Statics did not estimate any NO_2 removal in Areas A and E (Appendix A, Table A1). The Office for National Statistics' NO_2 estimates are significantly lower than those of i-Tree Canopy (i.e., from either version of the software). This outcome can be reasonably interpreted from the EMEP4UK model since, while trees remove NO_2 from the atmosphere, natural NO emissions from the soil under trees also exist, and these values balance out to a substantial extent [44]. Our findings are consistent with those of Jones et al. [44] who found that pollutant quantities assessed using the EMEP4UK model are roughly half those found in i-Tree studies. Furthermore, because the Office for National Statistics tool is based on a dynamic model, inhabitants of one area may benefit from pollutants absorbed in neighboring areas due to the nature of the model [43,44]. Additional pilot modeling, outside the purview of this study, can inform possible locations and vegetation parameters to maximize its impact for the least polluted conditions [8]. In the comparison of these results, consideration must be given to the fact that both tools do not consider pollution removal by building integrated vegetation (e.g., green roofs and green walls). In this regard, previous studies have shown that green roofs and green walls are effective to reduce pollution in streets [8]. Green walls, for instance, have been shown to reduce NO_2 levels at the street level by up to 40% and PM_{10} levels by 60%, according to researchers at Lancaster University [61]. It is also acknowledged that the impacts of vegetation on air quality at local scales (e.g., at the street level) are dependent on species composition and can be beneficial or negative. However, both tools were unable to model this level of detail [43,62].

The average monetary value of the Office for National Statistics is GBP 16.19 per person (i.e., the amount saved in healthcare costs) [44]. This result is higher than the UK average of GBP 15.39 per person. The i-Tree Canopy local data recorded a range between GBP $13,457 \pm 1444$ and GBP $32,252 \pm 1948$. Figure 6 shows the monetary value using i-Tree Canopy Version 7.1 with the UK average data and local data. There is a difference of GBP 2617 in area A while in area F it is only GBP 336 due to a lower tree canopy. The monetary values calculated using i-Tree Canopy Version 6.1 (i.e., using US average data) ranged from GBP 5486.59 ± 658.25 in area E and GBP $16,009.96 \pm 962.99$ in area C. However, these figures were estimated using the United States Environmental Protection Agency's BenMAP, which assesses the incidence of adverse health impacts and related monetary values caused by changes in NO_2, O_3, $PM_{2.5}$, and SO_2 concentrations [49]. Therefore, urban values were approximated using the national median externality values from the United States [49]. Contrarywise, i-Tree Canopy Version 7.1 as well as i-Tree Eco have been implemented using appropriate official values from the UK [63].

For the monetary values provided by GI, the two tools (i.e., i-Tree Canopy and the Office for National Statistics) offer a distinct benefit—differing valuation approaches [46]. The researchers from the Centre for Ecology and Hydrology calculate the health benefit from a change in air pollutant concentrations, whereas the i-Tree tools calculate damage costs per tonne of pollutant emitted [46]. The monetary value from the Office for National Statistics considers the costs of avoided health damage to people—i.e., the greater the number of people who benefit from pollution removal, the higher the value [43]. As a result, population density plays a significant role in final valuation. Moreover, the Office for National Statistics calculates avoided damage costs caused by NH_3 and PM_{10} within the parameters of $PM_{2.5}$, as this includes the aerosol fraction derived from NH_3 and $PM_{2.5}$ as the riskiest (albeit bottom end) component of PM_{10} [43]. As such, these assumptions make it difficult to compare the two tools. The comparison between the UK and the United States toolsets is further complicated by differences in the pollution levels of specific chemicals. This includes the degree of segregation between emission zones, forests, and receptor regions [46].

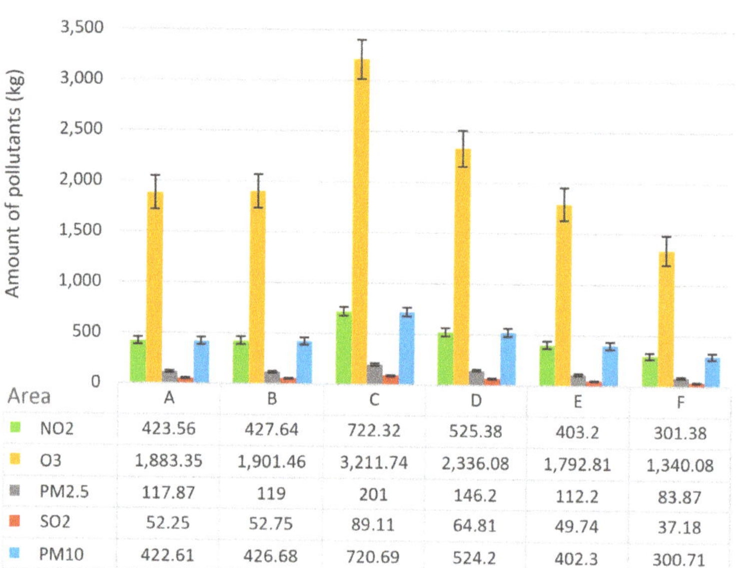

Area		A	B	C	D	E	F
	NO2	423.56	427.64	722.32	525.38	403.2	301.38
	O3	1,883.35	1,901.46	3,211.74	2,336.08	1,792.81	1,340.08
	PM2.5	117.87	119	201	146.2	112.2	83.87
	SO2	52.25	52.75	89.11	64.81	49.74	37.18
	PM10	422.61	426.68	720.69	524.2	402.3	300.71

Figure 5. Amount of each air pollutant removed annually using i-Tree Canopy Version 7.1 with local data (South West) (error bars represent SE, calculated from SEs of sampled and classified points).

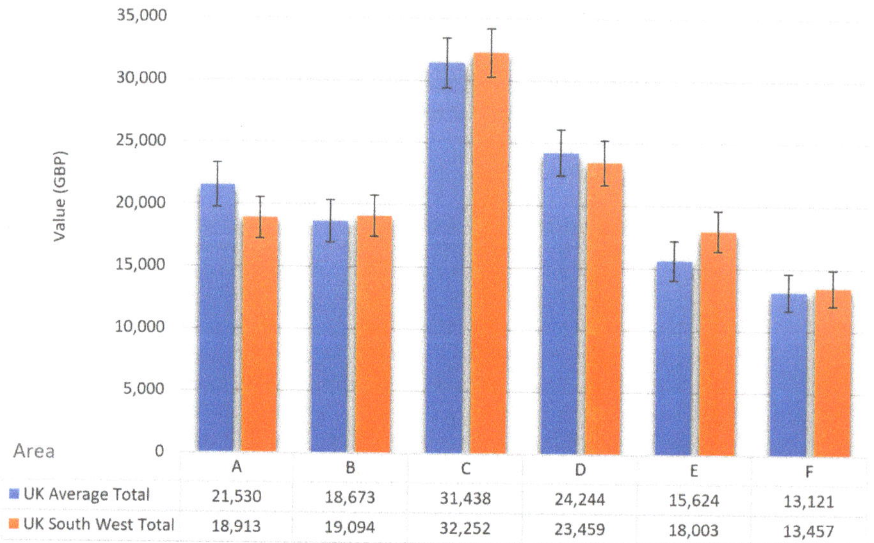

Area	A	B	C	D	E	F
UK Average Total	21,530	18,673	31,438	24,244	15,624	13,121
UK South West Total	18,913	19,094	32,252	23,459	18,003	13,457

Figure 6. Monetary values calculated using i-Tree Canopy Version 7.1 with the UK average and local data. Notes: currency is in GBP and rounded; UK air pollution estimates are based on these values in GBP/kg/year and rounded: NO_2 = GBP 0.19, O_3 = BGP 0.93, SO_2 = GBP 0.06, $PM_{2.5}$ = GBP 30.65, PM_{10} = GBP 33.71; local data air pollution estimates are based on these values in GBP/kg/year and rounded: NO_2 = GBP 0.11, O_3 = GBP 0.77, SO_2 = GBP 0.04, $PM_{2.5}$ = GBP 26.84, PM_{10} = GBP 33.71.

One key difference between the two tools is the i-Tree Canopy software allows researchers to define the project area at the beginning of the survey, while the Office for National Statistics just allows researchers to enter the postcode to look for the study area and they provide the kilograms of the pollutant removed per km^2. The precision of the result from i-Tree Canopy is based on researchers properly classifying each point into the correct cover type [62]. When the number of points augments, the accuracy of the survey increases. On the other hand, if insufficient points are input into the survey, SE increases. As stated in Section 2.3, 500 points were input into each area, i.e., within the suggested bounds of a proper i-Tree Canopy survey [62]. It is safe to suggest that some of the Google imagery provided when piecing together the survey may be of poor image resolution and may affect the decision of researchers during the input stage of the work [62].

3.3. Planting Strategies

In our study, we were only able to identify potential areas for future tree planting by combining the i-Tree Canopy and the Office for National Statistics results, using Google Maps as well as an online tree inventory (i.e., found at: https://bristoltrees.space/Locate/?latitude=51.47709&longitude=-2.58780, accessed on 3 June 2021). For example, to aid with the Office for National Statistics calculation for the postcode BS1-2 (i.e., area A where it recorded zero pollutant removal), city managers could think to increase the tree canopy as illustrated in Figure 7. However, both tools do not provide any information about soil and root space as they are not mainly designed for tree planting strategies. While the i-Tree Canopy's planar cover is valuable, it leaves the very essential vertical dimension unbound, and neither stem count nor tree-crown cover locates GI in the urban canyon three-dimensionally [10]. Several modeling studies reported that dense high-level canopy vegetation can lead to increased pollution concentrations inside street canyons by reducing turbulence, mi xing fresh air with polluted air, and trapping pollution at ground level [64–67]. As a result, it is important to consider tree interaction with local meteorological conditions and building arrangements in street canyons [65].

Figure 7. Possible areas for future tree planting: (**top left**) Fairfax Street, (**top right**) All Saints' Street, (**bottom left**) Nelson Street, and (**bottom right**) Union Street. Source: Google Maps.

It should also be noted that both tools do not provide information about species. However, species selection tools are available via i-Tree Species software which was created to assist urban foresters in choosing the best tree species for their needs (e.g., species' potential environmental services and geographic location). Using this additional toolset, users rank the importance of each desired environmental function provided by trees and the tool estimates the appropriateness of the tree species based on weighted environmental advantages of tree species at maturity [50]. In addition, a tree selection guide is available online which provides nearly 300 potential species for GI [68]. Both tools do not design or integrate tree planting applications; however, the i-Tree suite contains i-Tree Design and i-Tree Planting, but these only function in North America. In particular, i-Tree Planting is an online tool that calculates the long-term environmental benefits of a tree-planting project with a variety of trees and species. Its methods are based on the i-Tree Design and i-Tree Forecast tools [30]. Future research could develop a similar tool for Europe or a geographic information system (GIS)-like tool similar to the one developed by Wu et al. [69] that identifies suitable tree-planting locations by simulating the planting of large, medium, and small trees on plantable areas, with large trees taking precedence because they are projected to provide considerable benefits.

The suitability of each plant for each location, including tolerance of relevant stress and projected growth form, must be carefully considered when implementing robust and effective GI [67]. Lack of growth space, poor soil quality, light heterogeneity, pollutants, diseases, and conflicts with human activities, constructions, and pavements are all key issues for vegetation and green spaces in compact cities like Bristol City Centre [70]. If trees are well-managed and the correct trees are planted in the right areas, the ecosystem services they provide can greatly exceed the disservices, contributing to a city's or town's long-term sustainability and livability [71]. Tree initiatives should include recommendations on how to make public trees more resilient (e.g., promoting a broader species choice for public areas and ways to achieve greater size diversification) [71]. Recognizing that the potential for improving air quality through urban vegetation is limited, one important limitation to mitigating current air quality problems through vegetation is that the most polluted areas of cities have very limited space for planting, greatly limiting the potential for mitigation using these methods [72]. The benefits of an integrated policy that geographically isolates people from major pollution sources as much as possible (i.e., particularly transportation) and uses vegetation between the sources and the urban population are maximized [72].

4. Conclusions and Recommendations

Many developed nations are beginning to reduce pollution emissions and deposition through successful environmental policies [73–75]. At the same time, cities are implementing strategies to tackle air pollution; e.g., large scale urban afforestation projects are becoming more popular as a strategy to improve urban sustainability and human health [76]. In order to design and plan sustainable cities, landscape architects and urban planners need accurate metrics and indicators. In this communication, we have illustrated two free user-friendly online tools (i.e., the Office for National Statistics and i-Tree Canopy) using Bristol City Centre as an example. We found both tools are easy to use and communicate ecosystem services and monetary values. However, they produce different results due to the different methods that the tools incorporate. Our findings are in accordance with the conclusions of Timilsina et al. [77], who reported that the disparity in predictions by general models or average data have an impact on the estimation of ecosystem services. We also discussed the use of these tools for future tree planting strategies. According to Keith Sacre, co-founder of Treeconomics, the tree planting process should be strategized into several elements: "(1) creating a vision: what is wanted? (based on 'What is there now?'); (2) setting targets which are achievable and deliverable; (3) creating an action plan, comprising: where to plant, what to plant, how to plant, and what is needed to maintain?; and (4) monitoring and reviewing progress" [78]. This approach would potentially allow for "realistic and achievable targets to be set [and] suitable species to be selected" [78].

In terms of a key strength of the i-Tree Canopy tool, it allows users to simply photo-interpret Google aerial images to obtain statistically valid estimates of tree and other cover types, as well as evaluations of their uncertainty [30]. Random point sampling approaches such as i-Tree Canopy have the benefit of openly available data and software that may be used by a wide range of people [79]. Users of any random point sampling method, on the other hand, should be cognizant of the uncertainties involved with any urban tree cover estimate, especially if it is being used to track change [79]. i-Tree Canopy, as such, calculates several ecosystem services, e.g., avoided runoff, carbon storage and sequestration, and air pollutants removal; however, European users should be aware that i-Tree Canopy benefits (i.e., ecosystem services and monetary values) from selected locations are available only in Sweden and the UK. Using the i-Tree suite software, one can identify landscape features to predict other phenomena as well, e.g., wood tick presence [80]. Future research could explore the use of i-Tree Canopy to map edible green infrastructure as well as urban food provisioning ecosystem services [81]. The Office for National Statistics tool, on the other hand, provides meaningful urban metrics that highlight the linkages between GI and health which can improve health impacts through urban policies [82]. The Office for National Statistics website formulates its research from inputting postcodes and working from the preset-2015 dataset. The use of postcodes is a standard that many epidemiological studies utilize when pinpointing or narrowing in on specific phenomena (e.g., pollutants) [83–85]. This, in turn, parallels the tool's parameter structure and offers the prospect of measuring other performance or production elements via overlaid GIS. As an extension, we recommend the use of locals models for ecosystem services assessments [77]. Therefore, the Office for National Statistics uses landcover maps that are more appropriate for analyzing ecosystem services at the regional scale rather than the local scale since maps with limited resolution are unreliable for local studies unless additional data or fine-adjustments are supplied [86,87].

At length, both tools are excellent online resources which are easy to use, require little to no expert knowledge, and parallel a bottom-up concept, i.e., they are simplistic, fast, and trackable. A key difference, however, is that the i-Tree Canopy software specifically takes into account only tree (i.e., green) coverage while the Office for National Statistics considers the total environment (e.g., water, vegetation, etc.). This methodological difference would explain much of the discrepancy in results since Bristol is situated on the River Avon and water is considered a negative value in the Office for National Statistics methodology. However, results must be validated by fieldwork so future research could compare both tools using a stratified sample according to a rural-urban gradient [88–90]. Further case research would aid in better explaining if the toolsets could be integrated somehow (e.g., GIS) and if GI strategy can reliably be sought after if vast differences are present. Nonetheless, when factoring in an urban sustainable vision of designing green, urban-friendly cities, an uncertainty analysis should become a formal practice and necessary component of any modeling exercises, especially for models which aim to support "model transparency, model development, effective communication of model output, […] decision-making" [91] and policy formation. To further improve or develop new tools, researchers should also account for ecosystem disservices in order to assess the net benefits of GI [92,93]. To conclude, the results of this communication have updated the literature on the evaluation of GI tools in the UK [34] as well as provide a basis for the future development of a comprehensive online design tool that is site-specific for GI strategy and for the assessment of urban ecosystem services in Europe.

Author Contributions: Conceptualization, software, investigation, resources, data curation, writing—original draft preparation, methodology, validation, formal analysis, A.R. and W.T.C.; writing—review and editing, A.R. and G.T.C.; supervision, project administration, funding acquisition, A.R. All authors have read and agreed to the published version of the manuscript.

Funding: This research received no external funding.

Institutional Review Board Statement: Not applicable.

Informed Consent Statement: Not applicable.

Data Availability Statement: All datasets are available from the corresponding author on reasonable request.

Acknowledgments: The authors would like to thank the i-Tree team for clarification about the i-Tree model in the UK.

Conflicts of Interest: The authors declare no conflict of interest.

Appendix A

Table A1. Amount of each pollutant removed by trees using i-Tree Canopy Version 7.1 and Version 6.1 (kg), and the amount of each pollutant removed by blue and green infrastructure (i.e., urban trees and woodland, urban grassland, and urban water) using the Office for National Statistics (kg/km^2).

	Area	A	B	C	D	E	F
NO_2	i- Tree Canopy Version 7.1 UK	685.52	594.52	1000.97	771.93	497.45	417.76
	i-Tree Canopy Version 6.1 US	103.48	100.68	248.9	138.43	85.3	85.3
	i-Tree Canopy Version 7.1 UK South West	423.56	427.64	722.32	525.38	403.2	301.38
	Office for National Statistics	0	9.6	238.7	49.5	0	4.8
O_3	i- Tree Canopy Version 7.1 UK	2304.31	1998.43	3364.69	2594.77	1672.16	1404.28
	i-Tree Canopy Version 6.1 US	799.32	777.72	1920	1070	658.9	658.9
	i-Tree Canopy Version 7.1 UK South West	1883.35	1901.46	3211.74	2336.08	1792.81	1340.08
	Office for National Statistics	0	88	2685.8	526.5	−8.8	43.5
$PM_{2.5}$	i- Tree Canopy Version 7.1 UK	116.54	101.07	170.17	131.23	84.57	71.02
	i-Tree Canopy Version 6.1 US	40.83	39.73	98.22	54.63	33.66	33.66
	i-Tree Canopy Version 7.1 UK South West	117.87	119	201	146.2	112.2	83.87
	Office for National Statistics	0	0.5	125	34.6	−3.6	0.1
PM_{10}	i- Tree Canopy Version 7.1 UK	465.23	403.48	679.32	523.88	337.6	283.52
	i-Tree Canopy Version 6.1 US	226.86	220.73	545.7	303.51	187.01	187.01
	i-Tree Canopy Version 7.1 UK South West	422.61	426.68	720.69	524.2	402.3	300.71
	Office for National Statistics	0	0.7	193.9	55.5	−4.3	0.1
SO_2	i- Tree Canopy Version 7.1 UK	90.48	78.47	132.12	101.89	65.66	55.14
	i-Tree Canopy Version 6.1 US	50.87	49.49	122.36	68.05	41.93	41.93
	i-Tree Canopy Version 7.1 UK South West	52.25	52.75	89.11	64.81	49.74	37.18
	Office for National Statistics	0	12.3	311.9	124.5	28.9	7.8

References

1. Ritchie, H.; Roser, M. Outdoor Air Pollution. Available online: https://ourworldindata.org/outdoor-air-pollution (accessed on 20 June 2021).
2. Yim, S.H.L.; Fung, J.C.H.; Lau, A.K.H.; Kot, S.C. Air ventilation impacts of the "wall effect" resulting from the alignment of high-rise buildings. *Atmos. Environ.* **2009**, *43*, 4982–4994. [CrossRef]
3. Li, Z.; Zhou, Y.; Wan, B.; Chen, Q.; Huang, B.; Cui, Y.; Chung, H. The impact of urbanization on air stagnation: Shenzhen as case study. *Sci. Total Environ.* **2019**, *664*, 347–362. [CrossRef]
4. Śmiełowska, M.; Marć, M.; Zabiegała, B. Indoor air quality in public utility environments—A review. *Environ. Sci. Pollut. Res.* **2017**, *24*, 11166–11176. [CrossRef]
5. Peters, A.; Dockery, D.W.; Muller, J.E.; Mittleman, M.A. Increased Particulate Air Pollution and the Triggering of Myocardial Infarction. *Circulation* **2001**, *103*, 2810–2815. [CrossRef]
6. WHO World Health Assembly Closes, Passing Resolutions on Air Pollution and Epilepsy. Available online: https://apps.who.int/mediacentre/news/releases/2015/wha-26-may-2015/en/index.html (accessed on 15 July 2021).
7. Royal College of Physicians. *Every Breath We Take: The Lifelong Impact of Air Pollution*; Royal College of Physicians: London, UK, 2016.
8. Abhijith, K.V.; Kumar, P.; Gallagher, J.; McNabola, A.; Baldauf, R.; Pilla, F.; Broderick, B.; Di Sabatino, S.; Pulverenti, B. Air pollution abatement performances of green infrastructure in open road and built-up street canyon environments—A review. *Atmos. Environ.* **2017**, *162*, 71–86. [CrossRef]
9. Kumar, P.; Druckman, A.; Gallagher, J.; Gatersleben, B.; Allison, S.; Eisenman, T.S.; Hoang, U.; Hama, S.; Tiwari, A.; Sharma, A.; et al. The nexus between air pollution, green infrastructure and human health. *Environ. Int.* **2019**, *133*, 105181. [CrossRef]

10. Hewitt, C.N.; Ashworth, K.; MacKenzie, A.R. Using green infrastructure to improve urban air quality (GI4AQ). *Ambio* **2020**, *49*, 62–73. [CrossRef] [PubMed]

11. Russo, A.; Cirella, G.T. Urban Sustainability: Integrating Ecology in City Design and Planning. In *Sustainable Human–Nature Relations: Environmental Scholarship, Economic Evaluation, Urban Strategies*; Cirella, G.T., Ed.; Springer: Singapore, 2020; pp. 187–204.

12. Jim, C.Y.; Chen, W.Y. Assessing the ecosystem service of air pollutant removal by urban trees in Guangzhou (China). *J. Environ. Manag.* **2008**, *88*, 665–676. [CrossRef]

13. Cavanagh, J.-A.E.; Zawar-Reza, P.; Wilson, J.G. Spatial attenuation of ambient particulate matter air pollution within an urbanised native forest patch. *Urban For. Urban Green.* **2009**, *8*, 21–30. [CrossRef]

14. Dadea, C.; Russo, A.; Tagliavini, M.; Mimmo, T.; Zerbe, S. Tree Species as Tools for Biomonitoring and Phytoremediation in Urban Environments: A Review with Special Regard to Heavy Metals. *Arboric. Urban For.* **2017**, *43*, 155–167.

15. Brancalion, P.H.S.; Holl, K.D. Guidance for successful tree planting initiatives. *J. Appl. Ecol.* **2020**, *57*, 2349–2361. [CrossRef]

16. Pynnönen, S.; Haltia, E.; Hujala, T. Digital forest information platform as service innovation: Finnish Metsaan.fi service use, users and utilisation. *For. Policy Econ.* **2021**, *125*, 102404. [CrossRef]

17. Jones, B.A. Planting urban trees to improve quality of life? The life satisfaction impacts of urban afforestation. *For. Policy Econ.* **2021**, *125*, 102408. [CrossRef]

18. TDAG. *Trees in the Townscape: A Guide for Decision Makers*; Trees and Design Action Group: London, UK, 2012.

19. Mullaney, J.; Lucke, T.; Trueman, S.J. A review of benefits and challenges in growing street trees in paved urban environments. *Landsc. Urban Plan.* **2015**, *134*, 157–166. [CrossRef]

20. Bodnaruk, E.W.; Kroll, C.N.; Yang, Y.; Hirabayashi, S.; Nowak, D.J.; Endreny, T.A. Where to plant urban trees? A spatially explicit methodology to explore ecosystem service tradeoffs. *Landsc. Urban Plan.* **2017**, *157*, 457–467. [CrossRef]

21. Urban, J. *Up by Roots*; ISA: New York, NY, USA, 2008; ISBN 1881956652.

22. Jerome, G.; Sinnett, D.; Burgess, S.; Calvert, T.; Mortlock, R. A framework for assessing the quality of green infrastructure in the built environment in the UK. *Urban. For. Urban. Green.* **2019**, *40*, 174–182. [CrossRef]

23. Ellis, J.B. Sustainable surface water management and green infrastructure in UK urban catchment planning. *J. Environ. Plan. Manag.* **2013**, *56*, 24–41. [CrossRef]

24. Joint Core Strategy. *Green Infrastructure Strategy*; Joint Core Strategy: Gloucester, UK, 2014.

25. Natural England. *Green Infrastructure Strategies: An Introduction for Local Authorities and Their Partners*; Natural England: London, UK, 2008.

26. Russo, A.; Escobedo, F.J.; Timilsina, N.; Schmitt, A.O.; Varela, S.; Zerbe, S. Assessing urban tree carbon storage and sequestration in Bolzano, Italy. *Int. J. Biodivers. Sci. Ecosyst. Serv. Manag.* **2014**, *10*, 54–70. [CrossRef]

27. Daily, G.; Polasky, S.; Goldstein, J.; Kareiva, P.; Mooney, H. Ecosystem services in decision making: Time to deliver. *Front. Ecol. Environ.* **2009**, *7*, 21–28. [CrossRef]

28. Villa, F.; Bagstad, K.J.; Voigt, B.; Johnson, G.W.; Portela, R.; Honzák, M.; Batker, D. A Methodology for Adaptable and Robust Ecosystem Services Assessment. *PLoS ONE* **2014**, *9*, e91001. [CrossRef]

29. Brown, M.E.; McGroddy, M.; Spence, C.; Flake, L.; Sarfraz, A.; Nowak, D.J.; Milesi, C. *Modeling the Ecosystem Services Provided by Trees in Urban Ecosystems: Using Biome-BGC to Improve i-Tree Eco*; NASA: Washington, DC, USA, 2012.

30. Nowak, D.J. *Understanding i-Tree: Summary of Programs and Methods*; General Technical Report NRS-200; Northern Research Station, Forest Service, Department of Agriculture: Madison, WI, USA, 2020; 100p, [plus 14 appendices].

31. Roy, S.; Byrne, J.; Pickering, C. A systematic quantitative review of urban tree benefits, costs, and assessment methods across cities in different climatic zones. *Urban For. Urban Green.* **2012**, *11*, 351–363. [CrossRef]

32. Rogers, K.; Hansford, D.; Sunderland, T.; Brunt, A.; Coish, N. Measuring the ecosystem services of Torbay's trees: The Torbay i-Tree Eco pilot project. In Proceedings of the Urban Trees Research Conference, London, UK, 26–27 July 2021; Forestry Commission: Edinburgh, UK, 2011; pp. 18–26.

33. Raum, S.; Hand, K.L.; Hall, C.; Edwards, D.M.; O'Brien, L.; Doick, K.J. Achieving impact from ecosystem assessment and valuation of urban greenspace: The case of i-Tree Eco in Great Britain. *Landsc. Urban Plan.* **2019**, *190*, 103590. [CrossRef]

34. Natural England. *Green Infrastructure—Valuation Tools Assessment*; Natural England: Exeter, UK, 2013.

35. Greater London Authority. A Natural Capital Account for Public Green Space in London: How It Can Shape Future Policy and Decision-Making. Available online: https://www.london.gov.uk/sites/default/files/nca_supplementary_document.pdf (accessed on 14 July 2021).

36. UK Government. *The England Trees Action Plan. 2021–2024*; UK Government: London, UK, 2021.

37. Pincetl, S.; Gillespie, T.; Pataki, D.E.; Saatchi, S.; Saphores, J.-D. Urban tree planting programs, function or fashion? Los Angeles and urban tree planting campaigns. *GeoJournal* **2013**, *78*, 475–493. [CrossRef]

38. Bristol City Council. *The Population of Bristol*; Bristol City Council: Bristol, UK, 2019.

39. Climate Action. Friends of the Earth Bristol's One City Plan to Create an Urban Forest. 2021. Available online: https://takeclimateaction.uk/solutions/bristols-one-city-plan-create-urban-forest (accessed on 1 June 2021).

40. Consultants Air Quality. *Health Impacts of Air Pollution*; Consultants Air Quality: Bristol, UK, 2017.

41. Chan, W.T. *The Benefit of Green Infrastructure on Air Quality in Bristol*; University of Gloucestershire: Cheltenham, UK, 2019.

42. Garrett, J.; Connett, J. Campaigners Accuse Marvin Rees of Not Protecting Bristol's Mature Trees. Available online: https://thebristolcable.org/2021/02/campaigners-accuse-marvin-rees-of-putting-housing-above-climate-by-not-protecting-bristols-mature-trees/ (accessed on 14 July 2021).

43. Office for National Statistics UK Air Pollution Removal: How Much Pollution Does Vegetation Remove in Your Area? Available online: https://www.ons.gov.uk/economy/environmentalaccounts/articles/ukairpollutionremovalhowmuchpollutiondoesvegetationremoveinyourarea/2018-07-30 (accessed on 16 June 2021).

44. Jones, L.; Vieno, M.; Fitch, A.; Carnell, E.; Steadman, C.; Cryle, P.; Holland, M.; Nemitz, E.; Morton, D.; Hall, J.; et al. Urban natural capital accounts: Developing a novel approach to quantify air pollution removal by vegetation. *J. Environ. Econ. Policy* **2019**, *8*, 413–428. [CrossRef]

45. Vieno, M.; Heal, M.R.; Williams, M.L.; Carnell, E.J.; Nemitz, E.; Stedman, J.R.; Reis, S. The sensitivities of emissions reductions for the mitigation of UK PM2.5. *Atmos. Chem. Phys.* **2016**, *16*, 265–276. [CrossRef]

46. Jones, L.; Vieno, M.; Morton, D.; Cryle, P.; Holland, M.; Carnell, E.; Nemitz, E.; Hall, J.; Beck, R.; Reis, S.; et al. *Developing Estimates for the Valuation of Air Pollution Removal in Ecosystem Accounts: Final Report Office for National Statistics*; Centre for Ecology and Hydrology (CEH): Wallingford, UK, 2017.

47. Nowak, D.J.; Greenfield, E.J. Tree and impervious cover change in U.S. cities. *Urban For. Urban Green.* **2012**, *11*, 21–30. [CrossRef]

48. Russo, A.; Escobedo, F.J.; Timilsina, N.; Zerbe, S. Transportation carbon dioxide emission offsets by public urban trees: A case study in Bolzano, Italy. *Urban For. Urban Green.* **2015**, *14*, 398–403. [CrossRef]

49. Hirabayashi, S. i-Tree Canopy Air Pollutant Removal and Monetary Value Model Descriptions. 2014, pp. 1–11. Available online: https://www.itreetools.org/documents/560/i-Tree_Canopy_Air_Pollutant_Removal_and_Monetary_Value_Model_Descriptions.pdf (accessed on 10 June 2021).

50. Tree i-Tree Species. Available online: https://species.itreetools.org/ (accessed on 15 June 2021).

51. Trees of Bristol Bristol Tree Canopy Cover Survey. 2018. Available online: https://bristoltrees.space/trees/treecover-map.xq?fbclid=IwAR1QpV-CYSE6LYwZjd5yak_S3QZpoqBqILnj7_O3-SdzmTtBwwhbdg9dgcA (accessed on 16 June 2021).

52. Rogers, K.; Jaluzot, A. *Treeconomics Oxford i-Tree Canopy Cover Assessment*; Treeconomics: Oxford, UK, 2015.

53. Doick, K.J.; Davies, H.J.; Moss, J.; Coventry, R.; Handley, P.; VazMonteiro, M.; Rogers, K.; Simpkin, P. The Canopy Cover of England's Towns and Cities: Baselining and setting targets to improve human health and well-being. In Proceedings of the Trees, People and the Built Environment 3: Urban Trees Research Conference, Birmingham, UK, 5–6 April 2017.

54. Kenney, W.A.; Van Wassenaer, P.J.E.; Satel, A.L. Criteria and indicators for strategic urban forest planning and management. *Arboric. Urban For.* **2011**, *37*, 108–117.

55. Astell-Burt, T.; Feng, X. Association of Urban Green Space with Mental Health and General Health among Adults in Australia. *JAMA Netw. Open* **2019**, *2*, 1–22. [CrossRef]

56. Marselle, M.R.; Bowler, D.E.; Watzema, J.; Eichenberg, D.; Kirsten, T.; Bonn, A. Urban street tree biodiversity and antidepressant prescriptions. *Sci. Rep.* **2020**, *10*, 22445. [CrossRef]

57. Kondo, M.C.; Mueller, N.; Locke, D.H.; Roman, L.A.; Rojas-Rueda, D.; Schinasi, L.H.; Gascon, M.; Nieuwenhuijsen, M.J. Health impact assessment of Philadelphia's 2025 tree canopy cover goals. *Lancet Planet. Health* **2020**, *4*, e149–e157. [CrossRef]

58. Nemitz, E.; Vieno, M.; Carnell, E.; Fitch, A.; Steadman, C.; Cryle, P.; Holland, M.; Morton, R.D.; Hall, J.; Mills, G.; et al. Potential and limitation of air pollution mitigation by vegetation and uncertainties of deposition-based evaluations. *Philos. Trans. R. Soc. A Math. Phys. Eng. Sci.* **2020**, *378*, 20190320. [CrossRef] [PubMed]

59. Nowak, D.J.; Hirabayashi, S.; Bodine, A.; Greenfield, E. Tree and forest effects on air quality and human health in the United States. *Environ. Pollut.* **2014**, *193*, 119–129. [CrossRef]

60. Troy, A.; Morgan Grove, J.; O'Neil-Dunne, J. The relationship between tree canopy and crime rates across an urban–rural gradient in the greater Baltimore region. *Landsc. Urban Plan.* **2012**, *106*, 262–270. [CrossRef]

61. Pugh, T.A.M.; MacKenzie, A.R.; Whyatt, J.D.; Hewitt, C.N. Effectiveness of Green Infrastructure for Improvement of Air Quality in Urban Street Canyons. *Environ. Sci. Technol.* **2012**, *46*, 7692–7699. [CrossRef] [PubMed]

62. Tree. i-Tree Canopy Technical Notes. Available online: https://canopy.itreetools.org/references (accessed on 16 June 2021).

63. Sunderland, T.; Rogers, K.; Coish, N. What proportion of the costs of urban trees can be justified by the carbon sequestration and air-quality benefits they provide? *Arboric. J.* **2012**, *34*, 62–82. [CrossRef]

64. Janhäll, S. Review on urban vegetation and particle air pollution—Deposition and dispersion. *Atmos. Environ.* **2015**, *105*, 130–137. [CrossRef]

65. Jeanjean, A.P.R.; Buccolieri, R.; Eddy, J.; Monks, P.S.; Leigh, R.J. Air quality affected by trees in real street canyons: The case of Marylebone neighbourhood in central London. *Urban For. Urban Green.* **2017**, *22*, 41–53. [CrossRef]

66. Riondato, E.; Pilla, F.; Sarkar Basu, A.; Basu, B. Investigating the effect of trees on urban quality in Dublin by combining air monitoring with i-Tree Eco model. *Sustain. Cities Soc.* **2020**, *61*, 102356. [CrossRef]

67. Barwise, Y.; Kumar, P. Designing vegetation barriers for urban air pollution abatement: A practical review for appropriate plant species selection. *NPJ Clim. Atmos. Sci.* **2020**, *3*, 12. [CrossRef]

68. Hirons, A.; Sjöman, H. *Tree Species Selection for Green Infrastructure: A Guide for Specifiers*; Trees and Design Action Group: London, UK, 2019; ISBN 978-0-9928686-4-2.

69. Wu, C.; Xiao, Q.; McPherson, E.G. A method for locating potential tree-planting sites in urban areas: A case study of Los Angeles, USA. *Urban For. Urban Green.* **2008**, *7*, 65–76. [CrossRef]

70. Jim, C.Y.; Konijnendijk van den Bosch, C.; Chen, W.Y. Acute Challenges and Solutions for Urban Forestry in Compact and Densifying Cities. *J. Urban Plan. Dev.* **2018**, *144*, 04018025. [CrossRef]
71. Vaz Monteiro, M.; Handley, P.; Morison, J.; Doick, K. *The Role of Urban Trees and Greenspaces in Reducing Urban Air Temperatures*; Forest Research: Surrey, UK, 2019; ISBN 978-0-85538-984-0.
72. Defra Air Quality Expert Group—Defra, UK. Available online: https://uk-air.defra.gov.uk/research/aqeg/ (accessed on 20 June 2021).
73. Miranda, M.L.; Edwards, S.E.; Keating, M.H.; Paul, C.J. Making the environmental justice grade: The relative burden of air pollution exposure in the United States. *Int. J. Environ. Res. Public Health* **2011**, *8*, 1755–1771. [CrossRef] [PubMed]
74. Duvall, R.M.; Long, R.W.; Beaver, M.R.; Kronmiller, K.G.; Wheeler, M.L.; Szykman, J.J. Performance evaluation and community application of low-cost sensors for ozone and nitrogen dioxide. *Sensors* **2016**, *16*, 1698. [CrossRef]
75. Stevens, K.A.; Bryer, T.A.; Yu, H. Air Quality Enhancement Districts: Democratizing data to improve respiratory health. *J. Environ. Stud. Sci.* **2021**, 1–6. [CrossRef]
76. Yao, N.; Konijnendijk van den Bosch, C.C.; Yang, J.; Devisscher, T.; Wirtz, Z.; Jia, L.; Duan, J.; Ma, L. Beijing's 50 million new urban trees: Strategic governance for large-scale urban afforestation. *Urban For. Urban Green.* **2019**, *44*, 126392. [CrossRef]
77. Timilsina, N.; Beck, J.L.; Eames, M.S.; Hauer, R.; Werner, L. A comparison of local and general models of leaf area and biomass of urban trees in USA. *Urban For. Urban Green.* **2017**, *24*, 157–163. [CrossRef]
78. Sacre, K. i-Tree focus Tree planting strategies. *ARB Mag.* **2020**, *190*, 15–16.
79. Parmehr, E.G.; Amati, M.; Taylor, E.J.; Livesley, S.J. Estimation of urban tree canopy cover using random point sampling and remote sensing methods. *Urban For. Urban Green.* **2016**, *20*, 160–171. [CrossRef]
80. Omodior, O.; Eze, P.; Anderson, K.R. Using i-tree canopy vegetation cover subtype classification to predict peri-domestic tick presence. *Ticks Tick Borne Dis.* **2021**, *12*, 101684. [CrossRef]
81. Russo, A.; Cirella, G.T. Edible urbanism 5.0. *Palgrave Commun.* **2019**, *5*, 1–9. [CrossRef]
82. Prasad, A.; Gray, C.B.; Ross, A.; Kano, M. Metrics in Urban Health: Current Developments and Future Prospects. *Annu. Rev. Public Health* **2016**, *37*, 113–133. [CrossRef]
83. Grubesic, T.H.; Matisziw, T.C. On the use of ZIP codes and ZIP code tabulation areas (ZCTAs) for the spatial analysis of epidemiological data. *Int. J. Health Geogr.* **2006**, *5*, 1–15. [CrossRef]
84. Krieger, N.; Waterman, P.; Chen, J.T.; Soobader, M.J.; Subramanian, S.V.; Carson, R. Zip code caveat: Bias due to spatiotemporal mismatches between zip codes and US census-defined geographic areas: The public health disparities geocoding project. *Am. J. Public Health* **2002**, *92*, 1100–1102. [CrossRef]
85. Acevedo-Garcia, D. Zip code-level risk factors for tuberculosis: Neighborhood environment and residential segregation in New Jersey, 1985–1992. *Am. J. Public Health* **2001**, *91*, 734–741. [CrossRef]
86. Liu, O.Y.; Russo, A. Assessing the contribution of urban green spaces in green infrastructure strategy planning for urban ecosystem conditions and services. *Sustain. Cities Soc.* **2021**, *68*, 102772. [CrossRef]
87. Dales, N.P.; Brown, N.J.; Lusardi, J. *Assessing the Potential for Mapping Ecosystem Services in England Based on Existing Habitats*; Natural England: Newcastle, UK, 2014.
88. Baró, F.; Gómez-Baggethun, E.; Haase, D. Ecosystem service bundles along the urban-rural gradient: Insights for landscape planning and management. *Ecosyst. Serv.* **2017**, *24*, 147–159. [CrossRef]
89. Zardo, L.; Geneletti, D.; Pérez-Soba, M.; Van Eupen, M. Estimating the cooling capacity of green infrastructures to support urban planning. *Ecosyst. Serv.* **2017**, *26*, 225–235. [CrossRef]
90. Larondelle, N.; Haase, D. Urban ecosystem services assessment along a rural–urban gradient: A cross-analysis of European cities. *Ecol. Indic.* **2013**, *29*, 179–190. [CrossRef]
91. Lin, J.; Kroll, C.N.; Nowak, D.J. An uncertainty framework for i-Tree eco: A comparative study of 15 cities across the United States. *Urban For. Urban Green.* **2021**, *60*, 127062. [CrossRef]
92. Russo, A.; Escobedo, F.J.; Cirella, G.T.; Zerbe, S. Edible green infrastructure: An approach and review of provisioning ecosystem services and disservices in urban environments. *Agric. Ecosyst. Environ.* **2017**, *242*, 53–66. [CrossRef]
93. Von Döhren, P.; Haase, D. Ecosystem disservices research: A review of the state of the art with a focus on cities. *Ecol. Indic.* **2015**, *52*, 490–497. [CrossRef]

MDPI
St. Alban-Anlage 66
4052 Basel
Switzerland
Tel. +41 61 683 77 34
Fax +41 61 302 89 18
www.mdpi.com

Land Editorial Office
E-mail: land@mdpi.com
www.mdpi.com/journal/land